PHILONIS ALEXANDRINI DE ANIMALIBUS

STUDIES IN HELLENISTIC JUDAISM

Supplements to *Studia Philonica*

Edited by

Earle Hilgert and Burton L. Mack

Number 1

PHILONIS ALEXANDRINI DE ANIMALIBUS:

The Armenian Text with an

Introduction, Translation, and Commentary

by

Abraham Terian

PHILONIS ALEXANDRINI DE ANIMALIBUS:

The Armenian Text with an Introduction, Translation, and Commentary

by

Abraham Terian

SCHOLARS PRESS

Distributed by
Scholars Press
101 Salem Street
Chico, California 95926

PHILONIS ALEXANDRINI DE ANIMALIBUS:

The Armenian Text with an Introduction,

Translation, and Commentary

Abraham Terian

Library of Congress Cataloging in Publication Data

Philo, of Alexandria.
    Philonis Alexandrini de animalibus.

    (Studies in Hellenistic Judaism ; no. 1)
    English, Armenian, or Latin.
    "Supplements to Studia Philonica."
    Translation of: Alexandros.
    Bibliography: p.
    Includes indexes.
    1.Zoology--Pre-Linnean works. I. Terian,
Abraham, 1942-.   II. Studia Philonica.
III. Title. IV. Series.
QL41.P4612   591   81-836
ISBN 0-89130-472-X   AACR2
ISBN 0-89130-473-8 (pbk.)

Printed in the United States of America
1 2 3 4 5 6
Edwards Brothers, Inc.
Ann Arbor, Michigan 48106

TO

SARA MIRJAM

ARI ISAAC

SATU RUTH

SONJA ESTHER

TABLE OF CONTENTS

PREFACE

I was first introduced to the serious study of Philo during the
course of a doctoral seminar on the book of Hebrews. Eventually my
interest in his once lost works surpassed my eager quest for the
valuable comparative material necessary for a better understanding
of the New Testament book.

Philo is one of the most frequently quoted primary sources in the
religio-historical study of the New Testament. Unfortunately, however,
not all of his extant works are available in any one language. Moreover, neither of the two indices to his works is complete, for neither
includes his works extant only in "classical" Armenian (H. Leisegang,
Philonis Alexandrini opera quae supersunt, Indices, I-II [Berlin, 1926-
1930] and G. Meyer, *Index Philoneus* [Berlin, 1974]).

Not only biblical scholars but also historians of Greek philosophy,
early church historians, rabbinists, and medievalists are constantly
attracted to Philo in their quest for documentary sources, traditions,
and origin of ideas in his writings. Their verdicts, based on *partial
interest* in Philo, have left a detrimental mark on Philo scholarship
which has further suffered due to *partial use* of the works by Philo
scholars themselves. It is against such partial focusing on Philo
that E. R. Goodenough made the following statement: "We shall know
Philo only when we accept him as a whole" (*Introduction to Philo Judaeus*
[rev. ed.; Oxford, 1962], p. 19).

The present work is but a step in helping the scholar do justice to
Philo as a whole. The treatise *De animalibus*, extant only in Armenian,
was never before translated into any modern language. The reliability
of J. B. Aucher's Latin translation of Philonic and pseudo-Philonic
works extant only in Armenian (*Philonis Judaei sermones tres hactenus
inediti: I et II de providentia et III de animalibus* [Venice, 1822];
*Philonis Judaei paralipomena Armena: libri videlicet quatuor in Genesis,
libri duo in Exodum, sermo unus de Sampsone, alter de Jona, tertius de*

ix

tribus angelis Abraamo apparentibus [Venice, 1826]) has often been ques-
tioned. Moreover, this treatise was never thoroughly studied. The only
dissertation dealing with the subject, that by G. Tappe (*De Philonis
libro qui inscribitur* 'Αλέξανδρος ἢ περὶ τοῦ λόγον ἔχειν τὰ ἄλογα ζῷα:
Quaestiones selectae [Diss. Göttingen, 1912]), addresses itself to the
philosophical background and raises the question of sources.

    The desiderata in Philo scholarship are numerous indeed. Reliable
translations of the *De Providentia* and the *Quaestiones* are among the fore-
most needs. A more pressing need, however, is for a complete Armenian-
Greek and Greek-Armenian word index to the Philonic works extant only in
Armenian, an index based on equivalent terms found in Philonic works extant in both languages (cf. B. Reynders, *Lexique comparé du texte grec
et des versions latine, arménienne et syriaque de l' "Adversus Haereses"
de Saint Irénée, I: Introduction, index des mots grecs, arméniens et
syriaques, II: Index des mots latins*, Corpus scriptorum christianorum
orientalium, CXLI-CXLII, Subsidia, t. 5 et 6 [Louvain, 1954]).

    The task of editing, translating, and commenting on Philo's *De
animalibus* has been enormous and overwhelming. I need not mention the
undue pressures placed on my family during the long time spent on this
dissertation. My wife's patience and understanding permitted the long
hours of work, and her prayers--with those of the children--helped
lighten the burden.

    I wish to thank cordially Prof. Dr. Bo Reicke of the Faculty of Theo-
logy at the University of Basel for advising me to translate Philo's
De animalibus and to write a commentary on it at a time when I was thinking of another treatise. His sound guidelines, constant interest, and
immense patience made the completion of this work possible. I was fortunate to have had so able and amiable a source of strength. I also
wish to thank warmly the past and present directors of the Philo Institute, Drs. Robert G. Hamerton-Kelly, Burton L. Mack, and Earle Hilgert,
and Dr. John A. C. Greppin of Cleveland State University for their keen
interest in my work.

Berrien Springs, Michigan   Abraham Terian
May 9, 1978

ABBREVIATIONS

Titles of Philonic treatises are abbreviated in the forms listed in *Studia Philonica*, 1 (1972), 92; with some additions, they are as follows:

Abr	De Abrahamo
Aet	De aeternitate mundi
Agr	De agricultura
Anim	De animalibus
Apol Jud	Apologia pro Iudaeis
Cher	De Cherubim
Conf	De confusione linguarum
Congr	De congressu eruditionis gratia
Dec	De Decalogo
Deo	De Deo
Ebr	De ebrietate
Flacc	In Flaccum
Fuga	De fuga et inventione
Gaium	Legatio ad Gaium
Gig	De gigantibus
Heres	Quis rerum divinarum heres
Jos	De Iosepho
Leg All I-III	Legum allegoriae I-III
Migr	De migratione Abrahami
Mut	De mutatione nominum
Num	De numeris
Op	De opificio mundi
Plant	De plantatione
Post	De posteritate Caini
Praem	De praemiis et poenis
Provid I-II	De Providentia I-II
Quaes Ex I-II	Quaestiones et solutiones in Exodum I-II
Quaes Gen I-IV	Quaestiones et solutiones in Genesin I-IV
Quod Det	Quod deterius potiori insidiari soleat
Quod Deus	Quod Deus immutabilis sit

Quod Omn	Quod omnis probus liber sit
Sacr	De sacrificiis Abelis et Caini
Sobr	De sobrietate
Somn I-II	De somniis I-II
Spec Leg I-IV	De specialibus legibus I-IV
Virt	De virtutibus
Vita Cont	De vita contemplativa
Vita Mos I-II	De vita Mosis I-II

Authors and titles of Greek classical works are abbreviated in the forms given in H. G. Liddell and R. Scott, *A Greek-English Lexicon* (9th ed. by H. S. Jones, rev. with a supplement by E. A. Barber; Oxford, 1968), pp. xvi-xxxviii; and those of Latin classical works, with the exception of Pliny (the "Elder"), in the forms given in the *Oxford Latin Dictionary*, ed. A. Souter *et al.*, Fascicle 1 (Oxford, 1968), pp. ix-xx. Abbreviations of biblical and cognate literature follow the list in W. F. Arndt and F. W. Gingrich, *A Greek-English Lexicon of the New Testament and Other Early Christian Literature* (Chicago, 1957), pp. xxvii-xxviii. For the full titles, see Ind. VI and VII, pp. 323-331.

Other abbreviations:

Aucher	*Philonis Judaei sermones tres hactenus inediti: I et II de providentia et III de animalibus,* ed. J. B. Aucher (Venice, 1822).
Diels *VS*	*Die Fragmente der Vorsokratiker*, ed. H. Diels (5th ed. by W. Kranz; Berlin, 1934).
LCL	The Loeb Classical Library.
Migne *PG*	*Patrologia, Series Graeca,* ed. J. P. Migne *et al.* (163 vols.; Paris, 1857-).
PRE	*Paulys Realencyclopädie der klassischen Altertumswissenschaft,* ed. G. Wissowa *et al.*, Ser. I (A-Q), 48 vols.; Ser. II (R-Z), 18 vols.; 10 Suppl. (Stuttgart, 1894-).
SVF	*Stoicorum veterum fragmenta,* ed. J. von Arnim (4 vols.; Leipzig, 1903-1924).

I. INTRODUCTION

I. INTRODUCTION

A survey of the vast literature on Philo will not be attempted in this Introduction, which is primarily intended to acquaint the scholar with a little known Philonic work.[1] The treatise *De animalibus* (hereinafter abbreviated Anim) is one of Philo's works extant only in "classical" Armenian (hereinafter abbreviated Arm.), the original Greek (hereinafter abbreviated Gr.) having been lost at an early time. The first part of the Introduction deals with the Arm. corpus of Philo's works, the second with the treatise in particular.

A. THE ARMENIAN TRANSLATION OF PHILO'S WORKS

1. Contents

The Arm. corpus of Philo contains several of his works the Gr. original of which is extant: Abr, Dec, Leg All I-II, Spec Leg I 79-161, 285-

---

[1]See the critical surveys by W. Völker, *Fortschritt und Vollendung bei Philo von Alexandrien, Eine Studie zur Geschichte der Frömmigkeit,* Texte und Untersuchungen zur Geschichte der altchristlichen Literatur, XLIX (Leipzig, 1938) and H. Thyen, "Die Probleme der nueren Philo-Forschung," *Theologische Rundschau,* 23 (1955), 230-246. Among the general introductions to Philo, see J. Daniélou, *Philon d'Alexandrie* (Paris, 1958) and E. R. Goodenough, *An Introduction to Philo Judaeus* (2n ed.; Oxford, 1962). For the vast literature on Philo, see the following bibliographies: H. L. Goodhart and E. R. Goodenough, *The Politics of Philo Judaeus, Practice and Theory* with a *General Bibliography of Philo* (New Haven, 1938), J. Haussleiter, "Nacharistotelische Philosophen, 1931-1936," *Jahresbericht über die Fortschritte der klassischen Altertumswissenschaft,* 281-282 (1943), 107-116; L. H. Feldman, *Scholarship on Philo and Josephus (1937-1962),* Studies in Judaica,[I] (New York, n.d.), 1-25; G. Delling and R. M. Maser, *Bibliographie zur jüdisch-hellenistischen und intertestamentarischen Literatur, 1900-1970,* Texte und Untersuchungen zur Geschichte der altchristlichen Literatur, CVI (Berlin, 1975); E. Hilgert, "A Bibliography of Philo Studies, 1963-1970," *Studia Philonica,* 1 (1972), 57-71; "A Bibliography of Philo Studies in 1971 with Additions for 1965-70," *Studia Philonica,* 2 (1973), 51-54; "A Bibliography of Philo Studies, 1972-1973," *Studia Philonica,* 3 (1974-1975), 117-125; "A Bibliography of Philo Studies, 1974-1975," *Studia Philonica,* 4 (1976-1977), 79-85.

345, III 1-63, and Vita Cont. These were edited by F. C. Conybeare[2]
and with his help were utilized in L. Cohn's and P. Wendland's critical
edition of the Gr. text of these works.[3] The corpus contains four
other works of Philo extant only in Arm.: Quaes Gen I-IV, Quaes Ex I-II,
Provid I-II, and Anim,[4] and two fragments: a ten-page fragment on Gen
18:2a derivatively entitled De Deo, belonging to the large allegorical
commentary on Genesis, possibly to the first of the lost treatises following Mut and preceding Somn 1,[5] and a page-long fragment on the decad,
possibly part of the lost treatise Περὶ ἀριθμῶν mentioned in Vita Mos
II 115 and Quaes Gen IV 110 and alluded to elsewhere.[6] The corpus also
contains two pseudo-Philonic homilies: *De Jona* and *De Sampsone*, the

---

[2]*P'iloni Hebrayec'woy čaṙk' t'argmanealk' i naxneac' meroc' oroc'
Hellen bnagirk' hasin aṙ mez* ("Works of Philo Judaeus Translated by
Our Ancients the Greek Original of Which is Extant"), [ed. F. C. Cony-
beare] (Venice, 1892).

[3]*Philonis Alexandrini opera quae supersunt*, I-VI (Berlin, 1896-1915);
see especially I, lii-lvi. Conybeare was particularly interested in the
value of the Arm. text of Vita Cont for textual purposes; see his critical edition: *Philo about the Contemplative Life, or the Fourth Book of
the Treatise concerning Virtures* (Oxford, 1895).

[4]The Gr. fragments of these works, recovered from catenae and patristic writings by A. Mai, J. B. Pitra, C. von Tischendorf, P. Wendland,
and others, have been collected by J. R. Harris, *Fragments of Philo
Judaeus* (Cambridge, 1886); for later publications of Philonic fragments,
see R. Marcus, LCL Supplement II, 180.

[5]See L. Massebieau, "Le classement des oeuvres de Philon," *Bibliothèque de l'école des hautes études, Sciences religieuses,* I (Paris, 1889),
29; L. Cohn, "Einteilung und Chronologie der Schriften Philos," *Philologus Supplementband,* 7 (1899), 401; M. Adler, "Das philonische Fragment De Deo," *Monatschrift für Geschichte und Wissenschaft des Judentums,*
80 (1936), 163-170; and M. Harl, "Cosmologie grecque et représentations
juives dans l'oeuvre de Philon d'Alexandrie," *Philon d'Alexandrie,*
Colloques Nationaux du Centre National de la Recherche Scientifique, à
Lyon du 11 au 15 septembre 1966 (Paris, 1967), pp. 189-203. Cf. the
interpretation of this fragment by E. R. Goodenough, *By Light, Light:
The Mystic Gospel of Hellenistic Judaism* (New Haven, 1935), pp. 30-31
with that of H. A. Wolfson, *Philo: Foundations of Religious Philosophy
in Judaism, Christianity and Islam,* I (rev. ed.; Cambridge, Mass., 1968),
340-341.

[6]For an attempted reconstruction of this work, see K. Staehle, *Die
Zahlenmystik bei Philon von Alexandreia* (Leipzig and Berlin, 1931), pp.
1-18.

Hellenistic-Jewish authorship of which cannot be doubted.[7] Except for
the small fragment, the text of the works extant only in Arm. was published with a Latin translation early in the last century by J. B. Aucher.[8]

Excluding the fragments and the pseudo-Philonic homilies, the Arm.
titles of the works in the corpus and their Gr. equivalents are as
follows:[9]

Abr Փիլոնի կեանք իմաստնոյ

Φίλωνος βίος σοφοῦ

Anim Փիլոնի յաղագս բան ունել եւ անասուն կենդանեացդ

Φίλωνος περὶ τοῦ λόγον ἔχειν τὰ ἄλογα ζῷα

---

[7]H. Lewy, *The Pseudo-Philonic De Jona. Part I: The Armenian Text
with a Critical Introduction*, Studies and Documents, VII (London, 1936),
24. His English translation may soon be published by The Israel Academy
of Sciences and Humanities to complement *Part I*. C. Safrai, a graduate
student at The Hebrew University of Jerusalem, is currently editing *De
Sampsone* as part of her doctoral dissertation.

[8]*Philonis Judaei sermones tres hactenus inediti: I et II de providentia
et III de animalibus* (Venice, 1822)--hereinafter abbr. Aucher; *Philonis
Judaei paralipomena Armena: libri videlicet quatuor in Genesin, libri
duo in Exodum, sermo unus de Sampsone, alter de Jona, tertius de tribus
angelis Abraamo apparentibus* (Venice, 1826). The *Quaestiones* were translated into English with some reliance on Aucher's Latin translation by R.
Marcus, and were published as 2 supplementary vols. to the LCL edition
of Philo (Cambridge, Mass., 1953). No English translation of the Arm.
text of Provid I-II was ever attempted. Translations of the Gr. fragments of Provid II are readily available in translations of Eus. *PE*
vii. 21. 336b-337a (=§§ 50-51); viii. 14. 386-399 (=§§ 3, 15-33a, 99-
112). A translation of these fragments by F. H. Colson is part of vol.
IX of the LCL edition of Philo; a German version of Aucher's Latin
translation of Provid I-II is found in *Philo von Alexandria, Die Werke
in Deutscher Übersetzung*, VII, ed. W. Theiler (Berlin, 1964); and a
French translation of Aucher's Latin by M. Hadas-Lebel, *De Providentia
I et II*, constitutes vol. XXXV of Les oeuvres de Philon d'Alexandrie
(Paris, 1973). The work which comprises the subject of the present
study was never before translated into any modern language.

[9]Note that in all of these titles the Arm. maintains the word order
of the Gr. The Arm. titles corresponding with the Philonic De Deo and
the pseudo-Philonic *De Jona* (a page-long fragment bearing the same title
follows; translated by Aucher, *Paralipomena Armena*, p. 612; cf. Chrysos-
tom's *De paenitentia*, Hom. V [Migne *PG* XLIX, 310]) and *De Sampsone*,
respectively, are: Փիլոնի յաղագս զԱստուած, Յաղագս Յովնանու,
Առանց պատրաստութեան ի Սամփսոմն.

6

Dec	Փիլոնի յաղագս տասն բանիցն 10
	Φίλωνος περὶ τῶν δέϰα λογίων
Leg All I-II	Փիլոնի աստուածային ւերինացն այլարանութիւն
	Φίλωνος νόμων ἱερῶν ἀλληγορία
Provid I-II	Փիլոնի յաղագս յառաջախնամութեան/նախախնամութեան
	առ Աղերսանդրոս 11
	Φίλωνος περὶ προνοίας πρὸς ᾿Αλέξανδρον
Quaes Ex I-II	Փիլոնի այնցից որ յելան է խնդրոց եւ լուծմանց
	Φίλωνος τῶν ἐν ἐξόδῳ ζητημάτων καὶ λύσεων
Quaes Gen I-IV	Փիլոնի այնցից որ ի լինելութեանն խնդրոց եւ
	լուծմանց
	Φίλωνος τῶν ἐν γενέσει ζητημάτων καὶ λύσεων
Spec Leg I 79-161	Փիլոնի յաղագս քանանայից
	Φίλωνος περὶ τῶν ἱερέων
285-345	Փիլոնի յաղագս բագնին իրաց
	Φίλωνος περὶ τῶν τοῦ θυσιαστηρίου πραγμάτων
III 1-63	(see n. 10)
Vita Cont	Փիլոնի յաղագս վարուց կենաց տեսականի
	Φίλωνος περὶ βίου θεωρητικοῦ

The works extant only in Arm. may be remembered in pairs: the two
Quaestiones (Quaes Gen I-IV and Quaes Ex I-II) and the two treatises
ad Alexandrum (Provid I-II and Anim), as also the two fragments and the
two pseudo-Philonic homilies.

2. *Time and Place of Translation*

The establishment of the *terminus ad quem* for the Arm. translation
of Philo rests on the historian Elisaeus' use of the translation in the

---

10Under this title are included Spec Leg III 1-7, the page-long
fragment on the decad, Dec, and Spec Leg III 8-63.

11In some MSS this treatise is preceded by an anonymous scholium
on Philo, translated by Aucher, pp. vii-xi; cf. the traditions on
Philo cited by J. E. Bruns, "Philo Christianus: The Debris of a Legend,"
Harvard Theological Review, 66 (1973), 141-145.

latter part of the 6th century.[12]

The translation was among the first achievements of the so-called
Hellenizing School, which from *ca.* 570 - *ca.* 730 translated a number of
Gr. works.[13] These have been arranged into four groups representing
four successive periods of active translating: (1) the Τέχνη γραμματική
of Dionysius Thrax, a handbook on rhetoric belonging to Aphthonius, the
above cited Philonic and pseudo-Philonic works, Books IV-V of Irenaeus'
Adversus haereses and his *Demonstratio praedicationis evangelicae*--now
extant only in Arm., and the so-called Alexander Romance wrongly ascribed
in antiquity to Callisthenes; (2) the refutation of the creed of Chal-
cedon by the Monophysite Timothy of Alexandria (the "Cat"), the *Pro-
gymnasmata* of Aelius Theon, *Hermetica* ("To Asclepius"), Porphyry's
Εἰσαγωγή on Aristotle's *Categoriae*, the latter's *Categoriae* and *De interpretatione*, and Iamblichus' commentaries on Aristotle; (3) Plato's *Apologia*, *Euthyphro*, *Leges*, *Minos*, and *Timaeus*, the works of the Arm. Neopla-
tonist David, the collection and exposition of the stories ascribed to

---

[12]Elisaeus' use of Philo has been observed by Aucher, pp. iv-v; however, believing that Elisaeus' history of Vardan covers the Arm.-Persian
wars of 449-451 instead of those of 575, Aucher ascribed an early 5th
century date for the Arm. Philo. On the late date of Elisaeus, see B.
Kiwleserean, *Ełišē, k'nnakan owsoummasirowt'iwn* ("Elisaeus: A Critical
Study"), (Vienna, 1908) and the numerous articles by N. Akinean in
Handes amsorya (1931-1936) cited by Lewy, *The Pseudo-Philonic De Jona*,
pp. 9-16. Aucher was also misled by the spurius history of Movsēs
Xorenac'i, who purports to be an early 5th century writer and makes use
of Philo, p. iv; so was also Conybeare, "The Lost Works of Philo," *The
Academy*, 38 (1890), 32, *Philo about the Contemplative Life*, p. 115,
and in L. Cohn's "Prolegomena" to *Philonis Alexandrini opera*, I, lii-liii.
On the forgeries of Movsēs Xorenac'i, see Lewy, "The Date and Purpose of
Moses of Chorene's History," *Byzantion*, 11 (1936), 81-96.

[13]On the date of the founding of the Hellenizing School, see Lewy,
The Pseudo-Philonic De Jona, pp. 9-16, who follows N. Akinean, "Yownaban
dproc'ə," *Handes amsorya*, 46 (1932), 271-292; so also L. M. Froidevaux
in his edition of *Irénée de Lyon, Démonstration de la Prédication
apostolique*, Sources chrétiennes, LXII (Paris, 1959), 19-22 (like Philo,
Irenaeus was translated early by the same School); cf. K. Ter Mekerttschian and S. G. Wilson, "S. Irenaeus, Εἰς ἐπίδειξιν τοῦ ἀποστολικοῦ κηρύγματος, The Proof of the Apostolic Preaching with Seven Fragments: Armenian
Version Edited and Translated," *Patrologia Orientalis*, XII (Paris, 1919),
656, who suggest a *ca.* 612 date for the translation; H. Manandean,
Yownaban dproc'ə ew nra zargac'man šrǰannerə ("The Hellenizing School
and Its Development"), (Vienna, 1928), p. 107, dates the founding of the
School within a few years after the middle of the 6th century; elsewhere,
however, he has earlier dates, see pp. 223, 226.

Nonnus of Panopolis, the pseudo-Aristotelian *De mundo* and *De virtutibus et vitiis*, Eutyches' denunciation of the Nestorian doctrine of two persons or substances in Christ, and a number of anonymous works--possibly of Stoic origin; (4) the *Hexaemeron* of George of Pisidia, the *Phainomena* of Aratus, the *Historia ecclesiastica* of Socrates, Nemesius' *De natura hominis*, selections from Cyril of Alexandria, Gregory of Nyssa's *De hominis opificio*, and several mystic works attributed to Pseudo-Dionysius the Areopagite.[14]

The grouping is based primarily on the rendering of compound words in the Arm. As H. Manandean observes, there seems to be no serious effort in the earliest translations of the Hellenizing School to consistently render Gr. compounds with Arm. compounds. But in the later translations a mechanical imitation of Gr. compounds becomes more and more apparent, to the extent that newly compounded words become commonplace. In the latest translations of the School many of the new compounds stand out as *hapaxlegomena*.[15]

Unlike the biblical and other religious translations of the Golden Age (5th century), the predominantly philosophical translations of the Hellenizing School maintain the Gr. word order or syntax. The interlinear character of these translations is well explained by Lewy, who observes that these translations were primarily intended for the *trivium*--the three subjects taught in the schools of the late classical period as preliminary to biblical and theological studies: grammar in accordance with Dionysius Thrax, rhetoric, and introduction to dialectic, and that these translations facilitated the education of Arm. students seeking admission into the Byzantine schools in the metropolis of Gr. civilization of the time, where also the Hellenizing School must have been lo-

---

[14]Following S. Arevšatyan's ("Platoni erkeri Hayeren t'argmanowt'yan žamanakə" ["A propos de l'époque de la traduction en arménien des dialogues de Platon"], *Banber Matenadarani*, 10 [1971], 7-18) revision of Manandean's pioneering work: *Yownaban dproc'ə* ("The Hellenizing School"). For certain editions of the Arm. texts of these works, see Lewy, *The Pseudo-Philonic De Jona*, p. 11, n. 42, and pp. 16-18 of Arevšatyan's article.

[15]*Yownaban dproc'ə* ("The Hellenizing School"), pp. 86-255.

cated.[16] Such a thesis can be further substantiated by the fact that
the Arm. translations of voluminous writers like Plato, Aristotle, and
Philo are incomplete. No attempt was made by the Hellenizing School to
translate the complete works of such celebrities but only select works
which seem to have been used for tutorial purposes.

3. *Characteristics of the Translation*

The syntactical awkwardness encountered in the Arm. translation of
Philo is common to all the translations of the Hellenizing School. The
hardly intelligible syntax is due to the School's word for word rendering
which maintains the word order of the Gr.[17] This peculiarity was readily
recognized by Aucher,[18] Conybeare,[19] Lewy,[20] and Marcus,[21] all of whom
adopted some Gr. retranslation to resolve the ambiguities of the Arm.
text.

In his study of the Philonic works extant in both Gr. and Arm., Marcus
observes three classes of word renderings by the Arm. translator: (1) one
to one correspondences, (2) exact reproductions of compounds, and (3) one
word translating a number of Gr. synonyms or words of related but by no
means identical meaning.[22] Marcus, however, disregards three other

---

[16]*The Pseudo-Philonic De Jona*, pp. 15-16 (elaborating on Akinean,
"Yownaban dproc'ə" ["The Hellenizing School"], col. 285).

[17]For similar development in Syriac, see T. Nöldeke, *Kurzgefasste
syrische Grammatik* (Leipzig, 1898), p. ix; cf. the slavishly literal
and pedantic translations of the Heb. O.T. in the rival Gr. versions
and recensions of the LXX.

[18]p. ii.

[19]*Specimen lectionum armeniacarum, or a Review of the Fragments of
Philo Judaeus, as Newly Edited by J. R. Harris* (Oxford, 1889); "The
Lost Works of Philo," 32. Conybeare was the first to make use of the
Arm. version of Philo in textual criticism, see his edition of Vita
Cont: *Philo about the Contemplative Life*, especially p. 155.

[20]*The Pseudo-Philonic De Jona*, pp. 16-24.

[21]"An Armenian-Greek Index to Philo's *Quaestiones* and *De Vita Contem-
Plativa*," *Journal of the American Oriental Society*, 53 (1933), 251;
"The Armenian Translation of Philo's *Quaestiones in Genesim* (sic) *et
Exodum*," *Journal of Biblical Literature*, 49 (1930), 63-64.

[22]"An Armenian-Greek Index," p. 252.

classes of word renderings that emerge from lack of exact equivalency or import of meaning in the Arm. In such instances the translator was compelled to render a Gr. word with a number of Arm. words. Moreover, there are many instances where after once rendering a Gr. word by its Arm. equivalent, the translator elsewhere renders the same word by a number of words, often using *hendiadys* and sometimes *polysyndeton*. In like manner, one finds not only exact reproductions of compounds but also renditions of the same compounds in separate words.[23] The following examples, scores of which abound in any Philonic work extant in both languages, will suffice to show these additional classes of word renderings and the translator's treatment of words in general:[24]

αἰσχρός	Leg All I	ամաւթալի (61)
	Leg All II	ամաւթալի (32, 66)
		ամաւթանք (68, 70)
	Abr	ինչ համարեալ է ամաւթ (106)
	Dec	ամաւթալի (93, 115, 123)
		զարշելի (115)
	Spec Leg III	ամաւթալի գործ (24)
		ամաւթալի եւ անասուն (49)
		ամաւթ (51)
ἀκριβόω	Leg All I	ստուգել եւ ոշգրատել ինչ կարեմ (91)
	Abr	ոշմարտեալ ստուգեմ (2)
		ստուգեմ եւ ոշգրատել (167)
		ստուգեմ ոշմարտապէս (240)
	Dec	ոշգրատեմ (1, 18, 48)
		ստուգեմ (82)

---

[23]Consistantly exact reproductions of Gr. compounds characterize the latest translations of the Hellenizing School; see above, p. 8.

[24]References are limited to those works in Arm. the Gr. original of which is extant (excluding the fragments). Cf. B. Reynders, *Lexique comparé du texte grec et des versions latine, Arménienne et syriaque de l' "Adversus Haereses" de Saint Irénée, I: Introduction, index des mots grecs, arméniens et syriaques, II: Index des mots latins*, Corpus scriptorum christianorum orientalium, CXLI-CXLII, Subsidia, t. 5 et 6 (Louvain, 1954), and his *Vocabulaire de la "Démonstration" et des fragments de Saint Irénée* (Louvain, 1958). Note that Irenaeus was translated by the translator(s) of Philo, see above, p. 7 and n. 13.

Spec Leg I   * օչզրտագոյն ասեմ* (105)

             հաւաստի զեկուցանեմ (110)

Vita Cont   *ստուգեալ ասեմ* (14)

             օչզրիտ ստուգութեամբ կագմեմ ձեռագործ (49)

             կագմեմ (51)

ἀπόδοσις   Leg All I   *բացատրութիւն* (16)

             հատուցումն (99)

         Leg All II   *պատճառ* (14)

         Abr   *բացատրութիւն հատուցման* (88, 147)

             բացատրութեան հատուցումն (119)

             ենթադրութիւն (200)

ἀφθονία   Leg All I   *առատութիւն անչարակնութեան* (54)

         Abr   *առատութիւնք բերրցն* (134)

             առատութիւն մեծութեան (208)

         Dec   *առատութիւն եւ առաւելութիւն* (16)

             առատութիւն (17)

             աննախանձ առատութիւն (117)

         Spec Leg I   *բագմաշահութիւն* (141)

κυβόηλευω Leg All II   *խոտեմ եւ անգոսնեմ* (57)

         Dec   *նշաւակ խարդախութեամբ ապարտեմ* (3)

         Spec Leg I   *խոտեմ եւ անարգեմ* (326)

         Spec Leg III   *խանգարեմ* (11)

         Vita Cont   *խոտան եւ արհամարհ եւ անարգ գործ գործեմ*

                                                         (42)

κρατέω   Leg All I   *իշխեմ եւ տիրեմ եւ զաւրանամ* (73)

             ունիմ եւ ըմբռնեմ միշտ (100)

             բուռն հարեմ եւ հաստատահարեմ (106)

             իշխեմ եւ հաստատահարեմ (106)

         Leg All II   *ըմբռնեմ եւ տիրանամ* (29)

             ըմբռնեմ եւ տիրեմ (70)

             ըմբռնեալ իշխեմ (70)

         Abr[25]   *իշխեմ* (121)

             հասեալ կամ (151)

             հասեալ ըմբռնեմ (182)

---

[25]As Mangey observes, the word needs to be emended to κεχρῆσθαι in Abr 188.

		ունիմ եւ իշխեմ (220)
		բուռն ʃարեալ ունիմ զիշխանութիւն (243)
	Spec Leg I	յաղթեմ (312)
		ըմբռնեմ (343)
	Vita Cont	ըմբռնեմ (2)
συνέχω	Abr	միաբան եւ ամբողջ եւ ի միասին պաʃեմ (74)
	Spec Leg I	ամբողջ պաʃեմ (289)
	Vita Cont	ըմբռնեմ ի միաբանութիւն (63)
συνόλως	Leg All I[26]	միանգամայն (31)
	Leg All II	ընաւ իսկ ամենեւին (19, 71)
		ընաւ ամենեւին (46)
	Abr	միանգամայն (102, 141, 267)
	Dec	միանգամայն եւ առ ʃասարակ (21)
		առ ʃասարակ միանգամայն (51)
		միանգամայն (156, 171)
	Spec Leg I	միանգամայն (100)

No doubt the Arm. translator made inconsistent use of a lexicon, for
seldom does his choice of words fall outside a lexical pattern or appear
to have been based on recall. Certainly his choice of words was not always governed by context. He randomly chose words from among the alternate meanings provided in his manual. Sometimes he picked two or three
synonyms which he used either conjunctively or paraphrastically. When
picking two verbs, he often changed the first into a participle (but
no methodical rendering of participles is to be found in his translation--
even if one were to allow for the confusion resulting from the scribal
tendency in the Middle Ages toward abbreviating the participial ending
−եալ to −ել , the infinitival ending). His frequent use of auxiliary
verbs is all too obvious. But seldom does his "paraphrasing" obliterate
the lexical terms, which can be identified. Identifying the Gr. lexical
terms, however, could be difficult, since one Arm. word sometimes translates a number of Gr. synonyms or words of related but by no means identical meaning; e.g., ամաւթ, an equivalent of αἰσχρός, the first word
listed above, could also be a rendition of αἰδοῖον (Spec Leg I 83), and
միանգամայն , an equivalent of συνόλως, the last word listed above,

---

[26]Arm. omits the word in Leg All I 1.

could also be a rendition of ἀθρόος (Abr 42, 138), ἅμα (176), καθάπαξ
(36, 50), ὁμοῦ (45, 50, 136, 161, 246, 267), συλλήβδην (49), and ὅλως
(Leg All I 31).

Certain syntactical peculiarites are easily discernible in the Arm.
translation: (1) the placing of the verb after the subject instead of
before--as is customarily done in Classical Arm.; thus, one reads անա
սունն կենդանիքդ բանի բաժին ունին (Anim 1) instead of the more
familiar բանի բաժին ունին անասունն կենդանիքդ (the former is a
κατὰ λέξιν rendering of τὰ ἄλογα ζῷα λόγου μετέχουσι); and (2) the
placing of the attributive before the substantive instead of after; thus,
one reads զերիկեան բանն (Anim 1) instead of զբանն երիկեան (the
former is a κατὰ λέξιν rendering of τοὺς ἐχθὲς λόγους). Note the
attachment of the accusative prefix to a word with a genitive suffix,
an awkwardness resulting from the reversed position of the attributive;
e.g., զբանիցն խրախմանութիւն (Anim 3) instead of զխրախմանութիւն
բանիցն, զայլոցն լուրս (Anim 7) instead of զլուրս այլոցն, զկա
նանց տկարութիւն (Anim 11) instead of զտկարութիւն կանանց, etc.[27]

Awareness of the Gr. syntax underlying the Arm. translation of Philo's
works faded away in subsequent centuries. What most impressed the Arm.
scholiasts of the late Middle Ages was the obscurity of the language and
the challenge to convey its meaning. They accepted this as being the
fault of the readers, not of the author or the translator. For the
scholiasts Philo should be tackled only by men of profound ability.
While the would-be-interpreters were apt to quote Philo more than interpret him, some were eager to express their personal views rather than
those of Philo, others still were content with short catenae of Philonic
quotations.[28] As for the medieval and later scribes, they must have

---

[27]Cf. իմանալ զբանս Հանձարոյ (Golden Age), զՀանձարոյ բանս
իմանալ (Hellenizing School). For other observations, see Lewy, *The
Pseudo-Philonic De Jona*, pp. 16-19; Froidevaux, *Irénée de Lyon*, pp. 14-
19.

[28]Other than the occasional scholia in the MSS containing the works
of Philo (see below, pp. 15-21), scholia and catenae of Philonic quotations are found in about fifty MSS. About forty of these are at the
Maštoc' Matenadaran, Institute of Ancient Manuscripts in Erevan (Nos.
59, 99, 266, 437, 529, 538, 598, 631, 1053, 1138, 1426, 1480, 1701, 1843,

copied the seemingly obscure works of Philo with little or no understanding.[29]

4. *The Manuscripts*

There are two incomplete lists of Arm. MSS containing the works of
Philo: Lewy's synopsis in *The Pseudo-Philonic De Jona*[30] and R. P. Casey's
compilation in Goodhart's *Bibliography*.[31] Lewy selects for special consideration six MSS representing the best witnesses to the text of the
De Jona and enumerates another sixteen rather abruptly, without describing
their contents.[32] The list in Goodhart's *Bibliography* utilizes that of
Lewy, but abounds with errors wherever it attempts to supplement the latter.
Its coverage of the contents of the MSS is unreliable. So also is its
enumeration, which, incorporating both old and new numbers, sometimes given
separately, results in double-listing of some MSS.[33]

1897, 1919, 1931, 1960, 2059, 2151, 2369, 2679, 3197, 3207, 3276, 3506,
3860, 3937, 4150, 4199, 4381, 4612, 4765, 5254, 6036, 6353, 6354, and
10105). See J. Dashian's excursus on the scholia in *Katalog der armenischen Handschriften in der Mechitaristen-Bibliothek zu Wien* (Vienna,
1895), pp. 222-224. On the identity of certain scholiasts and their
place in the history of Arm. scholarship on Philo, see G. Grigoryan,
"P'ilon Alek'sandrac'ow ašxatowt'iownneri Hay meknowt'iownnerə" ("The
Armenian Scholia on the Works of Philo of Alexandria"), *Banber Matenadarani*, 5 (1960), 95-115.

[29]Lewy rightly observes, "Armenians who have not studied this subject
are rarely able to understand the syntactical forms of the Armenian
Philo." *The Pseudo-Philonic De Jona*, p. 19, n. 64.

[30]Pp. 4-8.

[31]Nos. 338-376, especially 338-363; 364-376 cover the Arm. scholia on
Philo (pp. 182-185).

[32]Because of his untimely death in 1945, the plan to provide a detailed
description of the then known twenty-two MSS was never realized; see *The
Pseudo-Philonic De Jona*, p. 4, n. 1. He was sponsored by the Preussische
Akademie der Wissenschaften to study these MSS and to prepare a critical
edition of the Arm. Philo.

[33]The *Bibliography* inadvertently omits Lewy's Etchmiatzin codex 513
(Erevan, Matenadaran 5239; see *The Pseudo-Philonic De Jona*, p. 6, n. 22).
Even the only complete coverage of the contents in No. 338 (Erevan,
Matenadaran 1500) is misleading; fols. 370-388 contain not only Provid

With the publication in 1965 and 1970 of the two-volume catalogue of
MSS in the Maštoc' Matenadaran, Erevan, Arm. SSR,[34] where more than half
of the Arm. Philo MSS are kept, several numbers had to be added to the
above cited lists--but none of greater textual significance than those
already known.[35]

Twenty-eight MSS are enumerated in the expanded list below,[36] with some
general information on those penned before the 17th century and with somewhat detailed description of the principal ones.

Erevan, Matenadaran 2101, dated A.D. 1223.

A miscellany in *bolorgir* or "round hand," penned by Step'anos (sic)
Aįt'amarc'i at the convent of Xoranašat. It consists of 205 paper folios,

---

I-II but also Anim--so also No. 339 (Erevan, Matenadaran 2101). No. 374
(Erevan, Matenadaran 2595) is a Philonic corpus and not a scholium.
No. 341 = Erevan, Matenadaran 2102; No. 343 cannot be accounted for;
No. 349 = Bzommar, Arm. Catholic Cloister 121; No. 352 = Venice, Mechitarist 1334; No. 354 should read: "a copy of No. 353"; No. 361 cannot
be accounted for; No. 375 = Bzommar, Arm. Catholic Cloister 194. Nos.
355 and 356 refer to the same MS, Istanbul, Arm. Patriarchate 69, dated
A.D. 1791; Nos. 357 and 359 refer to the same MS, Istanbul, Arm. Patriarchate 114, dated A.D. 1668 (Lewy has 18th century); Nos. 365 and 373
refer to the same MS, Erevan, Matenadaran 1897, dated 12th century
(scholium).

[34]O. Eganyan *et al.*, *C'owc'ak jeřagrac' Maštoc'i anvan matenadarani*
("Catalogue of the Manuscripts of Maštoc' Matenadaran"), I-II (Erevan,
1965, 1970). Unfortunately, the catalogue's coverage of the contents of
the MSS is inadequate. The hard work on a broader catalogue continues.

[35]In addition to the cited catalogue of the MSS in Erevan, see N.
Bogharian, *Grand Catalogue of St. James Manuscripts*, I-VII (Jerusalem,
1966-1974); M. Keschischian, *Katalog der armenischen Handschriften in
der Bibliothek des Klosters Bzommar* (Vienna, 1964); F. Macler, *Catalogue
des manuscrits arméniens et géorgiens de la Bibliothèque Nationale*
(Paris, 1908); S. Tēr-Awetisian, *Katalog der armenischen Handschriften
in der Bibliothek des Klosters in Neu-Djoulfa*, I-II (Vienna, 1970-1972).
B. Sarghissian, *Grande catalogue des manuscrits arméniens de la bibliothèque des PP. Mekhitaristes de Saint-Lazare*, I-II (Venice, 1914-1924);
Vol. III ed. by B. Sarghissian and G. Sarksian (Venice, 1966), goes as
far as No. 456 and does not cover the Philo MSS. Except for Macler's
compilation, all catalogues are in Arm., bearing bilingual titles.

[36]Erevan, Matenadaran 2059, dated 1599 and listed as *P'ilon Ebrayec'i,
Matenagrowt'iwnk'*, is not a Philonic corpus, but a commentary on selections from every work in the corpus.

measuring 17.2x12.5 cms. The pages are written in one column of 22-24
lines. The colophons of the scribe are found on fols. 85v, 204r.

It contains: Provid I-II (86r-156r), Anim (157v-190r).

Erevan, Matenadaran 5239, dated A.D. 1274.

A Philonic corpus in *bolorgir,* penned by Matt'ēos at the convent of
Aṙajnak'ar and owned by Esayi Vardapet. It consists of 231 paper folios,
measuring 17x13 cms. The pages are written in one column of 23 lines.
The colophon of the scribe is found on fol. 93. A later colophon is
found on fol. 85v (A.D. 1904). It is mutilated at the beginning and at
the end.

It contains: Quaes Gen beginning at IV 35 (1r-93r; has a missing
folio preceding fol. 7; fols. 86-92 have several missing and transposed
folios), Quaes Ex I-II (94r-164r), Spec Leg I 79-161, 285-345 (164r-194v),
Dec, Spec Leg III 8-? (194v-231v).

Lewy has the once new but long abandoned No. 513.

Erevan, Matenadaran 3932, dated A.D. 1275.

A Philonic corpus in *bolorgir,* penned by Barseł at the convent of
Ełiazar and owned by Nersēs Vardapet. It consists of 329 paper folios,
measuring 17.5x12.5 cms. The pages are written in one column of 22-23
lines. The colophons of the scribe are found on fols. 127v, 329v. Later
colophons are found on fols. 128v, 329v (14th century), 128v (A.D. 1867).

It contains: Quaes Gen I-III (1r-128r), Quaes Ex I-II (129r-193r),
etc. Reliance on a fragmentary microfilm copy would have led to over-
sights in citing the rest of the contents of this MS, which seem to be
identical with those of Erevan, Matenadaran 3935. Several striking
similarities between these two MSS may be noted.

Erevan, Matenadaran 1500, dated A.D. 1282.

A miscellany in *bolorgir,* penned by the scholiast Mxit'ar Ayrevanc'i
at the convent of Gełard and illuminated by one Sargis Grič'. It consists of 1189 paper folios, measuring 35.7x26.5 cms., and is written in
two columns with 52-59 lines to the page. The colophons of the scribe
are found on fols. 40r, 270v, 879r, 1041r, etc. The colophon of the
illuminator is found on fol. 723v. Later colophons are found on fols.
29v (A.D. 1604, 1690-1692), 393r (A.D. 1691), 499r (A.D. 1697). The
Philonic corpus covers fols. 370r-507v, with several folios missing

between 381-382.

It contains: Provid I-II 93 "they are in need" (370r-381v), Anim beginning at 14 "[human] voice" (382r-388r), Abr (388r-392v), Vita Cont (392v-401v), Leg All I-II (401v-412v), Quaes Gen I-IV (412v-466r), Quaes Ex I-II (466r-481r), Spec Leg I 79-161, 285-345 (481r-487v), III 1-7, Dec, Spec Leg III 8-63 (487v-497r), *De Sampsone* (497r-501r), *De Jona* (501r-506v), Deo (507r-v).

This is Lewy's A.

Venice, Mechitarist 1040, dated A.D. 1296.

A Philonic corpus in *bolorgir*, penned by the royal scribe Vasil by order of the Reubenite King of Arm. Cilicia, Het'owm II (A.D. 1289-1293, 1295-1297), who, wearied by the political issues of his day, abdicated in favor of his brother T'oros (A.D. 1293-1295) and withdrew to a monastery.   Political pressures and family intrigues brought him back to power only to retire again and to return to monastic life.   The MS consists of 318 parchment folios (636 pages), measuring 29.3 x 21 cms., and is written in two columns with 31 lines to the page.   The colophons of the scribe are found on pp. 504, 558, 610, 633-636.   There are several missing folios preceding pp. 1, 167, 219, 221, 223.

It contains: Quaes Gen beginning at I 49 "the senses of man" – IV 75 "authority and lordship over it, for ," 122 "not that he was born in the old age" – 242 "he passes over" (pp. 1-218), Quaes Ex beginning at I 10 "He thought it just and fitting" – 13, 18 "should [not] be defiled by any [thing]" – 21, resuming at II 44 "it is [ful]ly understood" (pp. 219-255), Spec Leg I 79-161, 285-345 (pp. 255-286), III 1-7, Dec, Spec Leg III 8-63 (pp. 286-334), *De Sampsone* (pp. 335-356), *De Jona* (pp. 355-382), Deo (pp. 382-386), scholia (pp. 386-389), Provid I-II (pp. 389-467), Anim (pp. 467-504), Abr (pp. 504-558), Leg All I-II (pp. 559-610), Vita Cont (pp. 611-632).

This is Aucher's A and Lewy's D.

Jerusalem, Arm. Patriarchate 333, dated A.D. 1298.

A Philonic corpus in *bolorgir*, penned and owned by Vahram Sarkawag of the twin churches of Astwacacin and Sowrb P'rkič' in the *Anapat of Armēn* (sic).   It consists of 391 paper folios measuring 25x19 cms., and is written in two columns with 33 lines to the page.   A lengthy colophon by the scribe and owner is found on fols. 390v-391v.   Later colophons are found on fols. 391v (15th-16th centuries), 317v (A.D. 1641).

It contains: Quaes Gen I-IV (1r-139v), Quaes Ex I-II (139v-179r), Spec Leg I 79-161, 285-345 (179r-196v), III 1-7, Dec, Spec Leg III 8-63 (196v-223r), *De Sampsone* (223r-235r), *De Jona* (235r-249r), Deo (249r-252r), Provid I-II (252r-295v), Anim (295v-316v), scholia (316v-317v), Abr (318r-347r), Leg All I-II (347r-374v), Vita Cont (375r-386v).

This is Aucher's B (*Paralipomena Armena*, p. 622; but on p. ii he cites the copy of this MS, Istanbul, Arm. Patriarchate 69, as his codex B) and Lewy's E.

Erevan, Matenadaran 3935, 13th Century.

A Philonic corpus in *bolorgir*, penned by Barseł and owned by Nersēs. It consists of 306 paper folios, measuring 17x12.5 cms. The pages are written in one column of 26 lines. The colophons of the scribe are found on fols. 120v, 181v, 277r. It is worn and illegible at the end.

It contains: Quaes Gen I-III (3r-120v), Quaes Ex I-II (120v-181v), Spec Leg I 79-161, 285-345 (182v-219r), III 1-7, Dec, Spec Leg III 8-63 (219r-255v), *De Sampsone* (255v-277r), *De Jona* (277r-302v), Deo (302v-306v).

Paris, Bibliothèque Nationale, Arm. 303, 13th Century.

An eleven-folio fragment of Provid I written in *bolorgir*.

Bzommar, Arm. Catholic Cloister 121, 13th-14th Century.

A Philonic corpus in *bolorgir*. It consists of 247 paper folios, measuring 17x12.5 cms. The first pages are written in two columns of 38 lines (fols. 1r-22v), the remaining pages in one column of 28-30 lines. It is mutilated at the beginning and, perhaps, at the end.

It contains: Provid beginning at I 84 "from meats" (1r-23v), Anim (23v-47r), Abr (47r-87v), Quaes Ex I-II (87v-141v), Spec Leg I 79-161, 285-345 (141v-167r), III 1-7, Dec, Spec Leg III 8-63 (167r-207r), *De Sampsone* (207r-224r), *De Jona* (224r-244r), Deo (244r-247v).

Erevan, Matenadaran 2104, dated A.D. 1318.

A Philonic corpus in *bolorgir*, penned by Mxit'ar and owned by Yovhannēs. It consists of 258 paper folios, measuring 17.2x12 cms. The pages are written in one column of 26-27 lines. The colophon of the scribe is found on fol. 164v. It is mutilated and illegible at the beginning and at the end.

It contains: Quaes Gen beginning at II 61 "overwhelmed by its prodi-
gal way of life"--III (2r-36r), Quaes Ex I-II (36r-84r), Spec Leg I 79-
161, 285-345 (84r-101v), III 1-7, Dec, Spec Leg III 8-63 (101v-132v),
De Sampsone (132v-144v), *De Jona* (144v-162r), Deo (162r-164v), Provid
beginning at I 5 "is in its nature"--II (165r-214v), Anim (214v-237r),
Leg All I-II 64 "has no part" (237v-257v).

This is Lewy's C (its dated colophon was not known to him).

Erevan, Matenadaran 2100, dated A.D. 1325.

A Philonic corpus in *bolorgir*, penned by Karapet Erec' and owned by
Kirakos Vardapet. It consists of 336 paper folios, measuring 16.6x12
cms. The pages are written in one column of 25-26 lines. The colophons
of the scribe are found on fols. 112v, 336r. Later colophons are found on
fols. 313r (14th-15th centuries), 2r (17th-18th centuries). It is slightly
mutilated at the beginning.

It contains: Quaes Gen beginning at IV 1 "the excellent and powerful
and sovereign ruler" (3r-112v), Leg All I-II (113r-159r), Provid I-II
(160v-230v), Anim (231r-264v), Abr (264v-313r), Vita Cont (313r-331r),
scholia (331v-335r).

Erevan, Matenadaran 2057, dated A.D. 1328.

A Philonic corpus in *bolorgir*, penned by Aṙak'el Kowsakron at the convent of Jermaɫbiwr and owned by Yakob Krakac'i. It consists of 309 paper
folios, measuring 24x16 cms. The pages are written in one column of 39-41
lines. The colophons of the owner are found on fols. 150r, 180v, 309r-v.
Later colophons are found on fol. 309v (A.D. 1409, 1881), where also a
seal impression is found with the letter   (Barseɫ Vardapet).

It contains: Provid I-II (1r-35r), Anim (35r-50r), Abr (50r-70r),
Quaes Ex I-II (70r-96v), Spec Leg I 79-161, 285-345 (96v-109r), III 1-7,
Dec, Spec Leg III 8-63 (109r-129r), *De Sampsone* (129r-137v), *De Jona*
(137v-148r), Deo (148r-150r), Leg All I-II (150r-172v), Vita Cont (172v-
180v), Quaes Gen I-IV (181r-309r).

This is Lewy's B (inadvertently, he has No. 2507).

Erevan, Matenadaran 2102, dated A.D. 1342.

A Philonic corpus in *bolorgir*, penned by T'or[os] and owned by Kirakos
Rapownapet and Yovhannēs Vardapet. It consists of 350 paper folios,
measuring 16x12 cms. The pages are written in one column of 24-25 lines.

The colophons of the scribe are found on fols. 317r, 341v, 346v-347r
(margin), and of the recepient and his intern on fol. 154r. It is slightly mutilated at the beginning and more so at the end.

It contains: Quaes Gen beginning at I 3-III (3r-154r), Quaes Ex I-II
(156r-232r), Spec Leg I 79-161, 285-345 (232r-264r), III 1-7, Dec, Spec
Leg III 8-63 (264r-317r), *De Sampsone* (318r-341v), *De Jona* 1-47 "as the
Scripture says," 78-103 "on an elevated place."

This is Lewy's F and a companion codex to Erevan, Matenadaran 2100.

Erevan, Matenadaran 2058, 14th Century.

A Philonic MS in *bolorgir*, owned by Kostandin K'ahanay. It consists
of 59 paper folios, measuring 24.7x17 cms. The pages are written in one
column of 29 lines. The colophon of the owner is found on fol. 59v.

It contains: Quaes Gen IV (2r-59v).

Erevan, Matenadaran 4275, 14th Century.

A Philonic corpus in *bolorgir*, penned by Karapet, Step'annos, and
Vardan. It consists of 329 paper folios, measuring 17x12 cms. The pages
are written in one column of 22 lines. The colophons of the scribes are
found on fols. 205v, 265v, 281v. It is mutilated at the beginning and at
the end.

It contains: Provid beginning at I 61-II (1v-56r), Anim (56r-89v),
Abr (90r-137v), Leg All I-II (137v-186r), Vita Cont (186r-205v), Quaes
Gen I-III 55 (206r-329v).

Venice, Mechitarist 1334, 14th Century.

A Philonic corpus written in two parts, on paper. The first part is
written in *bolorgir*, penned by Vardan, a pupil of the scholiast
Yo[v]hannēs (elsewhere spelled Yohanēs) [Erznkac'i] (*ca.* A.D. 1250-1326);
the second part is in *nodrgir* and belongs to an unknown hand of the 18th
century. The colophons of the scribe follow Vita Cont and Leg All II.

This is Aucher's C, *Paralipomena Armena*, pp. ii, 622. The same MS
was collated by Conybeare in his edition of that part of the Arm. corpus
of Philo the Gr. of which is extant, *P'iloni Hebrayec'woy cark'*, p. 288,
as well as in his edition of Vita Cont, *Philo about the Contemplative
Life*, p. 155. In these two editions it carries the sigla B and D,
respectively.

Jerusalem, Arm. Patriarchate 1331, 14th Century.

A Philonic corpus in *bolorgir*, penned by Nersēs and owned by Vardan Vardapet. It consists of 409 paper folios, measuring 16.5x12 cms. The pages are written in one column of 26 lines. The colophons of the scribe are found on fols. 145r, 191v. Later colophons are found on fols. 409v (A.D. 1610), 409r (A.D. 1891). It is mutilated at the end.

It contains: Provid I-II (1r-68v; a missing folio substituted, a number of folios transposed), Anim (68v-98v), Abr (98v-145r), Leg All I-II (145r-191v), Vita Cont (191v-210r), Quaes Gen I-IV 156 "to know everything" (212r-408v).

Ispahan, Arm. Monastery 186, 16th Century.

A fragmentary text of Leg All I-II.

The remaining ten MSS listed below have little or no independent value for text critical purposes since, for the most part, they depend on surviving exemplars. The first five of these belong to the 17th century, the following four to the late 18th century, and the last to the early 19th century.

Erevan, Matenadaran 1672 (contains Provid I-II, Anim)

Erevan, Matenadaran 2103 (contains Leg All I-II)

Erevan, Matenadaran 2379 (contains Provid I-II, Anim)

Erevan, Matenadaran 2056 (copy of Erevan, Matenadaran 1500)

Istanbul, Arm. Patriarchate 114 (Philonic corpus)

Jerusalem, Arm. Patriarchate 157 (copy of Jerusalem, Arm. Patriarchate 333)

Erevan, Matenadaran 2595 (copy of Jerusalem, Arm. Patriarchate 333)

Istanbul, Arm. Patriarchate 69 (copy of Jerusalem, Arm. Patriarchate 333)

Paris, Bibliothèque Nationale Arm. 159 (copy of Venice, Mechitarist 1040)

Venice, Mechitarist 253 (partial copy of Istanbul, Arm. Patriarchate 69)

Special consideration should be given to the principal MSS selected after collating the 13th-14th century witnesses to the text of Anim as a sample of the text of Philo (collated against the text of Venice, Mechitarist 1040). The authorities thus represented as preserving a more valuable text contain the whole of the Arm. version of Philo and date from before the middle of the 14th century.[37] These findings are

---

[37]Jerusalem, Arm. Patriarchate 1331 is the chief example of a late 14th century adulterated text. It is either of vulgar derivation or substantially independent. On the one hand, it abounds with conflate readings and, on the other hand, it shows signs of careless copying through its excessive orthographic errors.

equally apparent in Lewy's textual criticism of the pseudo-Philonic *De Jona*. The six MSS he selects for establishing the text of the *De Jona* date from before the middle of the 14th century and are basically the same ones selected for establishing the text of Anim.[38] Of these, four (A--Erevan, Matenadaran 1500, B--Erevan, Matenadaran 2057, D--Venice, Mechitarist 1040, E--Jerusalem, Arm. Patriarchate 333) contain the whole of the Arm. version of Philo, one (C--Erevan, Matenadaran 2104), grossly mutilated at the beginning and at the end, may not have lacked a treatise or a book other than Quaes Gen IV, and the other (F) is a complete Philonic corpus in two codices: Erevan, Matenadaran 2100--when considering the text of Anim; Erevan, Matenadaran 2102--when considering the text of the *De Jona*.

Considering the possible insufficiency of the samples (the texts of Anim and of the *De Jona*) and the desire to extend the collation to the whole corpus, including those works the Gr. of which is extant, the question of MS relations cannot be discussed in detail. However, certain observations leading to some tentative conclusions could be made. As for Lewy's pronouncements, although they are based primarily on the text of the pseudo-Philonic *De Jona*,[39] they seem to apply to the entire text of the Arm. Philo, of which corpus and textual tradition the *De Jona* is an integral part.

Traces of grammatical reworking in the earliest witnesses suggest that the archetype must have been a derived representative of some self-conscious recensional activity and that it had undergone orthographical changes due to Mediaeval Arm. phonology and morphology. Lewy observes: "All the manuscripts are proved by their common omissions and mistakes to be derived from a single archetype which had already suffered considerable textual corruption."[40] While there can be no doubt about the derivation of the surviving witnesses from a single archetype, Lewy's qualifying statement needs some modification. The grammatical reworking and orthographical changes spoken of above indicate neither "considerable

---

[38]Lewy, *The Pseudo-Philonic De Jona*, pp. 4-8.

[39]*Ibid.*

[40]*Ibid.* p. 7.

textual corruption" nor extensive alterations of the text as might be
implied. Moreover, he sees the "common omissions and mistakes" in the
light of his Gr. reconstruction of the *De Jona*. His overly critical observations may be balanced by those of Conybeare in the critical edition
of Vita Cont[41] and by further comparisons between the Gr. and Arm. texts
of other Philonic works extant in both languages.

The textual relationships obtaining between the six principal MSS
indicate a split into two recensions: α and β. The sigla assigned to
these MSS in the following description of their stemmatic relationship
have been adopted from Lewy, who rightly remarks: "The distinction between these two recensions should not be pressed too strongly, for though
as a rule the readings of β are preferable to those of α, no fixed rule
can be laid down."[42]

Recension α (ABC)

The representatives of this recension reveal further stages of non-
extensive editorial activity, seen mostly in A. Consequently, they are
to be regarded as somewhat inferior to those which have not undergone
later reworking. In spite of its inferiority, however, α preserves--now
and then--readings superior to those of β.

A--Erevan, Matenadaran 1500 (dated A.D. 1282).

The unsupported readings of A usually represent recensional activity
rather than haphazard variation due to careless copying--even though, as
Lewy observes, "it is the most careless of all the MSS of Philo."[43] Its
learned scribe, the famed scholiast Mxit'ar, attempted to make certain
problematic readings intelligible, hence its unreliability. According
to Lewy, "his text can be regarded as a would-be scholarly recension."[44]
Nonetheless, behind this recensional activity lies a valuable text

---

[41]*Philo about the Contemplative life, passim.*

[42]*The Pseudo-Philonic De Jona*, p. 7.

[43]*Ibid.*

[44]*Ibid.* This is why critics tend to prefer the work of a naive to
that of a sophisticated scribe. While Mxit'ar transmits valuable texts
of a number of biblical and other religious translations of the Golden
Age (5th century), his texts of Philo and other translations of the
Hellinizing School show signs of grammatical reworking.

attested by the readings of B and occasional agreements with E against D.
The often illegible handwriting may be resolved by a cautious reading of
Erevan, Matenadaran 2056, a 17th century copy of this MS.

B--Erevan, Matenadaran 2057 (dated A.D. 1328).

This MS is an excellent representative of the main body of MSS of α
and perhaps superior to the preceding representative. It descends from
a MS like the one which possibly lies behind A; however, it contains a
few readings, often in the margin, indicating a measure of reliance on β.
It thus stands somewhere at the beginning of a textual development characterized by correcting the readings of α by β.

C--Erevan, Matenadaran 2104 (dated A.D. 1318).

As between the two principal MSS, A and D, C shows, as usual, little
preference. It agrees first with one and then with the other, or endeavors to conflate the readings of the two. Nonetheless, the readings
of C contribute to the improvement of the text of the α recension.

Recension β (DEF)

The representatives of this recension derive from a textual tradition
characterized by marginal emendations which survive as such in MSS D, E,
and others. The recension encompasses the great majority of the extant
MSS, including the oldest, Erevan, Matenadaran 2101, dated A.D. 1223.
These are basically void of later reworking and whether they can properly be classified as a recension is therefore open to question. Several of the marginal emendations of β appear as conflate readings in α,
and thus seem to antedate the recensional activity reflected in α.

D--Venice, Mechitarist 1040 (dated A.D. 1296).

No doubt this MS has the best preserved text of the Arm. Philo. The
following is Lewy's testimony: "This is the main representative of β
and is the best manuscript of Philo, not only for its beauty but also
for the value of its text." He goes on to warn against using it un-
critically, since "the superiority of β to α is relatively small."[45]
It serves as the base text for Aucher's edition.

E--Jerusalem, Arm. Patriarchate 333 (dated 1298).

This MS was copied from an exemplar closely resembling that of D.

---

[45]*Ibid.*, p. 8.

Its scribe, like the scribe of D, reproduced the marginal readings of his model quite faithfully (Lewy thinks the scribe emended the text, generally in the margin, from a MS like D).[46] It is most useful in checking the peculiar errors of D, especially in instances where it agrees with A against D. Such instances, however, are extremely rare.

F--Erevan, Matenadaran 2100 (dated A.D. 1325).

Erevan, Matenadaran 2102 (dated A.D. 1342).

This corpus consists of two parts (each containing the works not included in the other) and was penned for the same recipient, known as Kirakos Vardapet in A.D. 1325 and Kirakos Rabownapet (Archimandrite), with an associate named Yovhannēs Vardapet, in A.D. 1342.[47] Lewy assigns the siglum F only to the second of these two codices which contains the *De Jona*. Although he fails to see the relation between the two codices, he keenly observes in a comment on the second: "formerly a complete Philonic corpus," and adds: "this often belongs to α, though not so frequently as C does to β."[48] The same could be said of the first part.

B. THE TREATISE *DE ANIMALIBUS*

1. *The Speakers*

While there can be no doubt about the identity of Philo[49] and his apostate nephew, Alexander, better known as Tiberius Julius Alexander,[50]

---

[46]*Ibid.*

[47]According to Goodhart's *Bibliography*, No. 341, the two parts were penned by the same hand.

[48]*The Pseudo-Philonic De Jona*, pp. 5, 8.

[49]The few known facts about Philo's personal life have so often been repeated in encyclopedic articles that it would be superfluous to repeat them here.

[50]He was born probably early in the reign of Tiberius (*ca.* A.D. 15), after whom he was named. He seems to have taken part in the Alexandrian Jewish embassy to Gaius in A.D. 39/40 (§ 54). Upon the accession of Claudius in A.D. 41, and possibly due to the mediation of Agrippa I, he entered Roman service and was known as Epistrategos of the Thebaid by A.D. 42. He became Procurator of Judaea under Claudius (*ca.* A.D. 46-48) and Prefect of Egypt under Nero (A.D. 66-70). He secured Alexandria for Vespasian and became Chief of Staff under Titus during the siege

the identity of Philo's interlocutor, Lysimachus, has been confused. The confusion arises from certain parts of the introductory and transitory dialogues between Philo (§§ 1, 75) and Lysimachus (§ 72), where both of them refer to Alexander as "our nephew." This common reference to Alexander by Philo and Lysimachus led to an erroneous identification of Lysimachus as a brother of Philo, taking him to be either Alexander the Alabarch, Philo's notoriously wealthy brother and father of Tiberius Julius Alexander,[51] or another younger brother.[52]

Such an identification of Lysimachus distorts his true identity clearly stated in § 2, where, speaking of Alexander, he says: "He is my uncle [lit., mother's brother] and my father-in-law as well. As you are not un-

---

of Jerusalem (A.D. 69-70). There are a number of good studies on Alexander that deal with the passages in Josephus, the Egyptian inscriptions and papyri, and the Roman authorities. See [A.] Stein, "Julius, 59," *PRE*, 19 (1918), cols. 153-157 and *Die Präfekten von Ägypten in der römischen Kaiserzeit* (Bern, 1950), pp. 37-38; A. Lepape, "Tiberius Iulius Alexander, Préfet d'Alexandrie et d'Egypte," *Bulletin de la Société Royale d'Archeologie d'Alexandrie*, 8 (1934), 331-341; E. R. Goodenough, *The Politics of Philo Judaeus: Practice and Theory* (New Haven, 1938), pp. 64-66; E. G. Turner, "Tiberius Iulius Alexander," *The Journal of Roman Studies*, 44 (1954), 54-64 (excellent discussion); V. Burr, *Tiberius Iulius Alexander*, Antiquitas: Abhandlungen zur alten Geschiehte, I (Bonn, 1955) (excellent documentation); J. Schwartz, "Note sur la famille de Philon d'Alexandrie," *Mélanges Isidore Lévy*, Annuaire de l'institut de philologie et d'histoire orientales et slaves. XIII (Bruxelles, 1955), 591-602; and J. Daniélou, *Philon d'Alexandrie*, pp. 11-39.

[51]It is to be suspected that from a mistaken interpretation of these passages, the name Lysimachus has been added to the name of Alexander the Alabarch in the 11th century Ambrosian MS of Josephus (*AJ* xix. 276; cf. xviii. 159-160, 259-260; xx. 100-103; *BJ* v. 205); see M. Pohlenz, "Philon von Alexandria," *Nachrichten von der Gesellschaft der Wissenschaften zu Göttingen*, Philologisch-historische Klasse, Neue Folge I, 5 (1942), 413. A. Schalit notes: "Der zweite Name Λυσύμαχος in A 19 276 ist zweifelhaft." "7 'Αλέξανδρος [Λυσύμαχος] Alabarch von Alexandrien," *Namenwörterbuch zu Flavius Josephus*, A Complete Concordance to Flavius Josephus, ed. K. H. Rengstorf, Supplement I (Leiden, 1968), p. 8. This erroneous association of the names was common in the last century and at the turn of this century; see, e.g., Stein, "Julius, 59," cols. 153-157.

[52]G. Tappe, *De Philonis libro qui inscribitur* 'Αλέξανδρος ἢ περὶ τοῦ λόγον ἔχειν τὰ ἄλογα ζῷα: *Quaestiones selectae* (Diss. Göttingen, 1912), pp. 4-5. This equally misleading identification has influenced a host of scholars (including Wendland, Tappe's major professor, and his associates) down to the present; see, e.g., H. Leisegang, "Philon," *PRE*, 39 (1941), cols. 1-50, especially 6-8; Schwartz, "Note sur la famille de Philon d'Alexandrie," p. 594.

aware, his daughter is engaged [lit., promised by an (betrothal) agreement] to be my wife." Aucher translates these lines correctly: *"Avunculus enim est, ac simul socer: quoniam non es nescius, quod filia ejus mihi juxta suam etiam promissionem desponsata uxor est."* However, being puzzled by Philo's and Lysimachus' calling Alexander "our nephew," he leaves the question of relation unresolved.[53] Tappe, who insists that Lysimachus is a brother of Philo and of the Alabarch, on the authority of a certain Andreas declares the text of this passage to be corrupt and goes on to provide the following translation: *"Avunculus enim sum ac simul socer: quoniam non es nescius quod filia mea ei iuxta meam etiam promissionem desponsata uxor est."*[54] But there is no basis for these forced emendations. Moreover, if Lysimachus is a brother of Philo and of the Alabarch (Alexander's father) and at the same time Alexander's "mother's brother," then the Alabarch would have married his own sister.

The absolutely clear relationship indicated in § 2 must stand and the references to Alexander as "our nephew" by Philo (§§ 1, 75) and Lysimachus (§ 72) must be explained--especially the latter. In a note to the translation of § 72 two possible solutions have been proposed: (1) Just as Philo in §§ 1 and 75 refers to Alexander as "our nephew," using a plural of modesty, likewise Lysimachus in § 72 refers to Alexander as "our nephew" out of respect to his interlocutor, Philo. Note that Lysimachus' reference to "Alexander, our nephew" comes immediately after addressing Philo as "honorable" (ὦ τίμιε Φίλωνε), an address used earlier in § 2.[55] (2) A possible corruption of "your nephew" to "our nephew," either in the Gr. ὁ ἀδελφιδοῦς ὑμῶν to ὁ ἀδελφιδοῦς ἡμῶν or in the Arm. եղբւորորդին ձեր to եղբւորորդին մեր, may be suspected in § 72 (the possibly confused letters are underlined).

---

[53]P. xi, n. 1.

[54]*De Philonis libro qui inscribitur* Ἀλέξανδρος, pp. 4-5.

[55]Aucher renders the first instance as "o venerate Philo" (§ 2) and the second as "o honoratissime Philo" (§ 72). The dialogues portray Philo as an old man and Lysimachus as a much younger man.

The genealogical table below is based on this treatise of Philo (§§ 1-2, 72, 75) and the passages in Josephus.[56]

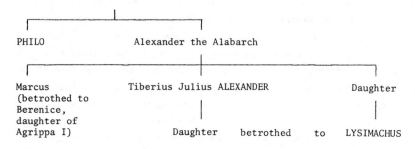

Considering the social status of the family of Lysimachus and his relation to Alexander, it is very likely that he too pursued a political career. He might be the same Julius Lysimachus mentioned in P. Fouad I 21, 8, whose three representatives appear among the nine magistrates with Tuscus the Prefect at the tribunal in the Great Atrium to hear grievances from veterans (dated Sebastos 7 of Nero's 10th year [September 5, A.D. 63]).[57]

2. *Authorship and Date*

The Philonic authorship of Anim cannot be considered apart from that of Provid I-II. Philo's opponent throughout these "dialogues" is his apostate nephew, Tiberius Julius Alexander. There is also a thematic relationship between Provid I-II and Anim. The latter treatise deals with a certain aspect of providence and thus complements the theme of the

---

[56]*AJ* xviii. 159-160, 259-260; xix. 276-277; xx. 100-103; *BJ* v. 205 (cf. 44-46; ii. 220, 223, 309, 492-498; vi. 236-243, on Alexander's career in Roman service).

[57]E. Balogh and H. G. Pflaum, "Le 'concilium' du Préfet d'Égypte. Sa composition," *Revue historique de droit français et étranger*, 30 (1952), 123, identified with Alexander the Alabarch; Turner, "Tiberius Iulius Alexander," p. 56, n. 17, suggests that he might be the nephew of Alexander. For the contents of the papyrus, see U. Wilcken, ed., *Archiv für Papyrusforschung und verwandte Gebiete*, XIV (Leipzig and Berlin, 1941), 174-175; W. L. Westermann, "Tuscus the Prefect and the Veterans in Egypt (P. Yale Inv. 1528 and P. Fouad I 21)," *Classical Philology*, 36 (1941), 21-29.

former.[58] The Alexandrian origin of both treatises is certain: Provid
II 55 and Anim 13 and 28 refer to Alexandria in Egypt,[59] and Anim 7 alludes to a mixed assembly of Romans and Alexandrians. Moreover, both
titles have the testimony of Eusebius, who cites them conjointly in his
list of Philo's works.[60] Consequently, a defense of the authorship of
one of these treatises invariably becomes a defense of the authorship of
the other. P. Wendland's[61] outstanding demonstration of the Philonic
authorship of Provid I-II may rightly be claimed for Anim--even though
it is devoid of any clear evidence of Jewish authorship.[62] Wendland
gives overwhelming evidence of the genuineness of Provid I-II by showing
philosophical, linguistic, and stylistic affinities between these books
and the rest of the works of Philo.[63] Likewise, the numerous parallels
between Anim and the Philonic passages cited in the Commentary[64] suggest
more than just a common literary heritage or use of sources.

The parody of an opponent's imagined speech is a common literary device.
Interestingly enough, a selection of Alexander's arguments (Provid II 3)
is introduced by Eusebius as a statement by Philo himself of the objections
which opponents might adduce.[65] There can be little doubt about Alexan-

---

[58]Providence and the question of animal intelligence are linked in Provid
I 9 (*SVF* II 577), 51-53, 67 (*SVF* II 1146), 70. For more on this thematic relationship, see Hadas-Lebel, *De Providentia I et II*, p. 52, and below, pp. 51-52.

[59]Cf. Quod Omn 125, Ἀλεξανδρία πρὸς Αἰγύπτῳ.

[60]*HE* ii. 18. 6; cf. Jerome *De viris illustribus* 11.

[61]*Philos Schrift über die Vorsehung, ein Beitrag zur Geschichte der
nacharistotelischen Philosophie* (Berlin, 1892), written primarily in
response to Massebieau, who doubted the Philonic authorship of Provid I,
"Le classement des oeuvres de Philon," pp. 87-90.

[62]The Jewish authorship of Provid I-II is clearly indicated in II 107
(cf. I 22, 84). On the possible biblical background of Philo's dialogues,
see below, pp. 46-48.

[63]See also Hadas-Lebel's French edition of the *De Providentia I et II*,
pp. 36-38, 357-361.

[64]See Ind. V, pp. 317-321.

[65]*PE* viii. 14. 386, Κατασκευάζει δὲ τὸν λόγον τοῦτον τὸν τρόπον. . . .
Ταῦτα εἰς ἀνασκευὴν καὶ μυρία ἄλλα πλείω τούτων εἰπών, ἑξῆς ἐπιλύεται
τὰς ἀντιθέσεις διὰ τούτων.

der's cherishing the thoughts attributed to him in Provid I-II and Anim,[66] for in answering him Philo finds himself in a predicament and, in search for answers, he sometimes contradicts himself.[67] Some of the questions raised by Alexander in these dialogues are not even dealt with in Philo's replies.[68] The objections may thus be considered to genuinely belong to Alexander. The composition of the treatises, however, may be taken with fair certainty to be Philo's.[69] The single authorship of Anim is to be seen in its structure, patterned after the first part of the Platonic *Phaedrus*.[70]

There are two datable events recorded in this treatise: the one in § 27, the other in § 54. The celebrations spoken of in § 27 were given by Germanicus Julius Caesar probably in A.D. 12, when he entered on his first term of consulship. The account, however, is taken from a literary source used also by Pliny and Aelian.[71] Some time, therefore, must be allowed for the period between the event and its literary description on the one hand and for the period between that literary composition and its use by Philo on the other hand. The embassy to Rome spoken of in § 54 is presumably the Alexandrian Jewish embassy to Gaius Caligula in A.D. 39/40. This delegation of five was headed by Philo himself and probably included his brother, Alexander the Alabarch.[72] It now seems that Tiberius Julius Alexander accompanied his uncle and, perhaps, his father on this delicate

---

[66]See comment on §§ 1, 8, and 74-75.

[67]Cf. Provid II 32 and 102; 105 and 110.

[68]See Colson's note to the Eusebian line between Provid II 3 and 15 (LCL IX, 458). In Anim, several of Alexander's arguments are answered not in specific terms but in sweeping generalities.

[69]Colson makes this observation: "Philo was able to manipulate, even if he did not entirely invent, the part which Alexander plays, and he does not seem to have treated his opponent fairly" (LCL IX, 449).

[70]See comment on § 1 and App. III, pp. 265-271.

[71]See comment on § 27.

[72]He was imprisoned on Gaius' orders and later released by Claudius. J *AJ* xix. 276 suggests, though it does not positively require, the arrest to have taken place in Rome.

mission described in Philo's Gaium and Josephus' *AJ* (xviii. 257-260).
The second of these two datable events is to be taken as determining the
terminus post quem and not the first as is generally supposed. Thus, the
terminus post quem of Anim has to be advanced by about thirty years--if
not more, since two other accounts in Anim can be dated by way of their
datable parallels in Pliny.[73]   The dates are A.D. 48 and 47 respectively.
We may thus conclude that Anim was composed *ca.* A.D. 50.[74]

Several other internal evidences indicate a late date for Anim. The
introductory and transitory dialogues (§§ 1-9, 72-76) portray an old man
conversing with a young relative, Lysimachus, who twice addresses Philo
as "honorable" (ὦ τίμιε Φίλωνε, §§ 2, 72; Philo's old age may also be
construed from a possible allusion to his poor eyesight in Provid II 1
and a reference to one of his several pilgrimages to Jerusalem in 107).
Alexander had a daughter, probably in her teens, betrothed to her cousin,
Lysimachus (§ 2). Granting that Alexander was born early in the reign of
Tiberius (*ca.* A.D. 15), whose *nomen* and *praenomen* he bears, was married
in *ca.* A.D. 35 at the age of twenty, he would by *ca.* A.D. 50 have been in
his mid-thirties, with a teenage daughter. Moreover, Alexander seems to
have held some public office (§§ 3-4) and, probably, was beyond Philo's
reach.[75] His apostasy, spoken of by Josephus in *AJ* xx. 100, that "he did
not persevere in his ancestral practices" (τοῖς πατρίοις οὐ διέμεινεν
ἔθεσιν), is reflected not only in the questioning of divine providence in
Provid I-II and, possibly, the Judaeo-biblical, anthropocentric view of
the cosmos in Anim but also in specific cases: as in § 31, where, contrary to Jewish dietary laws, he cites oysters as fit for food (he cites
the hare with animals fit for food in Provid II 92).

---

[73]See comment on §§ 13 and 58. It may be that in both cases Pliny has
adapted the same source(s) used also by Philo.

[74]E. Schürer observes that the book belongs to Philo's later works,
the embassy to Rome being already contemplated, *Geschichte des jüdischen
Volkes im Zeitalter Jesu Christi*, III (4th ed.; Leipzig, 1909), 685; so
also Pohlenz, "Philon von Alexandria," pp. 412-415. Turner, in due consideration of the events of Alexander's life, observes: "Philo cannot
have written these dialogues before A.D. 40-50." "Tiberius Iulius Alexander," p. 56.

[75]See comment on § 8; cf. Y. Amir, "Philo Judaeus," *Encyclopaedia
Judaica*, 13 (1971), cols. 410-411.

The following life sketch of Alexander should prove to be of some help in establishing the tentatively drawn dates for his birth, marriage, and betrothal of his teenage daughter.[76] The events mentioned in Anim are italicized.

Date	Event	Approx. Age
ca. A.D. 15	was born	--
ca. A.D. 35	was married	20
A.D. 39/40	*participated in the Alexandrian Jewish embassy to Rome*	25
A.D. 41	entered Roman service	26
A.D. 42	Epistrategos of the Thebaid	27
A.D. 46-48	Procurator of Judaea	31-33
ca. A.D. 50	*daughter betrothed to Lysimachus*	35
A.D. 66-70	Prefect of Egypt	51-55
A.D. 69	proclaimed Vespasian Emperor before the Alexandrian troops	54
A.D. 70	Chief of Staff under Titus during the siege of Jerusalem	55

The only and indirect reference to Alexander's age is found in *BJ* v. 46, where Josephus remarks on Alexander as a counsellor in the exigencies of war (σύμβουλος ταῖς τοῦ πολέμου χρείαις)[77] during the siege of Jerusalem by Titus in A.D. 70: "he was well qualified both by age and experience." When Josephus' *AJ* appeared in about A.D. 93/94 with the offensive remark on Alexander's renegadism (xx. 100), Alexander might have been either dead or politically inactive.

The references to Alexander as "the young man" (ὁ νήπιος, §§ 8, 75) are not to be taken literally as denoting age but rather metaphorically or derogatorily, denoting inexperience or ignorance.[78] After all, time must be allowed for Alexander's maturity and familiarity with the authori-

---

[76]See the authorities cited above, n. 50, especially Turner's article.

[77]Elsewhere Josephus describes Alexander's position as πάντων τῶν στρατευμάτων ἄρχων κριθείς and πάντων τῶν στρατευμάτων ἐπάρχοντος (*BJ* v. 45; vi. 237). For the rest of Josephus' references to Alexander, see above, n. 56.

[78]See comment on § 8.

ties and the arguments he is made to cite in Provid I-II and Anim. Note
that in Provid II he is referred to as ὁ ἀνήρ (1) and addressed with such
Socratic terms as ὦ γενναῖε (31, 62), ὦ θαυμάσιε (55), ὦ φίλε (56), etc.

The arguments for the traditionally held view that Philo wrote the
philosophical works Aet (if that is indeed Philonic), Quod Omn, Provid
I-II, and Anim before he undertook the expository works on the Pentateuch,[79] have been repeatedly challenged.[80] Proponents of the traditional
view observe in his works a progression from a purely philosophical orien-
tation to a more theological one and thus attempt to construct a chrono-

---

[79]See, e.g., P. Wendland, "Philos Schrift περὶ τοῦ πάντα σπουδαῖον
εἶναι ἐλεύθερον [Quod Omn]," *Archiv für Geschichte der Philosophie*, 1
(1888), 509-517; Cohn, "Einteilung und Chronologie der Schriften Philos,"
pp. 389-391, 426, cites the four works in the given order, as does also
W. von Christ, *Geschichte der griechischen Litteratur*, II, Handbuch der
klassischen Altertums-Wissenschaft, VII (6th ed.; Münich, 1920), 627-631.
With the publication in 1912 of Tappe's dissertation, *De Philonis libro
qui inscribitur* Ἀλέξανδρος, especially pp. 3-6, the placing of Anim at
the very beginning of Philo's works tends to become customary. W. Bousset,
*Jüdisch-christlicher Schulbetrieb in Alexandria und Rom. Literarische
Untersuchungen zu Philo und Clemens von Alexandria, Justin und Irenäus*,
Forschungen zur Religion und Literatur des Alten und Neuen Testaments,
Neue Folge VI (Göttingen, 1915), 137-148, ascribes the "dialogues" with
Alexander to the days of Philo's philosophical training. Leisegang,
"Philon," cols. 6-8, likewise lists Anim first as does also A. Lesky,
Geschichte der griechischen Literatur (Bern, 1957-1958), p. 729; etc.

[80]Schürer observes that Anim belongs to Philo's later works, the
embassy to Rome being already contemplated, *Geschichte des jüdischen
Volkes*, III, 685. M. Adler, *Studien zu Philon von Alexandreia* (Breslau,
1929), points out a gradual departure from the biblical text in the works
comprising the allegorical commentary: whereas the earlier works show
close attachment to the text, gradually breaking into free composition, the
later works are almost altogether free from such attachment and more
philosophical. He thus suggests that Philo's philosophical writings are
of a later development (pp. 66-67). H. Leisegang, "Philons Schrift über
die Ewigkeit der Welt," *Philologus*, 92 (1937), 156-176, argues against
Wendland that Aet belongs to the later years of Philo's life, to the time
when he defended the Stoics on the notion of divine providence. Although
the Philonic authorship of Aet is doubtful, most of Leisegang's arguments
could be claimed for Provid I-II and Anim, which he elsewhere (see the
preceding note) places at the beginning of Philo's works. After establishing the identity of the speakers and the right date of Anim, Pohlenz argues
that both the philosophical and the exegetical works belong to the same
period, to the closing years of Philo's life, "Philon von Alexandreia,"
pp. 412-415; so also Turner, "Tiberius Iulius Alexander," pp. 55-56;
similar questions are raised by Hadas-Lebel, *De Providentia I et II*,
pp. 38-40.

34

logical order of the works, dividing them into three groups: philosophical, exegetical, and historical or apologetic. While this is not the
place to discuss the chronological order of the works, especially those
of the second and largest division which in turn is subdivided into three
groups or major works on the Pentateuch,[81] suffice it to say that the
divisions are acceptable, but their sequence is not.

The often quoted passage seemingly favoring the traditional view of
Philo's growing out of the philosophical writings into theological maturity and exegetical writings, Congr 73-80, has been misconstrued. The
passage, rightly understood, shows Philo's concurrent interest in philosophy and theology. Note that in the introductory and transitory dialogues
in Anim,"a purely philosophical work," he calls himself an "interpreter"
(ἑρμηνεύς, §§ 7, 74)--if by that he is to be understood as an interpreter
of Scripture. Moreover, in a passage never before considered, Provid II
115, Philo speaks of philosophy as his life interest (in response to
Alexander's asking for time to hear more from Philo): "I always have time
to philosophize, to which field of knowledge I have devoted my life; however, many and diverse yet delightful duties that would not be fair to
neglect summon me." This conflict between political duties and personal
endeavors expressed at the end of a philosophical treatise is reminiscent
of that expressed at the beginning of an exegetical treatise, Spec Leg
III 1-6.[82]

Having demonstrated that the dialogues with Alexander belong to the
closing years of Philo's life and that his philosophical interest concurs
with his theological interest, we are led to conclude that perhaps most
of Philo's literary career belongs to the closing years of his life, to
the period following the political turmoils described in Flacc and Gaium.[83]

---

[81]This was correctly perceived by H. Ewald, *Geschichte des Volkes
Israel*, VI (3d ed.; Göttingen, 1868), 257-312.

[82]Goodenough in his brilliant interpretation of Spec Leg III 1-6 shows
that "Philo's literary career as an interpreter of the bible was a function of his life *after* he had gone into political affairs, carried on as
a hobby or escape from politics," *The Politics of Philo Judaeus*, pp. 66-68;
cf. Colson's comment in LCL VII, 631-632.

[83]The bulk of his writings does not make this supposition impossible.
Hadas-Lebel cites the example of Cicero (*De Providentia I et II*, p. 39).
One may also cite the example of Plotinus.

3. Man and Animals in Philo's Cosmology

A discussion of Philo's cosmology is neither within the scope of this
short survey nor deemed necessary in view of the fine studies on the subject.[84] However, due consideration of Philo's cosmology in Provid I-II
and that put in the mouth of the opposition will no doubt add to our
understanding of the subject, especially as it relates to Anim. Suffice
it to say that his involved understanding of the universe, as gathered
from the rest of his works, embraces every cosmological thought found in
his response to Alexander in Anim. The complexity is mostly due to his
use of different cosmological traditions.[85] Of the elaborate structure
which constitutes the metaphysical pyramid of his understanding of the
universe we shall consider only the branch that deals specifically with
categories of beings discussed in this treatise. These categories are
capitalized in the following stemma based on a key passage: Agr 139
(SVF II 182).[86]

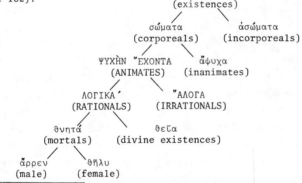

---

[84]The following are among the more outstanding studies: J. Gross,
Philons von Alexandreia Anschauungen über die Natur des Menschen (Diss.
Tübingen, 1930); H. Schmidt, *Die Anthropologie Philons von Alexandreia*
(Diss. Leipzig, 1933); U. Früchtel, *Die kosmologischen Vorstellungen bei
Philo von Alexandrien. Ein Beitrag zur Geschichte der Genesisexegese*,
Arbeiten zur Literatur und Geschichte des hellenistischen Judentums, II
(Leiden, 1968); and G. D. Farandos, *Kosmos und Logos nach Philon von
Alexandria*, Elementa: Schriften zur Philosophie und ihrer Problem-
geschichte, IV (Amsterdam, 1976).

[85]Früchtel, *Die kosmologischen Vorstellungen bei Philo*, pp. 184-186,
identifies three major traditions.

[86]Cf. Heres 209. For stemmatic outlines of Philo's cosmology and that
of Antiochus of Ascalon, Maximus of Tyre, and Cicero, see Früchtel, *Die
kosmologischen Vorstellungen bei Philo*, pp. 42-45; cf. Farandos, *Kosmos
und Logos nach Philon*, pp. 259-263.

In spite of the title of the treatise and the frequent references to animals, the work as a whole is basically anthropological; there must of necessity be a comparison between ἄλογον ζῷον and λογικὸν ζῷον.[87] The questions here raised are not simply on the nature of animals or their place in the scheme of things but especially on their relation to man. It is the anthropocentric view of the cosmos, that all things--including animals--were made for man's sake, that is challenged by Alexander and defended by Philo. Moreover, since each of the higher categories in the organic kingdom possesses all the properties of lower categories, the study of the orders inferior to man is also the study of a large part of human nature. We may perhaps advance this discussion if we bracket for the time being the treatise under consideration and go on to the rest of Philo's works.

The progression of corporeals (σώματα) proceeds from inanimates (ἄψυχα) to animates (ψυχὴν ἔχοντα or ἔμψυχα), from irrationals (ἄλογα) to rationals (λογικά), from mortals (θνητά) to divine existences (θεῖα) or immortals (ἀθάνατα). The human organism, of course, contains the properties of all lower organisms. Man, like animals, nourishes himself, grows, propagates his kind, moves about, and is endowed with the basic characteristic of the soul (ψυχή), which is sense-perception (αἴσθησις) with its dependent mental presentation and resultant impulse (φαντασία καὶ ὁρμή or, the first movement of the soul, τῆς ψυχῆς πρώτη κίνησις). In addition to this irrational aspect of the soul common to animals and men alike, the human soul has a rational element which constitutes man's advance beyond the animals. This element is reason (λόγος), which distinguishes humans as rationals (λογικά) from irrationals (ἄλογα).[88]

---

[87]Among other Philonic designations for man, note the following: βραχύτατον ζῷον: Heres 155; cf. Op 72; κάλλιστον ζῷον: Spec Leg I 10; III 108; cf. Jos 2; ἄριστον ζῷον: Spec Leg II 84; cf. Ebr 118; Vita Mos II 65; Spec Leg II 181; Aet 65; ἡμερώτατον ζῷον: Spec Leg I 295; Aet 68; cf. Heres 137; Vita Mos I 60; Dec 132, 160; Praem 92.

[88]Op 62, 66-67 (SVF II 722, 745), 73 (SVF III 372); Leg All I 30 (SVF II 844); II 22-23; Quod Deus 35-45 (SVF II 458); Plant 13; Heres 137; Somn I 136; Aet 75 (SVF II 459; cf. 714). Philo does not classify plants with animate beings; he places them among inanimates, making this distinction: inanimates have a cohesive principle described as "habit" (ἕξις), in addition to which, plants have an inner principle of growth described as "nature" (φύσις). The basic characteristic of the inanimate is lack of sense-perception.

Philo sometimes follows the Platonic tripartite division of the human soul (λογιστικόν, the rational part located in the head, θυμοειδές and ἐπιθυμητικόν, the irrational parts located in the chest and around the navel or diaphragm);[89] but more often he follows the Stoics' elaborations on its rational and irrational parts (λογικόν and ἄλογον) or masculine and feminine elements (ἄρρην and θῆλυς).[90] Like the Stoics, he speaks of the eight parts of the soul: the rational or sovereign part (τὸ ἡγεμονικόν) as the ruler of the irrational, which is divided into seven parts or faculties: the five senses and the organs of speech and reproduction.[91] He sometimes wonders whether the seat of the sovereign part is in the head or in the heart.[92] Occasionally, he follows the Stoics, who for the most part decide on the heart;[93] but more often he follows Plato in locating it in the head.[94] He also follows the Stoics' equation of the rational or sovereign part of the soul with the faculty of reason (διάνοια),[95] the fount of reason (λόγου πηγή),[96] the spirit (πνεῦμα),[97]

---

[89]Leg All I 70-73; III 115-116; Conf 21; Migr 66-67; Spec Leg IV 92-94; Virt 13; cf. *Tim.* 69E-71D; *Phdr.* 246A-256A; *R.* 439D; etc.

[90]Sacr 112; Spec Leg I 200; etc.

[91]Op 117; Leg All I 11; Quod Det 168; Heres 232; Agr 30 (*SVF* II 833); Mut 111; Quaes Gen I 75 (*SVF* II 832); six faculties in Abr 29, excluding reproduction.

[92]Sacr 136 (*SVF* II 845); Quod Det 90; Post 137; Somn I 32; Spec Leg I 213-214.

[93]Leg All I 59 (*SVF* II 843); II 6; Spec Leg IV 69; cf. *SVF* II 837, 879-880, 889, 898; III Diog 29-30.

[94]In addition to the passages cited in n. 87, see Quaes Gen I 5; Quaes Ex II 100, 124; cf. *Tim.* 90A.

[95]Op 135; cf. *SVF* I 202; III 306, 459.

[96]See comment on § 12.

[97]Spec Leg I 171; cf. Op 134, 139; see also Quod Det 80-85; Quod Deus 84; Plant 44; Fuga 133; Somn I 30; Mut 123; Aet 111; Gaium 63; Quaes Gen I 25; III 8, 22, 27, 49, 59; IV 1, 102; Quaes Ex I 4--on πνεῦμα relating to mind or soul; cf. also *SVF* I 137-138, 140, 521, 528; II 773-774, 778; etc.

or the mind (νοῦς)[98] in man. Moreover, his descriptions of the sovereign part of the soul also have much in common with those of the Stoics.[99]

In his interpretation of the creation of man in the image of God (Gen 1:27 and 2:7), Philo describes the human mind as being patterned after the divine mind (νοῦς), the divine word or reason (λόγος), the divine image (εἰκών), or the archetypal seal (σφραγίς) with which the mind of the earthborn man is impressed.[100] It is worth noting that he uses both Gen 1:27 and 2:7 to establish the God-like nature of the human mind and not, as it is sometimes supposed, to distinguish between a "heavenly," "image-like," or "ideal" man in 1:27 and an "earthly" man in 2:7. The rational or sovereign part of the soul is breathed upon the earthborn man.[101] Thus, man is made a rational being by divine agency; this is his natural constitution or characteristic (λογικὴ φύσις, λογικὴ ἕξις).[102] The creation of man in the image of God is often seen by him not only in the God-like nature of man's rational soul or mind but also in the human longing for the divine model.[103] Reason is man's special prerogative, whereby he is made superior to other animals.[104] Not only does Philo consider a child as a potentially rational being and unborn children as potentially separate, rational creatures[105] but he also speaks of the human semen as containing a rational principle.[106]

---

[98]Op 69; Vita Mos II 82; cf. S.E. *M.* ix. 93.

[99]See comment on §§ 10, 16, and 29.

[100]See the references to Philo's interpretation of Gen 1:27 and 2:7 in the comment on § 16.

[101]Spec Leg IV 123; cf. Leg All I 31-42; II 23 (*SVF* II 458); Quod Deus 47.

[102]Op 150; Leg All I 10; etc.

[103]From among the numerous discussions on Philo's conception of man's higher nature, see J. Giblet, "L'homme image de Dieu dans les commentaires littéraux de Philon d'Alexandrie," *Studia Hellenistica*, 5 (1948), 93-118.

[104]Quod Deus 45 (*SVF* II 458); Somn I 103, 113; Spec Leg II 173; etc.

[105]See comment on § 96 (*SVF* II 834).

[106]See comment on § 49.

Drawing upon Pl. *Tim.* 90A and 91E, Philo comments on the human posture
and that of animals.[107] The erect posture of the human body, with the
head, the brain, or the eyes in the most elevated position, is taken as a
mark of divine favor allowing man to contemplate the operations of the
heavens.[108] Man's stock is near akin to God;[109] reason is the bond of
this divine-human relationship, "For everyone who is left forsaken by
reason, the better part of the soul, has been transformed into the nature
of a beast, even though the outward characteristics of his body still retain their human form."[110] Philo thus speaks of two kinds of men: (a)
rational and (b) admixture of irrational and rational.[111]

To Philo creatures differ not only in their habitats, sizes, and qualities or formations (ποιότητες),[112] but also in their souls (ψυχαί)
and perceptions (αἰσθήσεις), the least elaborate of which has been alloted
to the race of fish.[113] The soul of the fish is said to have been given
to serve as salt, to keep it from rotting.[114] Worms are the vilest of
living creatures.[115] Others surpass man in bodily endowments: strength
or vigor, sense-perception, health, etc.[116] These are "the bodily good

---

[107]See comment on § 11.

[108]Plant 20; cf. Leg All I 38-39.

[109]Op 77; Abr 41; Vita Mos I 279; Spec Leg IV 14; Dec 134; etc.

[110]Spec Leg III 99; cf. I 295; IV 103; Abr 33. Philo, however, draws a
marked distinction between unreasoning men and animals: "Unreasonableness is of two kinds. One is the unreasonableness that defies convinc-
ing reason, as when men call the foolish man unreasonable. The other is
the state from which reason is eliminated, as with the unreasoning animals" (Sacr 46 [*SVF* 375]; cf. Quod Det 38).

[111]Leg All I 31-32; Heres 56-57; Fuga 71-72; etc. (see the references
to Philo's interpretation of Gen 1:27 and 2:7 in the comment on § 16).

[112]Op 63.

[113]*Ibid.* 65; cf. 68.

[114]*Ibid.* 66 (*SVF* II 722). In Cic. *ND* ii. 160 (*SVF* II 1154), a passage
attributed to Chrysippus, the same is said of the pig's soul.

[115]Vita Mos II 262.

[116]Provid II 20; Post 161; Agr 115; Somn I 49; Abr 266; Virt 146; etc.

things" (τὰ σωματικὰ ἀγαθά) or "virtues of the body" (αἱ περὶ σῶμα ἀρεταί),
which Philo distinguishes from "virtues of the soul" (αἱ περὶ ψυχὴν
ἀρεταί), the cardinal virtues.[117] The latter depend on knowledge
(ἐπιστήμη), which animals do not have.[118] Although one finds among animals some vague demonstrations of courage, self-restraint, justice, and
the like, they are not to be taken for real--the real attributes belong
to human souls.[119]

Since the inherent properties of all things in the universe--animate
and inanimate are maintained by nature (φύσις), all their peculiar characteristics are termed "natural." The divine word or reason (λόγος θεῖος
or, more often, λόγος) supervenes not as the opponent of these "natural"
characteristics but as the artificer or craftsman in charge of them.[120]
Time and again in this treatise Philo attacks the claim for reason among
the brutes with the argument that their "sagacity" is "natural" or instinctive, not intelligent.[121] He characterizes nature as the helper and
manager of all creatures and the designer of their mechanism.[122] "As
nature has fortified other living creatures each with appropriate means
of guarding themselves . . . so has she given to man a most strong redoubt
and impregnable fort in reason."[123] No doubt Philo was influenced by the
common notion of *natura creatrix*;[124] as a rule, however, his use of the

---

[117]Post 159; Sobr 61; cf. Leg All II 80; Quaes Gen IV 11 (*SVF* III 207).

[118]See comment on § 30.

[119]See comment on § 95 (*SVF* II 730).

[120]Op 36; Cher 36; etc. See Früchtel, *Die kosmologischen Vorstellungen bei Philo*, pp. 57-61; cf. *SVF* II 1132-1140. For more on the "natural" characteristics of plants, animals, and man, see comment on § 78 (*SVF* II 732).

[121]See comment on § 92 (*SVF* II 733).

[122]See comment on § 68.

[123]Somn I 103; see comment on § 11; cf. §§ 29, 71, 100, where nature is characterized as the giver of reason to man.

[124]See H. Leisegang, "φύσις. 2. natura creatrix," *Philonis Alexandrini opera quae supersunt, Indices*, II (Berlin, 1930), 837-838.

term "nature" is to be identified with the divine agency in things.[125]
Philo's interchangeable use of the terms God (θεός) and Nature (φύσις),[126]
as of the terms "divine law" (νόμος θεῖος) and "natural law" (νόμος
φύσεως),[127] does not necessarily imply the equation of the two--as in
Stoicism.[128] Although Philo's cosmological views owed much to the Stoics,
especially to their understanding of nature or universal reason as manifested in the animal creation, their conception of God as material and
immanent in the world was unacceptable to Philo, to whom God was wholly
transcendant and immaterial.[129]

Man is the only being on earth who is conscious of his ends.[130] He
alone among mortals has "the ability to reason of his own motion."[131]
Free will, God's "most peculiar possession," bestowed upon man, makes him
"accountable for what he does wrong with intent and praised when he acts
rightly of his own will," since his actions "stem from a volitional and
self-determining mind, whose activities for the most part rest on deliberate choice" (προαιρετικαῖς).[132] Certainly "involuntary acts done in
ignorance do not count as sin."[133] Likewise children are not to be held

---

[125]See §§ 11, 29, 68-69, 71, 80, 86, 92, 95, 97, 100; cf. Sacr 98;
Post 5; Heres 114-116; Fuga 170-172; Vita Cont 70; etc. On special uses
of the term φύσις in Philo, see Goodenough, *By Light, Light*, pp. 48-71;
Wolfson, *Philo*, I, 332-347.

[126]See Leisegang, "φύσις. 3. θεός est φύσις vel hac voce intellegitur,"
Indices, II, 838-839.

[127]See comment on § 48.

[128]*SVF* I 152-177.

[129]Leg All I 91; Migr 179-181; etc; cf. *SVF* I 532; II 774; etc. From
among the extensive discussions on Philo's conception of God, see the
authorities cited in the comment on § 10. For the Stoic views on nature
or universal reason as manifested in the animal creation, see *SVF* II 708-
772.

[130]Conf 178.

[131]Op 149.

[132]Quod Deus 47; cf. 49-50; Quod Det 11; Gig 20-21; Somn II 174.

[133]Leg All I 35; cf. III 210 (*SVF* III 512); Quod Deus 90; Agr 179;
Fuga 65-76; Spec Leg III 120-123.

responsible for the wrong they do before reaching an accountable age.[134] As for "plants and animals, neither praise is due if they fare well, nor blame if they fare ill, since their movements and changes in either direction come to them from no deliberate choice or volition of their own" (ἀπροαιρέτους).[135] "Unreasoning natures, because as they have no gift of understanding, they are also not guilty of wrongdoing willed freely as a result of deliberate reflection" (λογισμοῦ).[136] The enmity of wild beasts, unlike that of men, is actuated by natural antipathy without deliberation.[137] Suffice it to say that Philo's views on free will, deliberation, voluntary and involuntary acts, and responsibility for virtue and vice-- views broadly developed in Provid I-II--are reminiscent of Arist. *EN* 1109b30-1115a2, which became central in Stoic ethical thought.[138]

Philo's understanding of the impact of the mind (νοῦς) upon the senses (αἴσθησις) and that of the senses upon the mind is made sufficiently clear in his allegories on Adam and Eve, where he contrasts mental apprehension, symbolized by man, with sense-perception, symbolized by woman.[139] Allegorizing on the Genesis account of the fall of man, he shows how the superior mind, which in its rightful place of sovereignty controls the senses, can be influenced and deluded by them.[140] At such an instance the supe-

---

[134]See comment on § 96 (*SVF* II 834).

[135]Quod Deus 48.

[136]Conf 177.

[137]Praem 85.

[138]See H. A. Wolfson, "Philo on Free Will and the Historical Influence of His View," *Harvard Theological Review*, 34 (1942), 131-169, reprinted in his *Philo*, I, 424-462; D. Winston, "Freedom and Determinism in Greek Philosophy and Jewish Hellenistic Wisdom," *Studia Philonica*, 2 (1973), 40-50, and "Freedom and Determinism in Philo of Alexandria," *Studia Philonica*, 3 (1974-1975), 47-70; cf. J. M. Rist, *Stoic Philosophy* (London, 1969), pp. 219-232; A. A. Long, "Freedom and Determinism in the Stoic Theory of Human Action," *Problems in Stoicism*, ed. A. A. Long (London, 1971), pp. 173-199; A. -J. Voelke, *L'idée de volonté dans le stoicisme* (Paris, 1973).

[139]Leg All I 25; II 13, 38, 70; III 50, 200, 246 (*SVF* III 406); etc.

[140]Op 165-166 (*SVF* II 57); Quaes Gen I 46.

rior mind becomes one with the senses, "it resolves itself into the order of the flesh which is inferior, into sense-perception."[141] Although his works abound in contrasts between the heavenly, rational mind and the earthly, irrational senses,[142] he does not overlook the role of the senses in acquiring knowledge, which is dependent on sense-perception.[143] Sense-perception apprehends simultaneously with mind and gives it occasions of apprehending the objects presented to it.[144] Mind, helpless in itself, by mating with sense, comes to comprehend phenomena.[145] Thus, sense-perception is the handmaid of reason.[146] Moreover, the sensible and the mental must be so blended that "in the world of sense we may come to find the likeness of the invisible world of mind,"[147] or that by experiencing the world of sense we may get our knowledge of the world of mind.[148]

Closely related to Philo's basically Stoic understanding of the inter-action between sense-perception and mind, impression and impulse, are the two kinds of reason around which much of the discussion in this treatise centers: uttered reason or speech (λόγος προφορικός) and mental reason-ing or thought (λόγος ἐνδιάθετος).[149] The distinction between these two

---

[141]Leg All II 50; cf. III 185.

[142]Leg All III 108; Cher 70; Migr 206; Heres 185; etc.

[143]Cher 40-41; Heres 53; Congr 155. Such an elevated place for the senses in Stoic psychology must have aided opponents in arguing for the rationality of animals (see comment on § 13).

[144]Leg All I 28-30 (*SVF* II 844); III 56-58; Post 126 (*SVF* II 862).

[145]Cher 60-64; Leg All III 49; but cf. Heres 71; Virt 11-12; Quaes Gen III 3.

[146]Vita Mos II 81.

[147]Migr 105.

[148]Somn I 186-188; cf. S. E. *M.* viii. 56 (*SVF* II 88), that all elements of knowledge come either from sense and experience solely, or from sense and experience combined with reasoning; see also *SVF* II 850-860.

[149]This is not the place to discuss Philo's *logoslehre*, but only to introduce the reader to the kinds of reason distinguished in this trea-tise. Among the numerous works on the subject, see A. Aall, *Der Logos. Geschichte seiner Entwickelung in der griechischen Philosophie*

kinds of reason, implied in Plato[150] and Aristotle,[151] was emphasized by
the Stoics in their debates with the New Academy.[152] Subsequently, the
distinction was employed by Philo[153] and others.[154] To Philo, as to the
Stoics, mental reasoning or thought not only distinguishes man from animals but also engenders uttered reason or speech; that is to say, speech
is a manifestation of thought.[155] Animals are devoid of both of these
reasoning powers. Philo draws a marked distinction between the articulate human voice and the inarticulate utterances of animals which he
equates with the sound of musical instruments.[156]

Philo does not always hold animals in disdain. Man has much to learn
from animals. In discussing the honoring of one's parents he points out
that man is worsted by beasts. Lions, bears, and panthers can be tamed
as a result of their sense of gratitude to their keepers. Watch-dogs and
sheep-dogs can be "trained to know (οἶδε) how to return benefit for benefit"; they are faithful to death, and storks are exemplary for the care
the younger birds give the older ones.[157] He does not conclude that these

---

und der christlichen Litteratur, I (Leipzig, 1896), 184-231; L. Cohn,
"Zur Lehre vom Logos bei Philo," *Judaica, Festschrift zu Hermann Cohens
siebzigstem Geburtstage* (Berlin, 1912), pp. 303-331; M. -J. Lagrange,
"Le Logos de Philon," *Revue Biblique*, 32 (1923), 321-371; Farandos,
Kosmos und Logos nach Philon, pp. 231-275.

[150]*Tht.* 189E; *Sph.* 263E.

[151]*AP.* 76b24.

[152]*SVF* II 135, 233, etc.; for a fine discussion, see M. Pohlenz, "Die
Begründung der abendländischen Sprachlehre durch die Stoa," *Nachrichten
von der Gesellschaft der Wissenschaften zu Göttingen*, Philologische-
historische Klasse, Neue Folge I, 3 (1939), 151-198.

[153]Quod Det 40, 92, 129; Mut 69; Abr 83; Vita Mos II 127-130; Spec
Leg IV 69.

[154]See M. Mühl, *Der Logos endiathetos und prophorikos in der älteren
Stoa bis zur Synode von Sirmium 351*, Archiv für Begriffsgeschichte, VII
(Bonn, 1962).

[155]Philo uses the Stoic metaphor of a stream flowing from a fountain-
head in describing the derivation of the λ. προφορικός from the λ. ἐνδιάθετος.
See comment on § 12.

[156]See comment on §§ 98-99 (*SVF* II 734).

[157]Dec 113-117.

creatures have a share in reason since they are thus virtuous; on the contrary, he denies that animals could have virtue or vice since they have neither reason (λόγος) nor its dependent knowledge (ἐπιστήμη), which is the determining factor in all virtue.[158] Moreover, he finds in the Mosaic legislation special injunction of kindness to animals and plants; but this is interpreted anthropocentrically: through kindness to the irrational creation one learns to be kind to rational beings.[159]

Philo is at one with the Stoics in believing that the irrational creation exists for the sake of the rational: to the extent that wild animals are to exercise man's strength and tame animals his ability to lead people.[160] Man's dominion over the irrational creation is divinely sanctioned.[161] Man and animals are unequal (ἄνισοι); therefore, there can be no justice (δικαιοσύνη) obtaining between them: the rights of humanity do not extend to the unreasoning animals.[162]

In response to Alexander's arguments for the rationality of animals by way of demonstrating their virtues and vices, and his appeal for a moral and juridical relationship between man and animals (§§ 10-71), Philo maintains his usual, dogmatic position: animals are irrational and therefore inferior to human beings, the virtues are not relative but absolute, and free will is an attribute of the rational being. Hence, "to elevate animals to the level of the human race and to grant equality (ἰσότης) to unequals (ἀνίσοις) is the height of injustice (ἀδικίας)" (§§ 77-100, quoted from the last section).

---

[158]Leisegang, "ἀρετή II. 16. ἐπιστήμη, σοφία," *Indices*, I, 114-115.

[159]Virt 81, 116, 125-160; cf. Spec Leg II 69; IV 205-206, 226-229.

[160]Provid II 84 (*SVF* II 149), 91-92, 103; Vita Mos I 60-62; Spec Leg IV 119-121; cf. *SVF* II 1152-1167.

[161]Op 83-88, 148-150; Spec Leg II 69; cf. Leg All I 9; Mut 63; Quaes Gen I 18-23; also Abr 45; etc.

[162]Dec 61; Spec Leg IV 204; cf. Mut 226; Vita Cont 70; for the Stoic background, see the section below.

4. *The Biblical and Philosophical Background*

Notwithstanding the overwhelming Stoic influence reflected in Philo's description of the distance between man and animals, the Mosaic treatment of animals must be considered as the determining factor in moulding his thought. The Pentateuch makes it clear that man is master of the animal kingdom, that animals are subordinated to his purposes (note Gen 1:26 especially). It is to be noted that man as such, composed of body and soul, is made in the likeness of God, and that an aspect of this likeness is to be found in man's lordship over created things, whose relationship of subordination to him is analogous to that which he bears to his Creator. Philo insists in a number of passages (noted in the preceding section, especially those on Gen 2:7) that the likeness rests on man's receiving of the divine spirit, which he identifies with the rational faculty of the soul. Herein lies the basis for man's lordship over the irrational creation. In view of God's granting man dominion over the various creatures, it was most appropriate that Adam was privileged to name them (Gen 2:19-20).

Destruction of animal life--even for food--was out of harmony with sinless nature (Gen 1:29-30). However, animals were drawn into the consequence of original sin; they were killed to provide clothing for man and for sacrificial purposes (Gen 3:21; 4:4). Many of them became rebellious against and hostile to man; but man's dominion was reaffirmed after the flood (Gen 9:2). Man might now use flesh for his food, with the stipulation that it must be drained of its blood, because blood was regarded as the carrier of life. No restriction existed with respect to the type of animals that Noah and his family could eat (Gen 9:3-6). While this made it proper for man to kill animals for necessary food, he was not authorized thereby to indulge in needless slaughter for the sheer thrill of the hunt or to display personal prowess, as did Nimrod, whom Philo considers a rebel against God (Gen 10:9). A classification of animals is to be noted in the instructions to Noah to take with him into the ark seven of each clean animal and two of each unclean animal (Gen 7:2-3, 8-9). Since a flesh diet had not yet been authorized, this distinction between clean and unclean was probably determined upon the basis of what was acceptable for sacrifice (Gen 8:20). Dietary limitations based on this distinction belong to later legislation (Lev 11; Dt 14).

In connection with worship under the Mosaic law, cattle, sheep, and goats were among the creatures acceptable for sacrifice. Such animals were to be sound ones, and no castrated animal was admissable (Lev 22: 23-25). The use of blood for any purpose other than sacrifice was prohibited, "for the blood is the life," which must be given back to God or the earth (Lev 17:10-14; Dt 12:23-24). Worship of any representation of any animal or other created thing was strictly forbidden (Ex 20:4-5). In view of the needs of a pastoral, religious economy, the Mosaic law enjoined proper care of domestic animals (Ex 20:10; 22:30; 23:4-5, 12, 19; 34:26; Lev 19:19; 22:27-28; cf. Dt 5:14; 14:21; 22:1-4, 6-7, 10; 25:4). Philo overemphasizes the moral and juridical aspects of this.[163]

Of particular interest is the fact that there is not a word in this treatise that would suggest a Jewish context. The work lacks even the repeated symbolic use of animals[164] and the often encountered condemnation of Egyptian animal veneration.[165] In spite of this, the biblical background cannot be overlooked.

Leisegang[166] observes that by arguing for the rationality of animals Alexander is opposing not only the Stoic but also the Judaeo-biblical doctrine that only man is endowed with the rational spirit. We may add that by emphasizing the irrationality of animals and thereby sanctioning their unilateral use by humans, Philo is perhaps defending the anthropocentric view of the cosmos reflected in Gen 1:26-28; 2:19-20; 9:2--man's dominion over the irrational creation. One may likewise observe that within the context of Academic-Stoic polemic and without making use of Scripture in Provid I-II, Philo is perhaps defending the biblical view of teleology, including the necessity of evil for the ultimate good--as

---

[163]For the Philonic passages, see the corresponding references to the the biblical citations in Leisegang's *Indices*, I (Berlin, 1926), 29-43. For more on the biblical treatment of animals, see M. -L. Henry, "Tier," *Biblisch-Historisches Handwörterbuch*, III (Göttingen, 1966), cols. 1984-1987.

[164]For a list and references, see C. Siegfried, *Philo von Alexandria als Ausleger des Alten Testaments* (Jena, 1875), pp. 182-185.

[165]Ebr 95, 110; Vita Mos I 23; II 162; Dec 76-79; Vita Cont 8; Gaium 139, 163.

[166]"Philon," col. 7.

48

found in numerous accounts in Genesis and Exodus.[167] In view of these
observations, the dialogues with Alexander may be treated as apologetic
literature; indeed, the apologetic thrust of Provid II seems to have been
recognized by Eusebius.[168] Philo's dialogues thus seem to fall in line
with the rest of his works which are more or less colored by an apologetic overtone.

To add to the validity of the foregoing observations, we shall revive
a meritorious thesis put forward by Adler,[169] who observes in Philo's
works comprising the allegorical commentary a gradual departure from the
biblical text: whereas the earlier works (Leg All I-III, Cher, Sacr,
Quod Det, Post Gig, Quod Deus, Agr. Plant, Ebr) show close attachment to
the text, gradually breaking into free composition, the later works (Sobr,
Conf, Migr, Heres, Congr, Fuga, Mut, Somn I-II) are almost altogether
free from such attachment and more philosophical. Following his thesis
on Philo's method, Adler suggests that the philosophical writings are of
later development.[170] These concluding remarks can in turn be substantiated by our demonstration of the late date of Anim[171] and strengthened
by our elaborations on Leisegang's observations.

Before considering the philosophical background, we must stress the
fact that just as the Gr. word λόγος, which expresses both reason and
speech, must have contributed to the early belief that dumb animals
(τὰ ἄλογα ζῷα) lack the faculty of reason, so also mythological beliefs
(such as the primitive belief in metempsychosis, the mantic powers attri-

---

[167]Cf. A. A. Long, "The Stoic Concept of Evil," *Philosophical Quarterly*, 18 (1968), 329-343. One may sense the Stoic thought in Ro 8:28.

[168]*PE* viii. 14. 386; quoted above, n. 65. These works of Philo may
have influenced Christian apologists in the manner of responding to pagan
opponents, as seen in their appeal to Stoic Philosophy rather than to
Scripture (cf. Hadas-Lebel, *De Providentia I et II*, p. 75). As to the writings of the Fathers, of their vast *hexaemera* literature only the *Commentarius in hexaemeron* traditionally attributed to Eustathius of Antioch
(Migne *PG* XVIII, 707-794) shows reliance on Anim; the rest rely on Op and
Leg All I-III.

[169]*Studien zu Philon*.

[170]*Ibid.*, pp. 66-67.

[171]See above, pp. 30-34.

buted to animals, their congregating, flight, cries, and calls) must have
contributed to the early concept of rationality of animals. We shall not
delve with the pre-Socratic history of the arguments for or against the
rationality of animals, but shall content ourselves with reviewing the
immediate background.

Platonism and Stoicism have long been regarded as being of major importance in moulding Philo's thought, which nonetheless remains overly
religious in tone and determined by its Jewish outlook. The strength
of the Stoic influence is clearly seen in the responses to Alexander in
Provid I-II and Anim.[172] While Philo's frequent use of Stoic terms with
a meaning not at all Stoic may be true of several of his other works, it
certainly is not true of these treatises, for there are major arguments
as well as points of detail in which his thought, terminology, and phrase-
ology are explicable by the Stoic background. Note that selections from
about half of his responses to Alexander in these treatises are included
in von Arnim's *SVF*; certainly, there seem to be more Stoic views expressed
by Philo than von Arnim has admitted into his compilation. Notwithstanding the Stoic background, the structure of Anim owes to the Platonic
Phaedrus;[173] moreover, the arguments are developed with a wealth of illustrations drawn from epitomes covering much of the common mythology about
animals alongside many of Aristotle's observations.[174]

Although the views on the rationality of animals expressed in Alexander's discourse were anticipated by Plato[175] and Aristotle,[176] they were
doubtless taken over by Philo from the arguments used by the opponents of

---

[172]C. J. de Vogel, *Greek Philosophy*, III: *The Hellenistic-Roman Period*
(Leiden, 1963), 81, finds in Provid II a systematic exposition of the objections against providence and their refutation by the Stoics. The same
could be said of Book I and Anim.

[173]See comment on § 1 and also App. III, pp. 265-271.

[174]See below, on the sources, p. 55.

[175]*Tht*. 189E; *La*. 196E; *Sph*. 263E.

[176]Numerous examples of sagacious animals are provided in Book VIII of
HA, and even more in Book IX, which is generally attributed to Aristotle's
successors in the Lyceum.

the Stoics in the New Academy.[177] Chief among the opponents was Carneades
of Cyrene, who flourished as head of the Academy in the middle of the 2d
century B.C. The Academic origin of these arguments, which are reiterated by a host of classical writers[178] and attributed by Philo to Alexander, is well attested by Sextus Empiricus.[179] These and other questions
raised by Carneades and his predecessor, Arcesilaus, greatly challenged
the Stoic philosophers from Chrysippus to Posidonius--to whom the cumula-
tive Stoic reply may be traced.[180] Like most other arguments, those
against providence and the irrationality of animals led the Stoics to
create an arsenal of counter arguments which Philo, like later Stoics and
others, exploits.[181] His ingenuity in putting both the Academic criticism and the Stoic responses into good use appears in Provid I-II and
Anim, where he systematically supports the Stoic position against the
Academic criticism he attributes to Alexander.[182]

---

[177]Tappe, *De Philonis libro qui inscribitur* 'Αλέξανδρος, pp. 22-25,
49-54.

[178]Plutarch in three of his works: *De sollertia animalium* (*Mor.* 959A-
985C), *Bruta animalia ratione uti* (985D-992E), and *De esu carnium* I-II
(993A-999B); Aelian in *NA*; Oppian in *C.*; Celsus in Origenes *Cels.* (especially Book IV); Porphyry in *Abst.* (especially Book III); in certain of
the extant works of the Peripatetic Alexander of Aphrodisias: Περὶ
ψυχῆς I, Περὶ εἱμαρμένης; *et al.*

[179]*P.* i. 62-77.

[180]K. Reinhardt, *Poseidonios* (Munich, 1921), pp. 39-58.

[181]The Stoa provided Philo with a number of useful illustrations; two
of those can be traced to Chrysippus: that of the hound and that of the
peacock (see §§ 45-46, 84 [*SVF* II 726], 88 and comment; see also *SVF* II
1163, 1165; cf. *ibid.* 714-737). Note that Chrysippus wrote *On Providence*
(Περὶ Προνοίας, *SVF* III, App. II, p. 203) and Antipater of Tarsus wrote
On Animals (Περὶ ζῴον, *SVF* III Ant. 48). Among later Stoics and others,
note Cicero, in Book III of *Fin.*, which derives from traditional Stoic
sources, and Book II of *ND*, which provides the closest parallels to Provid II; Seneca, in Book I of *Dial.*, in the various *Ep.*, and in the preface to *Nat.*; Epictetus in his *Diss.*; and others, including Christian
apologists like Origenes in *Cels.*; *et al.*

[182]Numerous examples of his supporting the Stoa against the Academy
can be cited throughout his responses to Alexander. In defending divine
providence the distinction was made between God's primary works and the
secondary or consequential effects (Provid II 100); e.g., eclipses are
consequential effects, not God's primary intentions (79). Moreover,
Philo is at one with the Stoics in the fundamental position that the
irrational creation exists for the sake of the rational (see above, p. 45).

It is indeed difficult to discuss any aspect of Stoic doctrine without considering the interconnected mosaic of the whole philosophy--and that in the light of the fragmentary evidence. No doubt the related questions of providence, the nature of animals and their relationship to humans and the rest of nature are among the problems in Stoicism.[183] In a broad sense, the workings of providence range throughout the universe: from the majestic cycles of heavenly bodies to the minutest anatomical details of insects.[184] In support of their doctrine of providence the Stoics brought forward evidences of design throughout nature. They attributed to the workings of providence the peculiar characteristics of the various creatures: their inclinations or dispositions to move in a particular direction or act in a certain way as a result of some inherent quality or habit. They argued that the apparent evidences of reasoned action shown by animals are not due to reason but to their natural constitution; the universal, causal reason is at work and not that of the animal.[185]

As to the relationship of animals to humans, the Stoics argued that the irrational creation providentially exists for the sake of the rational, that animals were created for the sake of humans--just as humans were created for the sake of the gods.[186] This anthropocentric teleology, which finds its strongest proponent in Chrysippus,[187] characterizes the

[183]J. Christensen, *An Essay on the Unity of Stoic Philosophy* (Copenhagen, 1962), *passim*.

[184]Provid I 51-53; Epict. *Diss.* i. 16. 1-8; Origenes *Cels.* iv. 54; etc.

[185]*SVF* II 714-737, 988.

[186]*SVF* II 1152-1167; cf. the selections from Provid II in *ibib.* 1141-1150.

[187]Chrysippus went so far as to say that the pig was made more fecund than other animals in order to be a fitting food for man or a convenient sacrifice to the gods, horses assist man in fighting, wild animals exercise man's courage; the peacock is created for his tail and the peahen for accompanying symmetry, the flea is useful to prevent oversleeping, and the mouse to prevent carelessness in leaving the cheese about (Porph. *Abst.* iii. 20 [*SVF* II 1152]; Plu. *Mor.* 1044C-D [*SVF* II 1163]; Porphyry adds that Chrysippus' views were criticized by Carneades [*ibid.*]; Plutarch reiterates similar views in 1065B [*SVF* II 1181]: the serpent's venom and the hyena's bile are useful in medicine). Philo reflects similar views (see above, n. 160). Pohlenz thinks such anthropocentricism is alien to Greek thought and hence must be of Semitic origin (*Die Stoa. Geschichte einer geistigen Bewegung*, I [Göttingen, 1964], 100).

whole cosmology of the Stoics. It is well known that they studied the cosmos primarily to understand man's place in the realm of things.[188] They explained a creature's self-consciousness and relationship to its environment by formulating the doctrine of affinity or endearment (οἰκείωσις),[189] which was held as a natural principle of justice. Among animals this principle is seen in their longing for self-preservation, in their love for offspring, and even in the association of different species for their mutual advantage; moreover, it is attested through its opposite, the natural aversion or antipathy (ἀλλοτρίωσις) between certain species.[190] With humans, however, this relationship is so intimate and peculiar that it would be unjust to extend it to lower animals. The Stoics explained man's self-consciousness in terms of his rationality and his affective relationships, beginning, naturally, with relations according to propinquity and moving on to the rest of mankind.[191] To this fraternity of rationals as a *civitas deorum atque hominum* lower animals do not belong, for they are unequal in that they do not possess reason. Therefore, the Stoics maintained, there is no such thing as a justice which can obtain between humans and animals. Man cannot be charged with injustice when he makes unilateral use of animals, for it is to this end that animals are providentially made or naturally equipped.[192]

The Stoa became vulnerable to the attacks of the New Academy for cherishing these and other views, particularly those on the role of sense-perception--which is also shared by creatures without reason--in the

---

[188]D.L. vii. 88; cf. M. Ant. iv. 23; x. 6.

[189]S. G. Pembroke, "Oikeiōsis," *Problems in Stoicism*, ed. A. A. Long (London, 1971), pp. 114-149.

[190]See the passages cited under *De primo appetitu et prima conciliatione* in *SVF* III (178-189).

[191]In addition to the Ciceronian passages in the reference above, see the numerous other references in Pembroke's excellent article, "Oikeiōsis," especially pp. 121-132; cf. Mut 226; Vita Cont 70.

[192]See the passages cited under *Iuris communionem non pertinere ad bruta animalia* in *SVF* III (367-376) and those cited under *Animalia (et plantas) propter hominum utilitatem facta esse* in *SVF* II (1152-1167).

acquisition of knowledge.[193] The Academics argued that animals cannot
make use of sense-perception without some knowledge or understanding:
"the impact on eyes and ears brings no perception if the understanding is
not involved"; "all creatures which have sensation can also unterstand";
etc.[194] But it was not so much the rationality of animals that the opponents of the Stoics wanted to emphasize as the denial of the possibility
of knowledge or the existence of any positive proof or criterion for
truth. They wanted to maintain an attitude of suspended judgment and
thus utilized the question of rationality among the brutes.[195] To support their major arguments they often made use of material gathered by
the Stoics themselves.[196] The more the Stoics attempted to explain their
tenets, the more they exposed themselves to the criticism of the Academics.

Like Philo in Anim, the Stoics answered their critics by attributing
the workings of animals to the reasoning of nature (φύσις).[197] We regret
that it is not possible to discuss briefly that broad conception of
nature or universal reason inherent in the Stoic monism.[198]

5. The Sources

There are only a few acknowledgements of authorities in Anim: a
quotation attributed to Homer, "the poet" (§ 54); another to Hesiod
(§ 61); an anonymous verse, possibly of Orphic origin, attributed to
"the poets" (§ 47); a mention of Aesop by name (§ 46); and possible
allusions to Aristotle's followers: "men occupied neither with hunting

---

[193]SVF II 105-121; cf. Philo's understanding of the interaction between
sense-perception and mind (see above, pp. 42-43).

[194]Plu. Mor. 960D-961B.

[195]Cf.Pohlenz, Die Stoa, I, 37-63; II, 21-36; F. H. Sandbach, "Phantasia Kataleptikē," Problems in Stoicism, ed. A. A. Long (London, 1971),
pp. 9-21.

[196]See, e.g., the accounts on the spider and the hound (§§ 17, 45-46,
84 and comment).

[197]Cf. SVF II 714-737, 988 with §§ 77-100; Tappe, De Philonis libro qui
inscribitur 'Αλέξανδρος, pp. 38-49.

[198]On the totalizing value given to nature, see SVF II 1106-1186.

54

nor with inanimate things found in the forest" (§ 16); "the witnesses" (§ 61); and the "researchers into the history of animals" (§ 97).

Many of the illustrations used by Philo recur persistently whenever the question of animal intelligence comes into consideration in ancient authors.[199] Pliny in Books VIII and IX of his *NH*,[200] Plutarch in *De sollertia animalium* (*Mor.* 959A-985C), *Bruta animalia ratione uti* (*Mor.* 985D-992E), and *De esu carnium* I-II (*Mor.* 993A-999B), and Aelian in *NA*, use substantially the same illustrations, not always consecutively but scattered in various parts of their compilations. Philo's Anim is the earliest of these works whose dependence on common sources is readily discernible, especially in instances where they maintain a consecutive order of two or more illustrations. From this it would seem that in a number of places Philo has grouped together, more or less closely, passages derived from one and the same source.[201] Of the various writers, Plutarch in his *De sollertia animalium* (*Mor.* 959A-985C) provides the closest parallels to Anim, not just by way of illustrations but in subject matter as well.

M. Wellmann[202] rightly attributes the close agreement in many passages of these writers to the use of common sources, not to direct borrowing of the one from the other--though in certain passages Aelian seems to have

---

[199]A. Dyroff, "Zur stoischen Tierpsychologie," *Blätter für das bayerische Gymnasialschulwesen*, 33 (1897), 399-404; 34 (1898), 416-430; S. O. Dickerman, "Some Stock Illustrations of Animal Intelligence in Greek Psychology," *Transactions and Proceedings of the American Philological Association*, 42 (1911), 123-130.

[200]See the exhaustive list of authorities cited by Pliny in his outline of *NH*, comprising Book I; see especially the authorities used for Books VIII-IX, dealing with zoology (LCL I, 41-65).

[201]Tappe, *De Philonis libro qui inscribitur* Ἀλέξανδρος, pp. 22-38, 49-54.

[202]His source-analytical studies appear in a series of articles: "Sostratus, ein Beitrag zur Quellenanalyse des Aelian," *Hermes*, 26 (1891), 321-350; "Alexander von Myndos," *ibid.*, 481-566; "Juba, eine Quelle des Aelian," *ibid.*, 27 (1892), 389-406; "Leonidas von Byzanz und Demostratos," *ibid.*, 30 (1895), 161-176; "Aegyptisches," *ibid.*, 31 (1896), 221-253; "Pamphilos," *ibid.*, 51 (1916), 1-64.

---

made use of Plutarch.[203] Wellmann traces the original sources to celebrated authorities in antiquity: Aristotle and his monumental *HA*; Alexander of Myndos, the most renowned of all ancient ornithologists; the
once notorious Sostratus, the authority on insects, lower animals, and
snakes; Theophrastus, whose botanical works were as important as Aristotle's works on animals; and Apollodorus, the leading authority on
poisons and their antidotes. Among later writers he cites King Juba of
Mauritania, whose curious accounts of Asian and African expeditions were
most fascinating to later naturalists and explorers.

Our concern is not so much with the ultimate authorities as it is with
the immediate sources. Philo, like the others, could not have drawn these
illustrations mentioning eighty-two fauna and flora specimens[204] from
original works but from secondary sources, as the stock illustrations
suggest. A common use of ever-increasing epitomes by Philo, Pliny, Plutarch, Aelian, and others is to be suspected. Compilations of epitomes, in
which the specialized works of earlier authorities and others appeared
side by side with folklore, legends, and comments of scholiasts, can be
traced from one lexicographer of the Alexandrian school to another, from
Aristophanes of Byzantium (*ca.* 257-180 B.C.), to Aristarchus of Samo-
thrace (*ca.* 217-145 B.C.), to Didymus, nicknamed Χαλκέντερος and Βιβλιο-
λάθας (*ca.* 80-10 B.C.), to Pamphilius of Alexandria (flourished after A.D.
50), and on to others. Among the latter's compilations, a great lexicon
of ninety-five books, Περὶ γλωσσῶν ἤτοι λέξεων, absorbed many previous
collections. It was used by Aelian and Athenaeus, and later abridged by
a succession of epitomators. The last stage of this process is seen in
the extant lexicon of Hesychius of Alexandria.

Of particular interest are the accounts of events which Alexander (or
rather Philo) claims to have witnessed but are common to other writers
and obviously drawn from literary sources. The story of the monkey driv-
ing a chariot (§ 23), claimed to have taken place "the day before yesterday," occurs in Ael. *NA* v. 26. Another "eye-witness" account, likewise

---

[203]See A. F. Scholfield's "Introduction" to Aelian's *NA* in LCL I,
xx.-xxiii; cf. Tappe's excursus in *De Philonis libro qui inscribitur
'Αλέξανδρος*, pp. 70-77.

[204]See Ind. II, pp. 295-300.

claimed to have taken place "the day before yesterday," is that of the chariot race (§ 58); it occurs in Pliny *NH* viii. 160-161. A similar adaptation from a literary source to a contemporary scene is found in the story of the Thracian falcons (§ 37), claimed to have been verified by Thracian visitors the Alexandrian met in his native land; however, the story is told less elaborately in Pliny *NH* x. 23 and somewhat differently in Ael. *NA* ii. 42 and can be traced to Arist. *HA* 620a33-620b5, the ultimate source. The ingenious adaptation and appropriation of such literary accounts are noteworthy; they cast doubt on the veracity of other sporting and theatrical events which Philo claims to have witnessed, such as at the gymnasium and the theater.[205]

Among other evidences of Philo's use of sources, one may note the following possible hints in the text: "One [example] ought to be cited and the rest passed by in silence" (§ 50); and again: "I know of many other things which can be passed by in silence" (§ 80). From these statements one may deduce that Philo had a document before him.

In the preceding section we have already mentioned Philo's structuring of Anim after the first part of the Platonic *Phaedrus* and discussed the Academic-Stoic controversies which furnished not only the arguments and counter arguments but also several of the illustrations utilized in this treatise and in Provid I-II.[206] The main Stoic authority relied on was probably Posidonius,[207] the possible main source of Book II of Cicero's *ND*, which has much in common with these works of Philo.

---

[205]See §§ 24 and 52; cf. Provid II 103; Ebr 77; Quod Omn 26, 141; see also A. Mendelson's discussion on whether Philo participated in the cultural life of Gr. Alexandria: "A Reappraisal of Wolfson's Method," *Studia Philonica*, 3 (1974-1975), 11-26.

[206]See above, pp. 49-53.

[207]Wendland, *Philos Schrift über die Vorsehung*, pp. 83-84, identifies Posidonius as a primary source; cf. his *Die philosophischen Quellen des Philo von Alexandria in seiner Schrift über die Vorsehung*, Programme, LIX (Berlin, 1892); so also Reinhardt, *Poseidonios*, pp. 362-365; and M. Pohlenz, "Tierische und menschliche Intellegenz bei Poseidonius," *Hermes*, 76 (1941), 1-13.

6. *Text, Translation, and Commentary*

Except for the title of the treatise and three short excerpts in patristic catenae,[208] this work of Philo is extant only in Arm. The work, to be sure, is complete.

The text used for this translation is D, MS 1040 of the Venetian Mechitarists, the *textus receptus* of Aucher. Fortunately, he based the 1822 *editio princeps* of Anim on this MS alone,[209] emending it in but few instances. Excluding his parenthesized and italicized additions, the following are his departures from the MS:

§ 16   հոգի] հոգ
§ 27   ծանրութիւն] ծանրութեամբ
§ 36   փախստեան] փախստան
§ 46   դուզնաքեա] դուզնաքեայ
§ 96   բբայց] բայց
       զերկս] զերկ

Only the emendation in § 27 is substantive. The one in § 16 and the last one in § 96 are unintentional, as gathered from the Latin translation (*animus* in one and *laborem* in the other). The rest are orthographical. Except for the inadvertent departures, Aucher's emendations have been retained.

The following additional emendations have been made:

§ 27   հանդերձեալ] համբարձեալը
§ 37   մարտակցացն   2°] om
§ 38   բադիդիցն] փադանցացն
§ 66   մատակարարանին] մատակարարականին
§ 95   կադամբ] կադամբբ

---

[208]See App. II, p. 263.

[209]According to Lewy (*The Pseudo-Philonic De Jona*, p. 2), the text of Aucher's second volume is based on D, "but he frequently corrected it from the other [E] without noting the fact." According to Marcus (LCL Supplement I, vii), Aucher's second volume is based chiefly upon three MSS and in part upon two others. Both Lewy and Marcus seem to be overcritical. Although Aucher boasted of other MSS (see *Paralipomena Armena*, pp. ii, 622), his text is that of D, *seldom* relying on others.

The emendations in §§ 27 and 38 are derived from marginal readings.
The first of these is erroneously parenthesized by Aucher as *ամբարձաւէր;*
the second is supported by classical parallels cited in the Commentary.
The one in § 37 is *duplus* and the last two are orthographical. Moreover,
obvious interpolations have been identified in §§ 20 and 22 (enclosed in
brackets []).

One finds unnecessary but consistent modernization of spelling in
Aucher's text. Thus, *աւ* is changed to *o*, *հ* following *բ* is altered
to *է* in *թէ, էթէ, թէպէտ,* and *եւ* in *արբելր, արդելր, եղջելու,*
եղջելրուի is modified to *արբիլր, արդտօր, եղջիլու, եղջերուի.*

Since Aucher's punctuation of the text does not always indicate logical pauses, it has often been disregarded. And although sometimes his
section divisions are abrupt, they have been adopted with reluctance--
except for departures at the juncture of §§ 41-42 and 45-46. The section
enumeration is that of C. E. Richter's edition.[210]

It may be said with no disparagement of textual criticism that for all
practical purposes Aucher's edition of the Arm. text is fairly reliable.
After pointing out its few flaws, there is really no reason for not accepting it as a definitive text, especially since the substantive variants of the five other MSS[211] do not yield important results for the text.
It is reprinted here with the Latin translation (App. I, pp. 211-262).

There is really not much to be done with the Arm. text of Philo. Because of the syntactical ambiguities,[212] no attempt has been made to
emend the text beyond obvious corruptions of words. Certainly, there can
be no thorough emendation of a translated text, which leaves much of the
lost original to be desired, without thorough retranslation. In view of
the tendencies of the Arm. translator(s) of Philo, a reconstruction of

---

[210]*Philonis Iudaei opera omnia. Textus editus ad fidem optimarum
editionum*, VIII (Leipzig, 1830), 101-148, which has Aucher's Latin translation for the Arm.

[211]See above, pp. 23-25. The substantive variants in our collation do
not provide superior readings (all other variants, those which do not
affect meaning and are of orthographical and scribal character, have been
excluded).

[212]See above, pp. 9-14.

the Gr. original is impossible.[213] Partial retranslation may be under-
taken to the extent of clarifying extreme ambiguities.[214]

Aucher was aware of the peculiar and hardly intelligible syntax of the
Arm., resulting from a word-for-word rendering of the Gr.[215] The Latin
translation, however, betrays lack of a thorough understanding of the Gr.
syntax underlying the Arm. This is most noticeable in the translation of
passages where the meaning of the Arm. at first seems to be completely
obscure.[216] The resultant translation abounds not only with inconsisten-
cies but also with portions that are meaningless. Even in places where
Aucher seems to have rendered every word of the Arm., some sentences do
not convey the general sense. His translation of such "obscure" passages
also reveals strong tendencies toward amplification, thus leading to questions about not only the intelligibility of the translation but also its
reliability.[217] There can be no doubt that the Latin translation is

---

[213]Cf. Lewy, *The Pseudo-Philonic De Jona*, pp. 20-22. Lewy, however,
attempted to reconstruct the Gr. of the entire homily before concluding:
"any hope of a *complete* retranslation must be given up." The attempted
reconstruction of the *De Jona* is found in his archive at The Jewish
National and University Library, Jerusalem (Ms Var. 376-15).

[214]As is done by Marcus in his edition of the *Quaestiones* (LCL Supplements I-II).

[215]Aucher, p. ii. He must have prepared for his personal use an Arm.-
Gr. word index, much of which was utilized in the exhaustive *Nor baṙgirk'
haykazean lezowi* ("New Dictionary of the Armenian Language"), I-II, ed.
G. Awedik'ean, X. Siwrmēlean, and M. Awgerean [J. B. Aucher] (Venice,
1836-1837).

[216]At the outset of his translation of Anim, Aucher admits: "The
entire introduction of this dialogue [§§ 1-9] is very obscure. Learned
readers may find an alternate way of clarifying it," p. 124; for the
Latin, see App. I, p. 214.

[217]See my translation notes, pp. 67-108. When questioning Aucher's
Latin translation, Tappe consulted a certain Andreas (sic), *De Philonis
libro qui inscribitur* Ἀλέξανδρος, pp. 3, 5, 14, 42, as did Hadas-Lebel
with M. l'Abbé Mércier, *De Providentia I et II*, pp. 12, 22 (her work,
primarily, is a translation of the Latin, transmitting many of Aucher's
errors); cf. Colson's remarks on Provid (LCL IX, 449, 541, 546). Lewy
rightly observes: "He [Aucher] tried at all costs to read a clear sense
into the Armenian. When this could not be done directly he freely para-
phrased the ideas, and did not hesitate to adopt fanciful improvisations
which had scarcely any connection with the Armenian text." *The Pseudo-*

marred by the deficiencies in Aucher's mastery of Gr. as well as his insufficient knowledge of Hellenistic-Jewish and classical literature.[218] Nonetheless, one also appreciates the exact sense of many Arm. passages conveyed in the Latin.

Quite a few of the syntactical problems have been solved by Gr. retranslation, some of which is given in the translation notes and in the Commentary. The rendering of somewhat obscure words and phrases is also based on Gr. retranslation. The reliability of the retranslation is to be seen in that all Gr. equivalents to Arm. words are derived from Philonic works extant in both languages.[219] It is not to be assumed, however, that the Arm. and Gr. equivalents are consistently the same in those works of Philo extant in both languages. In choosing the Gr. equivalent to an Arm. word, the relative frequency of all its different renderings and their occurrences in conjunction with particular words in Philo have been considered.

No doubt some corruptions still remain; these have been translated without drastic departure from the text and their literal meanings have been given in the translation notes. Similar notations were made when a

---

Philonic De Jona. Marcus points out scores of errors in the Latin throughout his edition of the *Quaestiones* (LCL Supplements I-II, which, although imperfect, abound with improvements upon Aucher's translation). In "The Armenian Translation of Philo's Quaestiones in Genesim (sic) et Exodum," *Journal of Biblical Literature*, 49 (1930), 62-63, Marcus remarks: "A comparison of Aucher's Latin with the Armenian, in those passages for which the Greek is extant, will show that in every case the Armenian is closer to the Greek than Aucher's Latin to the Armenian," and goes on to give a number of examples. See also his "Notes on the Armenian Text of Philo's *Quaestiones in Genesin*, Books I-III," *Journal of Near Eastern Studies*, 7 (1948), 111-115.

[218]Note that he fails to identify--at times even to recognize--any of the quotations in Anim: §§ 28, 33, 47, 54, 61; cf. Colson's comment on Provid II 100 (= § 50, LCL IX, 546).

[219]See Ind. III, pp. 301-305. In indexing the Arm.-Gr. equivalents, I relied in part on an unpublished and incomplete index prepared by H. Lewy (Jerusalem, The Jewish National and University Library, Ms Var. 376-13); cf.the sampling by Marcus, "An Armenian-Greek Index." To maintain consistency in the retranslation and translation, I indexed all Arm. terms of scientific, philosophical, or general significance with all their Gr. equivalents, checked Leisegang's *Indices*, I-II, for their usage and G. Meyer's *Index Philoneus* (Berlin, 1974) for additional references, and often followed the renderings of Colson and Whitaker in the LCL edition.

literal translation failed to convey the thought of the original and also
when the English idiom was taken into consideration. Aside from these
exceptions, a literal translation was maintained throughout. Several
factors contributed to such a literal translation: (1) the problematic
state of the text; (2) the quest for Philonic and classical parallels;
(3) checking on Aucher's Latin translation; and (4) making this first
translation of Anim into a modern language readily translatable into
other languages. Great care has been taken lest there be any paraphrasing; also, lest alien thoughts emenating from classical parallels be
introduced in places where no such thoughts are embedded in the original.

Except for auxiliary verbs, adverbs, personal pronouns, and articles
employed in rendering participles, all supplied words have been italicized. Occasionally, tenses have been modified and certain verb forms
changed to noun forms and vice versa. Unfortunately, it still seems
necessary to remind even learned reviewers that one may depart from uni-
formly translating *pwʼuqʰ* (γάρ) by "for," *bι* (καί) by "and," or *pwjg*
(δέ) by "but," and that sometimes the translator may omit either of the
above words--or an article--as the most idiomatic way of translating it.
Suffice it to say that the translation was pursued with all the awareness of translation problems.

The difficulty of identifying the fauna and flora in Anim need not be
emphasized.[220] Parallel accounts in Aristotle, Pliny, Plutarch, and
Aelian helped in some of the more difficult identifications, especially
in instances where a name is not a transliteration of the Gr. and is rare
or *hapaxlegomenon* in the Arm. corpus of Philo. Sometimes Irenaeus'
Adversus haereses,[221] which also was translated by the Hellenizing
School,[222] helped in the identification. Much help was derived from
D'Arcy W. Thompson's glossaries of Gr. birds and fishes.[223] Much is also
indebted to the various editors and translators of the works in the LCL

---

[220]See Ind. II, pp. 295-300.

[221]Reynders, *Lexique comparé*, I-II.

[222]See above, pp. 7-8.

[223]*A Glossary of Greek Birds* (rev. ed.; London, 1936) and *A Glossary
of Greek Fishes* (London, 1947).

series which were often used.

The Commentary is not exhaustive. Primarily, it is an aid for a better understanding of the treatise as a whole. Its strength lies in the coverage of Philonic and classical thoughts paralleling those in Anim, especially those emanating from the Academic-Stoic controversies. Its brevity leaves room for amplifications, especially for inclusion of references to the vast *hexaemera* literature of the Fathers--even though their cosmology depends more on Philo's Op and Leg All I-III than on Anim.[224] The *Commentarius in hexaemeron* attributed to Eustathius of Antioch is the only known work which shows definite reliance on Anim.[225]

7. *Synopsis*

The treatise falls into two parts, each preceded by a dialogue between Philo and his interlocutor, Lysimachus (§§ 1-9, 72-76). Alexander's discourse on the rationality of animals, purportedly read in his presence, comprises the first part (§§ 10-71) and Philo's refutation of Alexander's premise, the second (§§ 77-100).

Alexander begins his polemical discourse with sweeping denouncements of man's appropriation of reason to himself (§§ 10-11). He attempts to show among the brutes instances of the προφορικὸς λόγος, the reason which finds utterance and expression (§§ 12-15). Then follows a lengthy argument for their possession of the ἐνδιάθετος λόγος, the inner reason or thought (§§ 16-71). Some talking and singing birds constitute the examples for the first kind of reason; but the examples for the second kind

---

[224]Special mention must be made of some of the more relevant works: Clem. Al. *Strom.*, Origenes *Homily on Genesis*, *Cels.*, Basil of Caesarea *Homilies on the Hexaemeron*, Gregory of Nyssa *Apologetica in hexaemeron* and *De opificio hominis*, Nemesius of Emesa *De natura hominis*, and, to some extent, Theodoret of Cyrrhus *De Providentia*. There is also a lost work of Tatian, *De animalibus*, cited among the lost books of early Christian literature in E. J. Goodspeed, *A History of Early Christian Literature*, revised and enlarged by R. M. Grant (Chicago, 1966), p. 198.

[225]Migne *PG* XVIII, 725, on the polyp, cf. § 30; 728-729, on the crocodile, the Thracian falcons, and the peacock, cf. §§ 50, 37, and 88; 740-741, on bears, bulls, boars, and deer, cf. §§ 51 and 33; 741, on the Cretian goats, cf. § 38; 745, on the snake fight, cf. § 52; and 748, on ants, cf. § 42.

of reason are far more numerous, including not only such stock examples
as spiders, bees, and swallows (§§ 16-22), but also several performing
animals (§§ 23-29) and a large number of others in whom demonstrations
of virtues and vices are believed to exist (§§ 30-71). The four cardinal
virtues are emphasized (§§ 30-65): wisdom or prudence (φρόνησις, §§ 30-
46), self-restraint (σωφροσύνη, §§ 47-50), courage (ἀνδρεία, §§ 51-59),
and justice (δικαιοσύνη, §§ 60-65). The exact opposites of these virtues
are discussed under the vices (§§ 66-70): lack of self-restraint
(ἀκολασία, §§ 66-67), cowardice (δειλία, §§ 68-69), injustice and foolishness (ἀδικία, ἀφροσύνη, § 70).

Philo, haphazardly and with some lack of kindness to his opponent,
argues that animals do not possess reason. He ascribes the seemingly
rational acts of animals to the reasoning of nature (§§ 77-100).

By emphasizing the rationality of animals, Alexander argues for a
moral and juridical relationship between man and animals (§ 10, his opening remarks). This Philo rejects by insisting that there can be no equality between man, who is privileged with reason, and animals devoid of it
(§ 100, his concluding remarks).

II. TRANSLATION

ON WHETHER DUMB ANIMALS POSSESS REASON[1]

(1) *PHILO:* You remember the recent[1] arguments, Lysimachus, which Alexander, our nephew,[2] cited in this regard, that not only men but also dumb animals possess reason.

(2) LYSIMACHUS: Admittedly, honorable Philo, some differing opinions have been amicably brought to the speaker[1] three times since then,[2] for he is *my* uncle,[3] and *my* father-in-law as well. As you are not unaware, his daughter is engaged[4] to be my wife. Let *us* resume the discussion[5] of this long, difficult, and wearisome *subject*[6] and its absurd interpretation[7] which does not appeal to me since it affects the clear light by distorting the obvious evidence.[8]

---

[1]Aucher, following the Gr. title cited by Eusebius (see App. II, Frag. 1), adds "(Dicebat Alexander)."

(1)   [1]Lit., "of yesterday."

[2]Lit., "our brother's son," the pl. form of the pron. denoting modesty. Cf. §§ 2, 72, 75.

(2)   [1]i.e. to Alexander.

[2]Aucher renders, "unum est istud trium illorum, quorum inspectiones convincentes efficiunt, ut auditur iterum veniat ad eum, qui locuturus est." He construes the ablatival sense of the first part differently, adds "ut auditur iterum," and omits ᚠ𝑛𝑏𝑝𝑑𝑛𝑖[𝑏𝑡𝑢𝑤𝑑𝑝 (μετ' εὐνοίας).

[3]Lit., "mother's brother."

[4]Lit., "promised by an agreement."

[5]Aucher has "Dictorum vero recordatio," modified by the following two adjs. Gr. retransl. of the last word, καινότης.

[6]Aucher renders loosely, "qui non est omnino consuetus his rebus."   [7]Aucher renders, "maxime circa arduam quaestionem." Gr. retransl., πάλιν καὶ περὶ τῆς χαλεπῆς ἑρμηνείας.

[8]Aucher renders somewhat differently, "cujus evidentia, certitudinis lumen oriri fecisset, si agitata fuisset; et quasi circumversata

(3) PHILO: *With regard to* great assertions, it is agreed[1] that *one ought* to listen to them carefully,[2] for nothing else seems to be so helpful to good learning as to critically examine what the lecturer is emphasizing.[3] Had he truly wished to continue[4] learning, he would not have allowed himself to become occupied with other concerns. Tell me, why would he leave *his* other affairs and come *merely* to entertain a relative[5] with useless words designed to tickle the ears?[6] Such *an action* would be considered neither kind nor appropriate by that person who has already rejected his former courtesies.[7] Therefore do not[8] anticipate receiving a particularly significant response to your request.[9] You will not get very far[10] with *your* request.

(4) LYSIMACHUS: Is not *his* want of leisure,[1] Philo, the reason? You are not unaware of how many things are involved given relatives, the

---

ut imprimeretur, non superficie tenus audita" and adds the following note on the obscurity of the text: "Totus introitus hujus dialogi obscuritatem nimiam habet: eruditi lectores aliunde possunt declarationem aliquam adinvenire." The last phrase seems to be an idiom expressing dissatisfaction. Gr. retransl., οὐκ εἰς τὸ πρόσωπόν μου ἀκο<ύετ>αι.

(3) [1]Aucher renders, "Magna res certe est magistrum amare." Gr. retransl., <τῆς δὲ> μεγά<λης> εὑρέσ<εως> ὁμολεγεῖται.

[2]Lit., "eagerly."

[3]Aucher observes in a note, "si legas դիտել [ἐπισκέπτεσθαι] pro գիտել [γινώσκειν] tunc sensus erit, *attendere magistrum ea, quae vult suggerere.*"

[4]Aucher has "cupiat." Gr. retransl., ὑπομένῃ.

[5]Lit., "brother." The indef. pron. ոք (τις) adds to the nonliteral denotation of the word.

[6]Aucher renders լսելեաց (ἀκοαῖς) as "auditoribus," and in § 5, correctly, "audiendi." Elsewhere, in inflected forms, "ad aures" and "auribus," §§ 12, 15.

[7]Possibly an allusion to Provid I-II. Aucher, misconstruing the demonstr. pron. այնորիկ (ἐκεῖνος, whereby Philo refers to himself), renders, "At tu fortasse eam ob causam, ne videare facile opus peragere, similemque esse iis, qui prompte gratiam projiciunt in conspectum." Gr. retransl. of the last part, τὰς πρώτας χάριτας.

[8]Lit., "retract from."

[9]Aucher renders, "supplicationes rogantis majores expectans." He omits յորդորումն (ἡ προτροπή), here translated "response."

[10]Lit., "you will hardly succeed."

(4) [1]Or, as Aucher has, "Occupationes." Gr. retransl., ἀσχολία.

classes,[2] and community[3] affairs at home.[4]

(5)   PHILO: Since I know that you are interested,[1] *indeed* that you
are always eager to hear new things, I shall begin to speak if you will
keep quiet and not always interrupt my speech by making forceful *remarks*
on the same matter.

(6)   LYSIMACHUS: Such a restrictive order is unreasonable.[1] But
since it is expedient[2] to seek and to ask for instruction, your order
must be complied with. So here I sit quietly, modestly, and with restored humility as is proper for a student; and here you are seated in
front of *me* on a platform looking dignified, respectable, and erudite,
ready to begin to teach your teachings.

(7)   PHILO: I shall begin *to interpret*, but I will not teach,[1] since
I am an interpreter and not a teacher. Those who teach impart their
own knowledge[2] to others, but those who interpret present through accurate
recall the things heard from others. *And they do* not *do this* just to a
few Alexandrians and Romans--the eminent *or* the excellent, the privileged, the elite of the upper *class*, and those distinguished in music
and other learning[3]--gathered at a given place.

(8)   The young man[1] entered in a respectful manner, without *that* overconfident bearing that some have nowadays, but with a modest self-reliance that becomes a freeman--even a descendant of freemen. He sat down
partly for his own instruction and *partly* because of his father's con-

---

[2]Presumably the upper classes in the society.

[3]Or, "state." Lit., "of this city."

[4]Presumably the Alexandrian Jewish community. Aucher, supplying
the personal and possessive prons., which are not implied, renders,
"quot cognatorum, et classium, civitatisque me circumduxerint examina
negotiorum in ipsa mea habitatione."

(5)   [1]Aucher has "non recusare."

(6)   [1]Lit., "unphilosophical."

   [2]Arm., "it is expediant and profitable," a rendition of ἀνυσιμώ-
τατον. See App. II, Frag. 2.

(7)   [1]Lit., "but not teaching."

   [2]Or, "skills," "arts" -- whether applied or fine.

   [3]Lit., "philosophy."

(8)   [1]i.e. Alexander. Cf. § 75.

tinuous, insistent urging.[2]

(9) Eventually one of the slaves, who was sent to a place nearby,
brought the manuscripts.[1] Philo took *them and* was about to read.[2]

ALEXANDER[1]

(10) Fathers' *care* for little children, guardians' responsibility
for orphans, and men's protection of women are essential to fulfill the
needs of nature. Those who can give help should not hesitate. The foregoing, that is to say complementing one another's weakness, should be
extended not only to mankind but also to all animals. I am not unaware
that selfishness is a common evil that eclipses the light leading to
the knowledge of truth, that eats its cankerous way into the soul and
tears down its sovereign part. Furthermore, those who dislike instruction detest such a discourse and are bored by it.[1] But we who learn
what belongs to wisdom, being cleansed from filth, with pure[2] minds
follow pure[2] truth, which is the most precious possession God has ever
given to man.

(11) Just as men ignore the weakness of women--as is common in every
community[1] whether in times of war or peace--*and* subjugate them only to
themselves, considering the disadvantaged female sex unfit for state affairs,[2] so, I think, when humans saw all the dumb animals bending downward

---

[2]This section is ascribed to Lysimachus in all the MSS. The last
phrases are not altogether clear. Aucher renders somewhat differently,
"tum ob patris liberalitatem, quam usurpavit propter id, quod prius dictum fucrat," and, in a note, "Sive quod jam scriptis mandaverat."

(9) [1]i.e. Alexander's manuscript.

[2]Ընթեռնույր (ἀνεγίνωσκε) is rendered as a conative imperfect.
Philo must have been prevented by Lysimachus who reads Alexander's
treatise out loud. See §§ 72-74.

[1]Aucher supplies "(Sermo Alexandri de probanda brutorum animalium ratione)."

(10) [1]Lit., "such a discourse is detestable and boredom *is caused* by
such a discourse."

[2]Lit., "naked."

(11) [1]Or, "city."

[2]Or, "state management."

to the earth, whereas they themselves stood upright and erect upon the
ground, they differentiated between their own good attributes and the
condition of the dumb animals.[3] And since their minds were elevated as
well as their bodies, they held the earthly creatures[4] in disdain. Reason is the best of things that exist, but they attributed none of it to
animals.[5] Rather they appropriated it to themselves as though they had
received an irreversible reward from nature. Since man[6] possesses reason, he may refute a fallacy by disproving *it*: because of *his* ability
to learn *and his* desire to discern clearly, he will discover truth.[7]

(12) However there are two kinds of reason: the one located in the
mind is like a spring which issues from the sovereign part of the soul,
whereas uttered *reason* is like a stream which, in the natural usage,
courses over the lips[1] and tongue and on to the sense of hearing. But
although both *kinds of reason* appear to be imperfect in animals,[2] they
are none the less fundamental.[3]

(13) But certainly the abstract concepts which the mind apprehends
in connection with[1] sensation are perceived[2] through hearing. This will
be considered first. Indeed many *birds* learn to articulate by listening
for themselves or are inherently self-taught, and many are taught to
talk: such as crows and Indian parrots. I know of parrots brought into
the wealthy homes of Alexandria in Egypt, which reiterate in a loud voice
like schoolchildren. Once they learn the sound of an amorous kiss by

---

[3]Lit., "their condition."

[4]Lit., "those that seek the ground." Said also of plants, § 20.

[5]Lit., "to them."

[6]Lit., "he."

[7]Aucher renders, "Quod tamen et animans participationem habeat
rationis, contentiosam siquis deponat oppugnationem, facile per studium
disciplinae verum quaerentis assequetur exactissime." He substitutes
"animans" for ἱῡ (αὐτός or ἐκεῖνος) and mistranslates most of the rest.

(12) [1]Lit., "through the mouth."

[2]Lit., "in them."

[3]Lit., "the elementary and foundational things."

(13) [1]Lit., "which also reach."

[2]Or, "are experienced," "are received," etc.

demonstration, they greet with kisses and *can* respond to greetings by
name. They are an especial adjunct[4] to princes; they flatter kings,
emperors, those who are august, and others of similar status.[5]

(14) They have a habit of narrating the past[1] and provoking old ru-
mors by recall. It is said that the Macedonian kings, especially the
Lagides[2] of Egypt, had crows *which* mimicked the human voice so well
that *they would* come out *and* give a greeting,[3] such as, "Hail King
Ptolemy!" It is likewise possible for many others to manipulate the
tongue properly and distinctly--even for those that are not well domesticated. Certainly voice always leads to belief in hidden powers. Moreover, there are two uses of the voice:[4] the one is to talk and the
other is to sing. Now you have been told enough about the former; consider the latter. And it is not necessary to discuss any of those *birds*
whose ears are not altogether deaf but sound *to a certain extent.*

(15) Blackbirds,[1] turtledoves, and swallows not only twitter[2] but also sing in rhythmic tunes, making it possible to express and to write
words for the tunes. And if one should wish to study *them* by careful listening,[3] he should not hesitate to go for a while to nearby parks, where
there are indeed all kinds of musical birds and, *where*, despite *seeming*
competition, they sing together in harmony. *There* one could certainly

---

[4]Or, "adorning."

[5]Lit., "who are like them."

(14) [1]Aucher renders, "Isti quidem solebant initium facere verbi."
Gr. retransl. of the last part, ἀρχαιολογεῖν.

[2]i.e. the Ptolemies. [F.] Stähelin, "Lagos, 1," *PRE*, 23 (1924),
cols. 462-464. Arm. is translit. of Λαγεύδας.

[3]Aucher renders, "ut exeunti foras regi salutem dicere solerent."
He supplies "regi" and renders adjectivally the first of the two infs.:
ɦ ηnιɾu quɩ (ἐξέρχεσθαι), nηɾnɟu̇ ɯɯɩ (ἀσπάζεσθαι).

[4]Lit., "kinds of voice."

(15) [1]Marg., "i.e. the starling," a gloss.

[2]Lit., "are used to twitter."

[3]Aucher renders loosely, "suis ipse auribus discere."

discover from the very first attempt[4] at studying, that the utterances of some dumb animals have been discredited by men who are *both* mistaken *and* truly embraced by selfishness.[5]

(16) But of what use is it to be verbose about uttered reason and to disregard that which is in the mind? This should be investigated.[1] It is indeed foolish for hunters of wild boars and lions to explore *their* tracks and haunts and never investigate *their* intelligence which pertains to the rational soul.[2] However, *some men*,[3] occupied neither with hunting nor with inanimate things found[4] in the forest, enter groves, thickets, swamps, and fens to observe the different species of animals and to discover whether only the human mind was made after the *divine* image and received a great honor separate and distinct from that of other *creatures* or whether God gave a common advantage to all *creatures*.

(17) It seems to me that the friends[1] of truth are not at all concerned about the beautiful *and* do not acknowledge unanimously that the faculty of reason is implanted in every *creature* endowed with a soul. Would that someone would consider *this* clear proof. Is not the spider very proficient in making various designs? Have you not observed how *it works* and what an amazing thing it fashions? For who else works as hard at spinning or weaving? I am not talking ironically but comparatively. Who rates second in art? Even those[2] who from childhood are neither listless nor careless in their pursuit[3] *of art* are actually surpassed *or* outdone.

---

[4]Lit., "work."

[5]Aucher renders somewhat differently, "mendacii redargui tum se tum caeteros omnes, qui irrationalia esse animantia pro lubitu fingunt, nonnisi φιλαυτίᾳ idipsum suadente." Gr. retransl. of the last part, σφαλεροῖς ἀνδράσι τοῖς ὑπὸ τῆς φιλαυτίας ἀληθῶς ἀσπασαμένοις.

(16) [1]Or, "should be pursued."

[2]Aucher inadvertently emends ⲓⲛ𝑞ⲃ (ψυχή) to ⲓⲛ𝑞 (φροντίς), but rightly translates it "animus."

[3]Aucher supplies "nos philologi" in parentheses.

[4]Lit., "dwelling."

(17) [5]Or, "associates."

[2]Lit., "all those."

[3]Or, "imitation."

Taking a useless substance, as though it were wool, it fashions it in a very skillful *and* artistic manner. First it spins it very thin, as though by hand. Then by stretching and intertwining, spinning and weaving so wonderfully, it is capable of creating a fine *piece of* art to make an open space[4] look like lace. With enduring patience it weaves back and forth, wisely having in mind a lyre with its curves and circular shape--the circular is always more durable than the straight. A clear[5] proof is that when the wind blows from every direction and things gradually pile up one on top of another, it hardly ever becomes torn.

(18) There is another amazing thing about the spider which cloth makers are unable to imitate because they divide their skills. Those whose task it is to spin do not weave and those who weave do not spin. Moreover the spider[1] contains within itself all that it needs. It produces in a most perfect manner every single thing without the aid of a co-worker. How wonderful! It is graced not only with the ability to join *or* to fit together as in artistry but also, to be satisfying, with tools. It has *within* itself the skill and the tools with which every aspect of its work may be accomplished.

(19) Without rudders, sails, and men cooperating together, sailors cannot sail a ship unerringly to a destination. Likewise without food, drinks, and drugs, physicians cannot succeed in bringing about healing. Is it not also appropriate to mention that artisans, who are familiar with tools and have vigor, cannot practice[1] without these things? Although the spider appears to be an insignificant and useless animal, it never needs a tool to spin a perfect web. It is self-equipped to accomplish whatever it chooses to fashion. Indeed I am cognizant of painters and craftsmen and their meritorious endeavors to keep intact *and* to perpetuate their handiwork: paintings and statues, which have become part of themselves. Certainly they hate as irreconcilable enemies those who damage them. It is evident that the spider does the same. As *soon as*

---

[4]Lit., "the space of air."

[5]Lit., "great."

(18) [1]Lit., "it."

(19) [1]Lit., "exist."

its network is finished, it conceals itself like a spot in the center,
watching all about for flying insects[2] which might break the web. It
is wary of anything unexpected. If anything should happen, it comes
out of the loop immediately to take vengeance against the cause of the
damage. Then it mends the torn place.

(20) There is that which not only peasants but also great kings keep
--the bee. Its intelligence is hardly distinguishable from the contemplative ability of the human mind. During the spring season, when every
fertile plain and mound is in bloom, swarms of bees fly over orchards,
gardens, arbors, and green fields and hover over the sweet smelling
blossoms, flowerets, and buds to suck the dripping dew, especially
that of the thyme and of several vines[1] generally called honeysuckle.
By the wonderfully fashioned mechanism of their *bodies*, they trans-
form the dew into natural honey. It takes place in this manner: the
bee receives the dew as if it were a seminal substance,[2] becomes full,
and hurries to pour it out in a place where the poured out substance will
not be wasted. [But beehives are constructed with the aid of humans, for
whose consumption God has given for an inheritance as it were of the pro-
ducts of the earth: of plants and animals alike.][3] When it is *ready* to
pour, it makes cells[4] which are compact, of suitable size, and hard e-
nough. The dual structure of *this* creature[5] has a twofold significance:
there is that which is like the substance of the body--wax, and there is
that which is like the soul that dwells in it--honey. The wax after being filled with dew is tightly sealed to protect the processing[6] work
from exposure and to shield it against the attacks of inherently mali-

---

[2]Lit., "those that fly in the air."

(20) [1]Lit., "those that seek the ground." Said also of animals, § 11.

[2]Aucher renders, "seminale quoddam," and elsewhere, § 96, "seminales vires." Gr. retransl., τὸν λόγον σπερματικόν.

[3]Aucher fails to recognize the interpolation which runs counter
to Alexander's thesis and interrupts the flow of the thought.

[4]Lit., "form," "shape," etc.

[5]Lit., "animal."

[6]Lit., "incomplete."

cious animals. The hive[8] is like a walled palace near a city, thickly walled all around. Since it would be easy to attack an exposed site, the inner sections are made with narrow and intricate passages that are inaccessible *and* cannot be easily attacked.

(21) It is not only important to enclose *the hive* tightly, but behind the mighty wall the bee takes charge *as* a captain of the guard and a keeper of the wall. It seems to me that it holds a leading position, for it is very clearly seen waiting[1] at the gates and looking all around like one watching from an observation post. If its adversaries are inactive, it is likewise quiet; but when they attack, it is immediately incited to avenge. If the structure needs reinforcement, it adds buttresses from within, for fear that enemies might break in unexpectedly. When the bees[2] are aroused from their hiding place, immediately they come buzzing, ready to attack. With their stings raised they terrify *their enemies* because they employ all their defenses[3] when it becomes necessary to avenge.

(22) What about the swallow? Is not this creature[1] prudent in exercising foresight? With mud and clay, or anything it happens to find, it builds *its* nest in accordance with two things: it figures out[2] a suitable shape and dimension, and it takes into consideration the stability of the site. Fleeing from the menace of vultures,[3] it appeals to man first of all and seeks shelter like those who take refuge in temples. [Moreover, taking off from the earth, it soars very very high.][4] It lays a round, protective wall beneath the rafters and builds its nest within the suspended structure. [It is at

---

(21) [8]Lit., "That thing."

[1]Lit., "sitting."

[2]Lit., "they."

[3]Lit., "whatever it may bear."

(22) [1]Lit., "animal."

[2]Lit., "it measures."

[3]Lit., "of stronger ones."

[4]Note the scribe's marginal correction of վերհաս (ἄνω) to լերհաս (ὄρη). An earlier interpolation, unnoticed by Aucher, is to be suspected.

odds with mice[5] and chases *them*; consequently, as they say in Egypt,
"Cats are thin"[6] since cats feed on mice.] And the structure *of the*
nest is *more* perfect than any[7] of the most artful constructions of
man. It is mindful of making good preparations for its offspring *before*
it hatches *them*. However it does not cease to work after hatching, but
nourishes its young[8] with the food it gathers from all quarters, dividing it equally to each of them, *thus* preventing greed and squabbling.
By dropping *the food* into *their* mouths, it manifests the attachment of
a mother and the affection of a nurse or a midwife. It even takes the
precautionary and necessary care of eliminating the waste matter by
training its brood to turn around and to excrete outside, for fear that
the nest might fall because of accumulated weight. With single-minded-
ness it returns bringing their essential nourishment, retracing its way
without getting lost. Neither does it alight on another nest but diligently cares for its own.

(23) Not only do some of them acquire self-heard and self-taught
knowledge[1] but many are very keen in learning to listen to and to comply
with what their trainers command--whether by intimidation when they fail
or by an appealing reward when they succeed.[2] But what need is there to
mention the trained? Certain whelps are hunters because of their innate
courage and eating habits and attack huntsmen *whenever* they meet *them*.
Several of our domesticated *animals can* master learned skills. A monkey, which is the most unruly of all the animals, drove a chariot *and*
made quite a show at the theater the day before yesterday. Riding a
chariot drawn by four goats, it drove into the arena like a charioteer

---

[5]Aucher, incognizant of the interpolation, renders, "quo
armatur contra mures."

[6]Lit., "Cats have hair" or, "Cats are shaggy." Aucher renders,
"ita ut in Aegypto felis pilus dicatur" and explains in a note: "Aut
hirundinem ipsam, aut nidum suum hoc nomine appellari notat. Apud Plinium x. 85. haec tantum habentur. 'Aegyptiis muribus durus pilus, sicut
herinaceis.'"

[7]Lit., "all."

[8]Lit., "the creatures."

(23) [1]Aucher renders, "naturalis auditus ac instructionis habent ingenium." Gr. retransl., αὐτηκόου καὶ αὐτομαθοῦς ἐπιστήμης.

[2]Lit., "for a right thing."

78

fighting for victory. It waved the reins, lashed with the whip, screamed into the ears of the goats, and gave great and wonderful amusement to the spectators.

(24) On a long thin rope strung above the ground, the fawn can climb *and* run as though walking steadily on a flat surface *or on* a paved road. Whenever it thinks that it might fall, it counterbalances itself. Once I saw this very thing. *A fawn* came on stage dancing and playing ball with its forelegs. It seemed to be a marvelous act[1] on the part of *a creature* that has no hands with which to catch the ball.[2] But then when it mounted a round *platform*, it performed lengthier, greater, more wonderful, and amazing feats. Burning torches were thrust into each and every hole all around the rim of a wheel, *and* the wheel was turning increasingly fast. The fawn[3] stood in the midst of the device like one trapped in the flames. A tame ape[4] came walking about, inspecting the round *platform* over which he was put in charge. Then the *two* animals stood in readiness for the show. *The fawn* snatched some of the torches with its mouth by turning its head[5] to the right and lifted one up to give to the tame ape. Receiving it from the *fawn's* horns, as easily as from *one's* hands, he held it up *and swung it* around like a child carrying a lamp. He performed well and was careful not to let the flames burn anything beyond their confines. Those gathered around were astonished, and to me the performance seemed amazing. During the performance, he occasionally stared fearfully at the trainer, I think afraid something might fail; and once in a while he stared at the people to see[6] if they were enjoying themselves or if they were restless. When the show was over *and* his fears were dispelled, he was thrilled with exultant joy like a victorious gladiator. This was obvious from his excited jumping and from the sheer delight expressed in his eyes.

---

(24) [1]Lit., "thing."

[2]Lit., "that which is thrown."

[3]Lit., "it."

[4]Arm. is translit. of καλλίας, the meaning of which was unknown to Aucher who notes, "Vox ipsa Graeca videtur servata ab interprete; de qua studiosi lectores dijudicabunt."

[5]Lit., "the neck."

[6]Lit., "examining."

These *movements* were nothing less *than* common expressions of gladness. Meanwhile the fawn on its own was nodding its horns, which to the tame ape had seemed like hands.

(25) There are some *animals* that learn servile tasks, performing *their* work like servants, attentively *and* with courteous regard. The theater is full of these animals that together with their trainers are said to engage routinely in public shows for entertainment. But such is not funny; *it is* worthy of high regard.[1] Certainly this *performance* is laughable to fools whose eyes of the soul are blind,[2] but it is serious to the wise who are clear-sighted *and* have enlightened minds.[3] Therefore without derision consider diligently the wealth of their rational nature. Leaving suppositions aside, learn the truth. *And* you will be persuaded[4] anew that these creatures are[5] esteemed by God and respected and commended by the God-loving race of mankind.

(26) Now of the animals *previously* mentioned, some say that only the domesticated learn; they say that the wild are incorrigible. But even certain of the wild and unapproachable were brought to trainers and were tamed.

(27) Is not the genus of elephants, especially the Libyan, the wildest? Yet at the time when Germanicus Caesar celebrated the victories of his consulship *and* proclaimed that shows for entertainment be given in various places, Baebius did something great *and* altogether new.[1] He sent *as* a present a herd of elephants that had been accustomed to enjoy household foods and drinks. What more could be said of those that are of immense size? As soon as they entered the theater, they lined up in a row--as if so ordered by the silence of the audience. First the elephants knelt to-

---

(25) [1]Aucher has "studio virtutis dignum." Gr. retransl., σπουδῆς ἄξια.

   [2]Lit., "who have blinded the eyes of the soul."

   [3]Lit., "whose minds are enlightened *and* clear-sighted."

   [4]Lit., "they will be persuaded."

   [5]Lit., "this genus is."

(27) [1]Aucher renders, "elephantorum gens nonne ferarum maxime agretis est? Attamen ex Libya Baobas, Germanico Caesari undique ad pompan ludi apparatum disponenti, eo tempore quo de consulatu certamina proponebat, novam rem magnifice praestans." Լիբիացիրց (Λιβυϰῶν) undoubtedly refers to փիզց(ἐλεφάντων). Բաιπրաս (Baebius) is mistransliterated "Baobas."

gether, their faces down, doing obeisance to the victor. Then they stood up and raised *and* waved their trunks, indicating that they were greeting the people. After the spectators applauded, as though having received a double greeting they lowered their trunks, as *one* lowering *his* right hand. Soon afterward some of them sat down on specially re-inforced *and* pre-tested iron beds which would not break under the enormous weight of the elephants, while the others stood by like attending servants, constantly serving much of that which is used for merrymaking. Before long one of the young arose as from a wine table, pretending to stagger, and danced to the sweet music of men entertaining with trumpet and lyre. Meanwhile the others waved their trunks and from time to time trumpeted to the dancer, as though applauding. Then one came carrying a lantern. When they saw that, they knew[2] it was time to quit, *and* they stood up and began to leave. Nowhere else in the world could the appearance of staggering drunks be *so* demonstrated. Stumbling and walking unsteadily, they were hardly able to move *on* their feet while leaving. Like drunks they were roused and led about until they left the theater.

(28) It is told in Alexandria, Egypt, that once upon a time *an elephant* learned lettering to the extent that he could designate: "I myself wrote this." Such an animal had also learned erotic plays. As the ode says,

> An elephant was enamored of a camel:
> Like a child, kissed, hugged, and caressed it--
> Even indulged it with food.[1]
> Day and night refused to part with it,
> Save by command of the trainer of the beast.[2]

(29) Thus nature has placed a sovereign mind in every soul. In some it has a very faint imprint, an inexplicable and ill-defined form of an image; in others it has the likeness of a well-defined, very distinct, and fastidiously clear image; and still in others it is of an indistinct kind.

---

[2]Lit., "thinking."

(28) [1]Lit., "And put food in its mouth."

[2]Origin unknown.

But the deep[1] and distinct impressions are borne upon the image of man.

(30) In addition to what has been said, it could be stated without disparagement, that many other animals have wisdom, knowledge, excellent discerning, superior foresight, and all that is related to the intellect--those things which are called "virtues of the rational soul." Their reasoning *ability* is not deceptive; the evidence is very obvious: the expediency of providing the necessary food, *their* understanding, stratagem, and various ingenuities. Indeed the polyp makes its body look like a rock whenever it is encountered as prey. It evades the fish by pretending to be a rock, even when it is hit hard by the swimmer. As for what is called a crampfish, when it leaves a mysterious discharge on *other* fish which touch it, it can easily catch them, as with a hook, utilizing the force with which they resist. As for what is called a starfish, it very skillfully,[1] like a cook, prepares *for itself* a tasty and delightful *meal*. And it seems to me, when it feels something, it by nature produces heat and with courage attacks a larger *fish* which cannot withstand the scorch of the heat; *thus* it feeds without effort.[2]

(31) Creatures of water, air, and earth alike are endowed with abundant wisdom. None is deprived of wisdom, in spite of the regional and somatic differences that separate them. The oyster,[1] the flesh of which is edible, is an excellent example.[2] It lies in ambush between two shells, one on either side.[3] The shells tightly trap whatever falls in them. It is impossible to pry open what are called shellfish.[4] Whatever is snared is absorbed easily, since it dissolves by the warmth of the gullet. What keeps the shells tight is an elastic, *which*, when relaxed, releases the clasp, so that they open wide to engulf food.

---

(29) [1]Or, "firm."

(30) [1]Lit., "cleverly."

[2]Or, "painlessly."

(31) [1]Or, "Molluscs."

[2]Lit., "is a most perfect proof."

[3]Lit., "one on the one side and one on the other."

[4]Marg., պեղականք, a synonym of թունդք. Aucher has "Paelotes."

82

(32) Moreover *animals* do not panic when they are in great want *of food*.[1] They resort to contrived schemes against those animals whose scent they eagerly track *and* for which they lie in wait to ambush in deep dens or in the woods. Catching the scent which their nostrils draw from the air, they are attracted to hunt by the sense of smell,[2] to hide in lairs, and to wait to spring. *Then* when they charge from *their* ambush, they devour immediately.

(33) Some are so concerned about self-preservation that they turn away not only from the snares of other *animals* but also from the ingenious designs of man. It is said that after deer shed *their* antlers, they seek shelter in seclusion, in the heights of mountains *which* they ascend, in woods, in inaccessible cliffs,[1] and in uninhabited deserts. So when deer lose their protective ability, being disarmed as it were of their defensive weapon, they become dismayed at the very prospect of having trouble.[2] As if flight is their only resolve.[3] From this the saying derives:

> Where the deer shed *their* antlers
> Is truly the most solitary place.[4]

Because the wound at the roots gets painful after the antlers are shed, and lest flies should hover over *it*, feed *upon it*, and cause severe pain, they hide in the dense brushwood and the thick forests during the day, taking cover beneath the boughs and the leafy branches. At night, while the birds are at rest, they get up to graze fearlessly. When their antlers have budded again, they do not dare to come out even on an ordinary day, fearing that their antlers have not yet developed *fully*. For while they are tender, they might be cracked by the dazzling scorch

---

(32) [1]Aucher renders, "Nec tamen in solitudine (vel solicitudine) degentes extrema laborant cibi penuria." Gr. retransl. of the first part, Καὶ μὴν οὐκ ἀπορούμενα.

[2]Aucher renders loosely, "foetorem vero sentientes, nares dimovent."

(33) [1]Or, "steeps."

[2]Lit., "a troublesome thing."

[3]Lit., "that which alone remains before them."

[4]Cf. Arist. *HA* 611a26.

of the sun, as the potter's clay is by the fire. Not being convinced
yet that these are well-developed *and* sturdy, they strike their heads
against the trees, rubbing and butting, to test their antlers for
strength.[5]

(34) Not just one but all[1] species of animals have intelligent foresight which they exhibit especially when it is time to bring forth
young. There are some which head for the heights of mountains, others
which rummage the depths of the Scythian and Libyan deserts and others
which hide in the heavy underbrush of the woods. And you will find many
in rushy lands, where they are attracted to give birth. These *acts* together with the intelligent *and* precautionary[2] care shown toward their
young are indications of abundant wisdom.

(35) Indeed partridges[1] do an amazing thing which is applicable to
reason in every respect. First of all for fear of being found *and* cap-
tured by those who are after them, they keep changing the place where
they nest and feed. They flee cautiously from the many which lie in
wait to trap them. When they sense that some have come to catch their
brood, they come out to contend very skillfully. They escape by flitting
about slowly slowly and then in quick *and* short flights. Before long
they fly a *certain* distance. While they divert the fowlers' constant
watching, they allow time for their brood to escape. When they feel
they have flown a *certain* distance, they alight,[2] and then they fly in
the air in a contemptuous manner. Whereas the fowlers,[3] disappointed by
being mocked *and* eluded by deception, duly become frustrated.

(36) Those who report on the semelē,[1] which is a creature of the

---

[5]Lit., "whether they are firmly developed."

(34) [1]Arm. has a dittographical error.

[2]Or, "foresighted."

(35) [1]Aucher, as in § 80, has "Palumbes."

[2]Lit., "waiting on their heels."

[3]Lit., "they."

(36) [1]Unidentified. Aucher transliterates "Semela" and suggests the
possibility of being a rendition of σμίλη, σμιλίον, ζῶον μέλαν, or μύξον.
This possibility, however, is unlikely. The etymology of the word could
be traced to the Laconian σέμελος.

sea,[2] say that it employs some demonic tactics.[3] If this forewarning
fails to caution men about their various ill-contrived plans, then in
the end, they will find their pursuit unrewarding. Sensing the evil
scheme, *the fish* either bites off the hook or avoids it--no matter how
well-baited *or* food-like it might be. But if it advances to attack, it
will eat the line, accomplishing two favorable things: saving itself
and tormenting those who attempt to harm *it*.

(37) One would think that the story of the Thracian falcons is most
fascinating. When I heard it for the first time I did not believe *it*,
not until many natives of the province and some of the foreigners who
visit us--those simple-minded *and* unpremeditating men--[1] told whatever
they know about them. They said that these work concertedly, aiding the
fowlers by preying[2] swiftly upon the catch. They live[3] in the forests,
as is proper, in thick, tangled woods where a variety of birds is often
found, and cooperate with those who come to fowl, *especially* those who
approach cautiously and who do not refuse to share their reward, as is
only fair for comrades in arms.[4] What an example they set of mutual
aid! First of all,[5] the fowlers shake the trees. Because the weak and
frail birds cannot withstand the shaking, they fall from fear and try to
fly away. But the falcons swoop down and knock them to the ground by
pecking, preparing easy game for those who have come *to fowl*. When they
are forcibly thrown from the branches by the shaking, they flutter fearfully as a result of the pounce and are thrown into a state of utter
confusion and terror. *Then* they are easily caught, even by the fowlers'

---

[2]Or, "a marine fish." Lit., "of the nature of the sea."

[3]Aucher renders, "cooperatione." Gr. retransl., συντάξει<ς>.
The Arm. word for "cooperation" is *qnpδwկgnιβhιն*, which he renders
correctly in § 37.

[4]Or, "finding," "knowing," etc.

(37) [1]Aucher renders loosely, "homo simplex," and parenthesizes it
with the preceding clause.

[2]Lit., "they seize," "they gain," etc.

[3]Lit., "They have their habitation."

[4]Arm. has a dittographical error. Aucher renders, "socios a
sociis belli."

[5]Or, "Firstly, to begin with."

hands. Nevertheless the fowlers are glad to share the catch partly
to compensate *the falcons* for their cooperation and partly to invite
their aid for future expeditions.[6]

(38) *Animals* grow and abound in substantial wisdom; so much so that
when it comes to most essentials, they surpass man in wisdom. When one
of us becomes sick and does not know what treatment to apply, he calls
physicians, who recommend a cure which restores to health. But animals[1]
never need *to call on* another, because they have instinctively learned
the self-taught art of healing. When deer are pricked by venemous spiders, they consider hops for their remedy and are healed. In Crete when
a goat is hit by an arrow, it looks for what is called dittany, and after
the goat eats it, the wound heals easily.

(39) It is said that when tortoises devour[1] asps because of gluttony,
they eat marjoram for fear of dying; for only by this means can they be
relieved of the harmful venom. Those that do not believe in making an
immediate effort to search *for the plant,* most certainly suffer the
stated consequence. When a tortoise that has devoured[2] an asp discov-
ers[3] a marjoram plant nearby, it snatches it up with the roots as soon
as it sees *it.* But when it fails to find *any* to eat after venturing, it
dies.

(40) It has been observed that *animals* not only heal *their* existing
and impending diseases but also feign sickness lest they be compelled
to work. It is said that Aristogiton, one of those valiantly involved
in making Athens a prominent city, had a stubborn horse which, upon appearance in the hippodrome, pretended to be lame. But it gave up cheating when it was eventually detected by the breeders; moreover, it ceased
to be slow *when* beaten by them.[1] But the hypocrite was actually putting

---

[6]Lit., "hunting" or "fowling."
(38) [1]Lit., "none of them."
(39) [1]Lit., "having filled themselves with."
     [2]Lit., "having filled itself with."
     [3]Or, "sees." Aucher renders, "Videntes." Gr. retransl., ἰδοῦσα.
(40) [1]Lit. "by whose lashes."

on a pretense that was[2] a clear indication of *its* cleverness of mind.

(41) It is not superfluous to say that cattle, herds of oxen, and flocks of goats and sheep have reasoning minds. They guard against those beasts which are the wildest. They have no lack of knowledge--not just about eating, drinking, and mangers, but about everything that relates to their good or harm. Their minds are capable of discerning between many things and are cunning and rational *as well*.

As for house management, ours is ridiculous when compared with theirs.[1] (42) Have you not observed the ant? Though it is *but* a small animal physically, it *achieves* great deeds through its careful industry and mental powers. It makes *such* provisions during the summer that its nest is well stocked[1] when the heavy rains begin. Being cautious of the harmful cold and rains, it secludes itself at the end of autumn. When the wet winter arrives, being happily sheltered right where its harvest is stored, it consumes freely what it needs from the supply, and if it happens to run short of essentials, it comes out again to gather the necessary food for another season. It also escapes the grave misfortune of hunger. It cuts into two pieces the wheat, the barley, or whatever it happens to gather, to prevent sprouting *and* want of food.

(43) The day is not *long* enough to tell of similar concerns of other animals which are constantly engaged in gathering[1] food. Of all the *animals*--terrestrials, aquatics, and aerials--none ever lacks provisions, since they use foresight to take care of themselves. Unless they possess a reasoning mind, would they know how to distinguish between the useful and the useless, the worthwhile and the worthless? Their power of discernment *and* understanding[2] is well attested and closely related to the ra-

---

[2]Lit., "but." Aucher, punctuating differently, has "non aliunde."

(41) [1]Aucher renders, "ut proinde absurdum sit maxime, inter se conferre nostram et illorum providentiam." He supplies the initial resultant and intensive senses and translates մատակարարութեան (οἰκονομίας) as "providentiam." Elsewhere he translates the same word more accurately as "oeconomia" (§ 91) and "dispensatio" (§§ 65, 91).

(42) [1]Lit., "weighed to capacity."

(43) [1]Or, "preparing," "procuring," etc.

[2]Or, "learning."

tional soul.[3]

(44) However it is worth noting--*for* it is obvious--that they have
various ways of coping with opposites as well as of facing obstacles.
With regard to heat and cold, sweet and bitter, white and black, large and
small, or whatever inconveniences[1] *result* from these opposites, they set
their reasoning mind differently toward them, so as to make *them* pleasant
and agreeable. They long for that which produces pleasure[2] and flee
from that which is loathsome *and* painful. *Although* they are unable to
express[3] their mental conceptions because of their inarticulate tongues,
they conduct themselves with such abundant wisdom that they exhibit many
characteristics of speech. To the keenly perceptive there is something
more evident than voice--the truth which their actions reveal.

(45) A hound was after a beast. When it came to a deep shaft which
had two trails beside it--one to the right and the other to the left,
and having but a short distance yet to go, it deliberated which way
would be worth taking. Going to the right and finding no trace, it returned and took the other. Since there was no clearly perceptible mark
there either, with no further scenting it jumped into the shaft to track
down hastily. This was not achieved by chance but rather by deliberation of the mind. (46) The logicians call this thoughtful reckoning
"the fifth complex indemonstrable syllogism": for the beast might have
escaped either to the right or to the left or else may have leaped.

Similar traits and resemblances of speech can be explained by
man. But nothing has been heard from anyone[1] that is *as* significant and
accurate *as* Aesop's account in the best part of his works. Although animals lack the ability to speak, no fable is created but that it sub-

---

[3]Aucher renders somewhat differently, "Istorum vero peritia nimis certa est (apud animalia), et solidata disciplina, animae rationali
propria."

(44) [1]Or, "disagreements."

[2]Aucher, punctuating differently and supplying the first two
words, renders, "ut animans quisque velit ac appetat quod placet et
affert voluptatem ad quam properat." Gr. retransl. of the middle part,
ποιητικὸν ἡδονῆς.

[3]Lit., "to interpret."

(46) [1]Lit., "from the others."

stantiates their wisdom and knowledge in many *ways*. And often in prov-
erbs astute listeners among men are appropriately called foxes. These
animals have caused many to marvel at the wittiness[2] and various
schemings of their minds. Even the most stupid monkey is a very artful
gesticulator and playful enchanter; and after captivating at will, it
laughs at those whom it has outwitted.

(47) Yet there are many other accounts of how animals conduct themselves wisely, but those mentioned will suffice. It should be noted
that they exhibit[1] not only wisdom but also self-restraint. When they
need a word for something, they *merely* utter a sound. Do they indulge
in any feasts, winebibbing, drinking parties, or superfluous works of
cooks and bakers whose craft is to cater for the pleasures of our
wretched stomachs? We who swell with pride are not ashamed to open our
mouths, even more so to knit our brows or hold our necks high. We are
filled with irrational thoughts, have barren and uncultivated minds,
and are easily puffed up. As the poets say,

> We're thought to be of seed divine
> And of kinship closely akin.
> *But* when enslaved by food and drink,
> *Such* lowly things do make us fall
> From heav'n above to earth *beneath*.[2]

Moreover we ought to be ashamed even of our physical weakness. Certainly insatiability and lack of restraint cause minor diseases among the
young and incurable *ones* among the old.

(48) As soon as *animals* are born, they purposely adjust to simple
foods, they eat grass and are satisfied with little. They drink water
infrequently, about *once* every few days. Sometimes they *can* abstain
from everything for several months. Some of the hibernating *animals* do
not drink for a *whole* season. They are satisfied with cool air instead
of water and with the dampness of the dew *instead of* eating grass. *Oth-*
ers also *can* give up liquids and the pleasures below the belly *and* can

---

[2]Lit., "brilliance."

(47) [1]Lit., "possess " or "partake of."

[2]Reminiscent of the so-called Orphic poems.

become very temperate.[1] Some copulate in the spring, others in the
fall, and some abstain altogether for a year. The females succumb to
mounting only for the purpose of impregnation; then they run away from
the males, fulfilling the law of nature which has found out the corruption that results from the male joining the female in sexual indulgence.

(49) Is it not then proper that men should blush at their unleashed
sexual indulgence when it is compared with the chastity of these *animals?* For at what season of the year do we cease[1] from venereal pleasures and from practicing our customary and constant vices? There are
those who are not satisfied with their wives alone but look elsewhere
and stray to prostitutes, who do not keep their part unblemished but
exchange their honor for such things *as* disgrace and misery. Some young
men even defy death by choosing lust over life. They infringe on the
marital rights of others and do not *even* blush during trial before the
magistrates. They fear neither the present laws concerning adultery nor
the raging and inexorable wrath of husbands threatening with death,
having the freedom to kill, unhindered by inexperience in evil.[2] Some
resolve to wickedness and fall into such violent passion for unlawful
sexual indulgence that they commit sodomy. They disturb not only commu-
nities but also the very order of nature. However truth herself convicts them for transgressing unalterable law, for committing immoral
acts, for giving the seed to the immature, and for wasting and destroying the seed.

(50) There are other explicit examples of self-restraint among animals that invoke such emulation that those *men* who are shamelessly in-
satiable in sexual indulgence will immediately want to desist, if not at
once, then gradually. Not only among *animals* domesticated and reared[1]

---

[1]Aucher, omitting ꭒꮟꭇꭒꮃꭹꮟꮃꭵ (ἠρεμοῦντα), renders loosely,
"Ad haec prurigo illa ex humiditate orta, cupiditasque sub ventre existens nimium minor et exigua illis est."

(49) [1]Aucher renders, "homines ... abstinent." Gr. retransl., ἱστά-
νομεν.

[2]Aucher amplifies in rendering, "quibus ad majora scelera passio
inexperta vim facit ut homicidiam admittant."

(50) [1]Or, "fed."

by us but also among the other *species* there are those which appear to have self-restraint. There are[2] many clear examples to convince those who doubt, but one ought to be cited[3] and the *rest passed by* in silence. When the Egyptian crocodile--which is an anthropophagus and amphibious animal--is inclined to copulate, he diverts the female to the bank and turns her over, it being natural to approach *her* lying on her back. After copulating, he turns her over with *his* forearms. But when she senses the copulation and the impregnation, she becomes malicious in purpose and pretends to desire copulation once more, displaying a harlot-like affection and assuming the usual posture for copulation. So he immediately comes to ascertain, either by scent or by other means, whether *the invitation* is genuine or merely pretense.[4] By nature he is alert to hidden things. When the intent of the action is truly established by their looking into each other's eyes, he claws her guts *and* consumes *them,* for they are tender. And then unhindered by armored skin or hard and pointed spines, he tears her flesh apart.

(51) Enough about self-restraint. Is it not necessary to discuss courage? Which of the wild boars, the young bulls,[1] the elephants, or any of the other wild animals endowed with strength and vigor does not know how to wage quick counterattacks with a most amazing skill of warfare? Indeed I have seen them rushing to fight, stirring up dust with their feet like wrestlers, staring steadily and threateningly with red flashing eyes, and aiming their piercing horns with fiery wrath. Those who go hunting in the forests tell that when a boar[2] sees hunters, it stops *and* does not force its way[3] until it first whets its tusks at the nearest tree. Bears separate themselves *to entrap their pursuers.* They

---

[2]Lit., "There ought to be."

[3]Aucher amplifies in rendering, "verum satis erit unius mentione facta similiter argumentari."

[4]Lit., "perhaps opposition."

(51) [1]Aucher has "taurorum." He does not distinguish between *qnιw-pw\q* (μόσχος), which occurs here and in § 64, and *gnιʃ* (ταῦρος), which occurs in §§ 66, 80.

[2]Aucher has "aprum." He fails to distinguish between *խnq* (σῦς), which occurs here, and *վшɲwq* (κάπρος), which occurs above and in §§ 16, 70.

[3]Lit., "a foot."

break *their* necks and backs,[4] inflict very heavy blows, and entangle the feet by skillful ensnaring. They practice all the cunning tricks[5] which the drillers customarily teach; *but* it would be superfluous to mention them separately.

(52) I saw a python fight with an Egyptian cobra the day before yesterday. I was not only amazed but also persuaded to believe the truth about them. Do only rational beings[1] have abundant reason or are other living creatures also endowed? The venomous cobra advances to fight treacherously and deceitfully. It crawls, dragging and rustling its coiled tail and lifting up its head above its elongated body. It is high-breasted, has an expanded and swollen neck, *and* is full of wrath and venom. As some[2] would signal with *their* hands, *so* it moves its tongue back and forth, hissing very angrily. But when the python rises and attacks headlong to inflict *the cobra* from behind, just as it would move rapidly to assault any *other* hostile animal, *the cobra* lowers itself completely, lifting only the front part of its head to look carefully, and waits. It wonders[3] if the time has come to inflict the decisive blow and is altogether ready to strike back. When the cobra realizes that its plan of attack can be executed favorably,[4] it quietly chases after *the python and,* like a beast, suddenly strikes from behind. It grips its neck, overpowers *it,* and suffocates *it.*

(53) These animals conduct themselves *more* efficiently than centurions and tribunes,[1] and they do not need each other's adulation--since it is more expeditious to work without excitement. Their work is always

---

[4]Or, "loins." Aucher has "sub medio."

[5]Or, "skills."

(52) [1]Aucher has "quod rationis particeps esse." Gr. retransl., λογικαὶ φύσεις. He also disregards the question mark above *մ՛ի այն* (μόνος).

[2]Lit., "others."

[3]Lit., "Fearing."

[4]Lit., "are according to *its* will."

(53) [1]Aucher renders *զդասապետս եւ զդասապետս* (λοχαγοὺς καὶ ταξιάρχους) as "cohortis praefecti ac centuriones," and *ի դասապետ-աց եւ ի դասապետաց* (ὑπὸ λοχαγῶν καὶ ταξιάρχων), § 65, as "praefectis."

characterized by vigilance.[2] They never allow slackness, but press on
to valor.[3]

(54) When I went on an embassy to Rome, I often attended the contests of wild animals. They speed swiftly as *do* good runners, goading
themselves by lashing their flanks and ribs from either side with
their tails, just as charioteers are said to do with whips while rac-
ing--not only inflicting the pain of severe punishment but also prod-
ding[1] the legs. Taking note of this, the poet has truly said,

> With *its* tail it lashes its ribs and flanks from side to side,
> In order to rouse itself to fight.[2]

(55) Many *animals* have such a great longing for courage that after
fighting in a cowardly manner[1] instead of a hearty one obviously they
feel shame. It is said that when lions yield to defeat even once[2] out
of many confrontations, they leave, roaring softly and roaming here and
there, exhibiting evidence of shameful failure. They roam about with
their faces and voices *revealing* shock and embarrassment. *But* they con-
sent cheerfully to the chasening of *such* ignominy, *while* always desiring
relief. Then they return running as if from a starting point *in a race*.[3]

(56) The genus of horses is such a lover of honor that just as men
were rewarded when the Greeks *fought* the barbarians so also are horses
rewarded at the racecourse. And not only *is this true* in various minor
games whose importance[1] is trivial but also in the sacred games to
which people from every region gather for various purposes: some for
competition in the hope of victory, some in their desire *to witness* a
spectacle, and some to trade. Indeed those who most eagerly attend

---

[2]Aucher amplifies in rendering, "quo necesse est homines identidem ad opus impelli, quod acuit audientes."
  [3] Or, "power."

(54) [1]Lit., "for reminding" or "for prompting."

  [2]Hom. *Il.* xx. 170-171, οὐρῇ δὲ πλευράς τε καὶ ἰσχία ἀμφοτέρωθεν
μαστίεται, ἓε δ' αὐτὸν ἐποτρύνει μαχέσασθαι.
  (55) [1]Lit., "with a great lack of strength."
  [2]Lit., "in a confrontation."

  [3]Aucher renders ɟwnʂwʔwpwuʐ (ὁρμῆς) as "ex statio," and the
acc. pl. of the same word in § 57, correctly, "circi carceres."

  (56) [1]Lit., "thing."

the solemn gatherings consider *it* shameful if they have nothing to
relate on their arrival home.

(57)   At the opening of the grand spectacles of horse *racing* at the
Olympian, the Nemean, the Pythian, and the Isthmian *games*, the horses
attract most of the attention, in their unsurpassed meticulous upkeep,
the reckless expenditure of building stables, of feeding, *and* of hiring
trainers, who rouse the innate speed of the animals by tender care.
When the race is about to start, *the horses* sense it and head to the
starting points at once, whereupon they begin to trot, prior to assuming
a fast gait. Then when they are let out, they spurt and bolt in very
long[1] and rapid strides.

(58)   There was a wonderful show the day before yesterday, which drew
crowds of people. It was a four-horse-drawn chariot race of seven
courses. Naturally some were left behind, but the two fastest had a
close race.[1] The charioteers, *urged* by the eagerness of the spirit--as
they should be--if not by the impulses of the body, *merely* accompany them.
One *of the charioteers* was *so* carried away by the love of honor that he fell
unconscious. Being left without a charioteer, the horses[2] took charge
of themselves *and* resolvedly kept pressing on, *ever* pushing themselves
into a faster pace. They were knowingly demonstrating the exemplary
speed of a charioteer. They did not take steps in vain, such as needless swerving here and there from the course, but kept running in an on-
ward direction, staying steady in the track. Whenever they reached
those racing against them, they spurted forward and overtook them, suc-
ceeding reasonably in places where many fail. Determined to keep the
axle close to the circle, the horse on the left kept setting the line
lest its partner, and the others as well, should circle a wider radius.

(59)   There are other things told about the elephant, which is the
most ferocious animal. The oldest of these ancient tales is from Antioch. A chief of Asia[1] had bred a herd of elephants to assist in fighting against cavalry, because a horse is frightened by the trumpeting of
elephants *and* runs as if being goaded by fear. Those that were superior

---

(57)   [1]Lit., "long long."

(58)   [1]Lit., "were competing closely together."

      [2]Lit., "they."

(59)   [1]Or, "Asiarch."

in physical vigor and utilized their strength, he named after heroes. The highest honor was given to Ajax, for he was the most virtuous *elephant* of all, *both* in appearance and in performance. When they were to cross a river on the way to a fort, Ajax kept standing on the bank, afraid to cross the stream, I suppose. But those that were with him kept the line carefully *and* fearlessly, as it is proper. Thereupon the king summoned the breeders--it was time for immediate attention. He also set a silvery prize and high honor to the elephant which would cross before the others. Now when Patroclus began to run across the river, considering it commonplace crossing, he overtook all the others and was about to win the honor. But it was difficult for Ajax to be denied the high honor and it was difficult to be enthusiastic over an inferior rank. *Ajax,* the son of Telamon, would have succumbed in a short time, for he was enraged just as for the arms of Achilles. Thereupon he impatiently held the line, running and throwing himself headlong on toward the fort, refusing disgrace and craving for the honor. Would that man did the same: to aspire not only for courage but also for that which follows valor, *the achievement* of true happiness.

(60) Besides the mentioned virtues, some of the swimming, terrestrial, and aerial animals exercise justice.[1] The partnership of the marine pinna and the pinna-guard is obvious; indeed they feed together and share their food equally. The *Egyptian* plover *and the crocodile* do the same thing. The pilot fish have lasting agreements on partnership *and* trust with the weaker ones[2] *of their own kind.*[3]

(61) Among the birds the stork exhibits supreme justice by feeding *its* parents in return. As soon as it begins to fly,[1] it sets no other task above that of reciprocating the favors of those that fostered it.

---

(60) [1]Or, "have *a sense of* fairness."

[2]Or, "smaller ones."

[3]Aucher, disregarding his punctuation of the text, renders, "Idem facere et Trochilum ac Pompilum in confesso est apud omnes, qui minores iis comperiuntur in vita communi." Հասարակութիւն միարաունութեան (κοινωνία), which occurs twice in this section, is rendered "aequitas in societate" and "vita communi" respectively. Elsewhere, in an inflected form, it is rendered "communem societatem," § 93.

(61) [1]Lit., "the growth of the wing."

It is possible that some of the terrestrial *animals* behave like it. But since we cannot tell all of *their* instincts and can neither see nor ascertain *them*, we ought[2] to believe what the witnesses have revealed about these things *we have* not seen. For it is right that the universe should be composed not of some *only* of its parts but of them all. Moreover that *part* in which justice and injustice are found is preeminently *endowed* with reason, since both pertain to reason, which, just as is *imparted* to men, should likewise be imparted to those animals cited. The storks that feed their parents *and* give due consideration to the old, also punish. The pinna-guards, noted for sharing their food together with the pinnae, deal fairly with the injurious and the adverse, lest those named should be done away with.[3] As for the bees, all swarms of workers consider the drones detrimental to honeycombs *so* they kill them. For indeed as Hesiod has said,

They reap the toil of others into their own bellies.[4]

(62) It has been observed that animals which attain to the cardinal virtues abstain from eating flesh, whereas the Greeks are noted for having maintained the customs of the barbarians for a long time.[1] They plunge into a voluptuous life of hedonism soon after they are weaned. They *continually* seek new pleasures *and* fill the cities with the most illicit indulgences. Their principles are corrupt and their crimes abhorrent.[2] Thus a nation of educated people, who are zealous for the Pythagorean philosophy, are unable to shun the things of the body or to have any regard for chastity[3] or disease.

---

[2]Lit., "one ought."

[3]Aucher renders somewhat differently, "si soli consument dictum (cibum)."

[4]*Th.* 599, ἀλλότριον κάματον σφετέρην ἐς γαστέρ' ἀμῶνται.

(62) [1]Lit., "many a time."

[2]Lit., "unclean."

[3]Aucher renders կրաւնից ձուձկուլութեանն (τῆς ἐγκρατείας) as "sobrietatem religiosam" and in § 49, more accurately, as "continentiae."

(63) The body naturally serves the purposes of the soul not only through a tie of domestic relationships, which includes tame animals,[1] but also through reciprocal acts of kindness. For one who is loved loves in return, and all who have been helped wish to reciprocate.[2]

(64) Animals,[1] no less than men, show great--if not better--demonstrations of equality and justice. The leaders they appoint and the offices to which they designate them are never ignored. They follow with all willingness those whom they choose to be leaders. Eyes ascertain the veracity of what is said. A young bull[2] leads herds of cattle, a he-goat flocks of goats, and a ram flocks of sheep. An extremely large number of animals follow as irregular troops raised[3] by royal commissioners. This *sense of following* must be divinely imbued.

(65) Doubtlessly a swarm of bees designates a king, under whose guidance it serves with fear and trembling. *Otherwise* all his servants would get their due punishment. And when the swarm thrives and their numbers increase, as when a city gets overpopulated, they migrate to another place, setting out, as it were, *to establish* a colony. If any of those which were given authority fails to supervise, then, being shaken by anarchy--as much as if by ochlocracy--they become separated *and* are dispersed. I also have seen many birds in a row, flying under the leadership of some and following others. In the rivers, in the ponds, and in the seas, schools of fishes follow in a line as straight as a furrow *and* like soldiers lined up by centurions and tribunes.[1] Thus several species of animals possess the virtue of statesmanship, as is also demonstrated in their household management. These are obvious arguments for judgment[2] and foresight.

---

(63) [1]Aucher renders, "Etenim habent ex natura relationem aliquam corpus et anima, (animalis et homines,) non solum per familiaritatem cum iis, quibus cum solis assuefacta sunt." Gr. retransl., ἔχει γὰρ ἐκ φύσεως ἔργον τι σῶμα ψυχῇ·οὐ μόνον δι' οἰκειότητα, ἣν πρὸς τὰ ἥμερα.

[2]Lit., "to do a similar favor in return."

(64) [1]Lit., "They."

[2]Aucher has "Taurus." See § 51, n. 1.

[3]Or, "formed."

(65) [1]Aucher omits ԲԻ Ի ՀՄՄՊՄԲՄՄԳ (καὶ ταξιάρχων). See § 53, n. 1.

[2]Aucher renders, "moralis disciplinae." Gr. retransl., γνώμης.

(66) Moreover vices, no less than virtues, are indications of innate
reason. I have little need to show similarities of faults between man
and other animals: folly, lack of self-restraint, cowardice, injustice,
all that are related to these, and numerous common debaucheries. However animals have been left destitute of truth. Some of them are sly,
for example, wolves and foxes. Likewise he-goats and pigeons are lewd in
sexual relations. Indeed they possess a frantic ardor. The male *pi-
geon* has a habit of breaking the eggs and keeps hovering about while the
female sits on the eggs. When they are seized with wild passion, they
do not confine themselves to their own species. Instead they rush de-
viously into perversions because of insatiability. Men think in like
manner as they continually indulge in various foods and develope that
inevitable[1] wantonness. When their prostitutes are engaged by others,
they pursue new passions and are incited to do unusual things. It is
said that Pasiphaë, daughter of Minos of Crete, was enamoured of a bull
and gave birth to a monster, the Minotaur. Some men have similar insi-
dious lusts, like that ram which *lusted* after Glauce the harpist. Otherwise how would a different, odd looking, and monstrous creature[2] come
into being if not through promiscuous intercouse of unlike species?

(67) It is told in Achaia that once upon a time a dolphin fell pas-
sionately in love with a boy who had a very beautiful face. Many times
it carried *him* from the shore to the deep sea to ride on its back, on
the crests of waves. Many times it carried *him* deeper still *and* safely
brought *him* to the shore after covering a long distance. But when the
boy died in youth--as was his *lot* in life--the dolphin grieved over him
and gave up its *own* spirit when its tense respiratory tract was severed
because of the strain on its body.

(68) Some *animals* are so fearful that they succumb to the weakest
ones and are terrified by the shadow of the mighty. There are those
that hide fearfully in cavernous mountains and valleys and in thick forests and others in the thickets of woods in high places, turning their
faces *as* they watch here and there. Only of the flight of birds are

---

(66) [1]Aucher renders, "quibus de more dediti sunt." Gr. retransl.,
ἣν ἐπιτηδεύουσι.

[2]Lit., "thing."

they not afraid. Deer confirm my words. Apparently when their cowardice was first found out by nature, they were given large[1] defensive
weapons.[2] Unlike other animals which have two horns attached to their
heads, *deer* have many branches *on theirs*, growing here and there like
boughs from a *tree* trunk. Since they are in desperate need to defend
themselves, their armament has to be appropriately large. But what does
it profit cowards to be powerfully armed? Not even ornaments make ugly
women look beautiful. Therefore it is very shameful to make oneself
gaudy with ornaments.

(69) In a contest of fear, the hare, no doubt is next to the deer.
This animal always trembles and is terrified. It does not trust anything. I mean not *only* beasts, but also the *various* parts of nature.
It becomes frightened by the gurgle of gushing fountains, the splash of
the sea, the wind, the breeze, and by everything else on earth. It
hides from streams, shrubs, and dry leaves. Therefore it seems to me
that it was not in vain that the poets[1] thought[2] of the hare *as* a
crouching *creature* because of its cowering[3] and consternation. For this
reason nature, the helper of all, has given this animal an aid to make
up for *its* fearfulness *in* the ability to run swiftly. Its legs move *as*
swiftly *as* wings. Unlike *most* other *creatures*, it is not listless,
malodorous, or thick-skinned. Rather it is as soft as sponge; *and* because it takes in plenty of air, *its* paces rise very easily.

(70) Just as these *animals* are stricken with dread, others[1] have allowed themselves to grow shamelessly and insolently daring.[2] Is it
necessary to talk about the aggressiveness of wild boars, leopards, and
lions in being ready to act unjustly and to injure? Even the dog,

---

[1]Or, "numerous."

[2]Lit., "instruments."

(69) [1]Lit., "the genus of poets."

[2]Aucher has "vocat." Gr. retransl., δοκεῖν.

[3]Or, "crouching." Aucher notes a play on the words, "Λαγώς
enim *lepus* est, et λαθεῖν, *latere*."

(70) [1]Lit., "some."

[2]Aucher renders loosely, "Quantum vero isti pavore sunt pleni,
tantum nonnulli in impudentem audaciam feruntur."

though this animal is fed by man, charges at strange faces and is frantic and fierce. If it happens to see *someone* very far away, it howls
incessantly, and of course, barks as it attacks. When it gets close *e-
nough*, it charges, unmindful of all the injuries which it might suffer.
It assaults very furiously, staring with bloody eyes, its mouth full of
froth. Not only does it withstand rocks thrown at it, but it also defies the strikes of lances and arrows and keeps advancing in spite of
them.

(71) Now each has attained[1] its share. It is obvious that not only
men but also various other animals have inherited the faculty of reason.
Furthermore it is believed that they possess *both* virtues and vices. An
excuse is considerately made[2] for those who have neither heard such a
subject nor studied it on their own *and so* have remained in ignorance.
But to those who have been endowed by God[3] and natural agencies with
fundamentals of knowledge and who have been instructed orally, it is
fair to speak angrily[4] as to laggards and enemies of truth.

(72) LYSIMACHUS: These are *the matters*, honorable Philo, that Alexander, our nephew,[1] presented and discussed when he came in.

(73) PHILO: Wonderful Lysimachus; time is longer than life! These
matters may interest not only the peasants, but also those trained in
philosophy. Now it is not *as though* I was not taught the things referred to; in fact I was nurtured with such instructions throughout
childhood, on account of their certainty, intriguing names, and easy
comprehension. *And* it is not that I studied them thoroughly, but surely I do know them well. Nor are you ignorant, as expressed by the tone
of your voice and indicated by the constant nodding of your head. Since

---

(71) [1]Lit., "they have attained."

[2]Aucher renders, "contrarium omnino tuentur." Gr. retransl.,
ἀπολογίαν ἐστοχάσθαι.

[3]Aucher renders the indicative subjunctivally, "Deus concesserit."

[4]Or, "to scold."

(72) [1]Lit., "our brother's son." The relation so clearly stated in
§ 2 would suggest that the pl. pron. here denotes another form of modesty. A possible corruption of ꝺɓƿ (ὑμῶν) to ꝺɓƿ (ἡμῶν), either in
the Gr. or in the Arm., may be suspected. Cf. §§ 1, 2, 75.

you were listening to what was being read, what else would you need?[1] You seemed to be absorbed like bacchanals and corybants,[2] whose self-proclaimed revelations are not consistent with[3] the reports of researchers and interpreters. On the one hand, there is a diction[4] which results from the up and down movements of the tongue and terminates at the edges of the mouth; on the other hand, there is that which stems from the sovereign part of the soul *and*, through the marvelous employment of the vocal organ, makes sensible utterances.

(74) The affection of a father or of a mother for *their* children is unequaled. Even honest, wise, *and* knowledgeable parents blend with their words an indescribable affection when they relate their experiences[1] to those who listen. They add[2] quite a few nouns and verbs. That is fine and appropriate, you say.[3] But from the interpreter's point of view, I admire your method. You appeared to present the subject much as the author himself would have presented *it* by reading. It seems to me that you have not omitted anything.

(75) As for the recent questions which the young man raised, the new and diverse discoveries, and the terms *used* to delineate everything that is being disclosed, I am not persuaded by them[1] as the fickle-minded, whose habit is to be easily attracted by any fascinating thing. But I will thoroughly examine the truth, as one accustomed to do so, and will make it known to everyone after analyzing it critically. I must not always be impressionable to persuasive argumentation; otherwise what our nephew[2] has already written, which is contrary to sound learning, would be readily believed. If you want to concern yourself

---

[1]Lit., "seem to have."

[2]Aucher renders, "adinstar divinantium sacra quadam insania." Gr. retransl., ὥσπερ οἱ βαχχευόμενοι καὶ κορυβαντιῶντες.

[3]Or, "the same as."

[4]Lit., "prediction."

(74) [1]Lit., "discoveries."

[2]Or, "treasure," "heap," etc.

[3]Aucher, departing from the punctuation of the MS, attaches "inquis" to the preceding sentence.

(75) [1]Lit., "by the same."

[2]Lit., "our brother's son." See § 1, n. 2.

with these matters, I will discuss them right now; but if you want to wait, let us agree[2] to defer them to some other time.

(76) LYSIMACHUS: Do you not realize Philo, that I hold in low esteem all other duties for the sake of *my* love for learning and hunger for truth? If you wish to teach these matters now, I would be most pleased.

(77) PHILO: For my part Lysimachus, I am ready. But take care[1] that by such questioning and surmising we do not sin against the sacred mind. What about the spiders which are considered[2] to be weavers and others, the bees, which make honeycombs? They do this *neither* by means of skill in the arts nor by innate reason.[3] They have no *particular* accomplishment, to tell the truth, except that they work diligently.[4] For since art is an acquired *skill*, what accomplishment[5] is there when there has been no previously acquired knowledge which is the basis of the arts?

(78) Now for example, birds fly, aquatics swim, and terrestrials walk. Is this done by learning? Certainly not. Each of the *above* mentioned *creatures* does it by its nature. Likewise bees make honeycombs by nature, not by learning. Spiders also make their fine work of lace spontaneously.[1] If one wishes to be dissuaded from wrong *thinking* about creatures, he should go to look at the trees and notice every intricate detail. These too have many aesthetic features but no skill in the arts.

(79) Have you not seen the vine when it begins to bear fruit in the springtime? First it covers *the buds* with leaves. Then like a mother,

---

[2]Lit., "let us say."

(77) [1]Or, "bear in mind."

[2]Lit., "you consider."

[3]Aucher, following the punctuation of the MS, renders the sentence with a question mark.

[4]Aucher amplifies in rendering, "Siquidem horum omnino, si oportet verum fateri, admirabilis habenda diligentia, quae tamen non a disciplina deducta est."

[5]Aucher, as in the preceding note, has "disciplina." Gr. retransl., πόρος.

(78) [1]Aucher renders loosely, "per se edoctae subtilem."

it nurtures them gradually, makes them grow, and keeps changing the
sour grapes to sweet grapes until the fruit is fully ripe. Has it been
instructed in *this* care? Through its most amazing and wonderfully
functioning nature, it not only bears essential fruit but also elegant-
ly decorates its trunk with long layers of bark *and* twining sprays, *and*
puts forth green leaves on its branches of various shapes. Its appearance is lustrous and of marvelous aspect. Its comeliness strongly appeals not only to the eyes but also to the olfactory sense; for it
sends out a sweet odor of incense that fills the surrounding air with
a pleasant and pervasive fragrance.

(80) Much could be said about other plants; however I will refrain,
not because of indolence but for brevity, to avoid tiresome prolixity.
I know of many other things which can be *passed by* in silence.[1] I think
that by way of one analogy the *whole* matter is made clear. Thus we af-
firm that the invisible natures and all orderly and artful things are
wrought by rational beings, never by those that are soulless.[2] Yet not
only those but also the animals *previously* mentioned cannot do anything
through foresighted care and thinking; but whatever they do is involun-
tarily done through the peculiarity of their design. For example
partridges[3] fly away from kites and their brood is scared of reptiles;
and when one touches clams[4] they become hard as a rock and close up
tight as if held between two hands which can hardly be pried apart. Is
this by thinking? No, *but* by the peculiar designs of nature which ena-
bles every one to carry on *its* suitable work. Consequently, as soon as
their wings are grown, birds fly; and as soon as their horns are grown,
bulls strike and withstand those who come to fight. Likewise never
think that the scorpion raises *and* waives its stinger deliberately, so
as to inflict an immediate harm by its own choice. Such are the genera

---

(80) [1]Arm. has a rhetorical question, "Don't I know ...? " Aucher
omits Ո՞չ գիտեմ (οὐκ οἶδα;).

[2]Aucher renders somewhat differently, "Dicimus enim et de invisibilibus naturis, quod quum ordinatae et artificiosae sint, integrum
servant esse rationales, etiam illae, quae omnino carent anima."

[3]Aucher, as in § 35, has "palumbes."

[4]Aucher renders ծովային զատտակուրրդ as "marini murices,"
and զզատտակրբաղագն, § 84, as "collectoribus conchyliorum."

of snakes when they strike with their venomous power those who attack.
Do they know *anything* about what causes death to the victim? The providing of food, the healing of diseases, the hidden power of self-preservation which each of them receives in order to inflict injury on those
who attack, and all similar *provisions* are also designs[5] of nature.

(81) It is not by learning that the mouse scurries into its hole or
burrow. Is not even the youngest of them all eager to tunnel before it
is grown?[1] Likewise the brood of swallows turn around to eliminate the
waste matter. Which of the animals really has a sense of the beneficial
or ever does anything with regard?[2] As for that which is well done, is it
accomplished through foresight *or* in the course of the creatures' growth?

(82) It is difficult to think that animals behave with great fraudu-
lence and substantial wisdom. Certainly those who attest to such a wisdom do not realize that they themselves are utterly ignorant. And even
if they have any understanding it is very insignificant, superficial,
and dull. In regard to those who remain childlike, their grasp of knowledge is uncertain, wavering, and childish compared with the rational
mind of the mature.

(83) Even if the horse of Aristogiton was limping without having an
injured leg, it was not cheating. It *only* appeared to be cheating,
since cheating is rational. Perhaps it had a protracted pain in its leg
and was reminded to have a rest. Is not a quiet repose very often for-
gotten[1] until such a time when its memory is rekindled?

(84) Even the assertion of those who think that hounds track by
making use of the fifth mode of syllogism is to be dismissed. The same
could be said of those who gather clams[1] or any other thing which moves.
That they seem to follow a definite pattern is *only* logical speculation
on the part of those who have no sense of philosophy, not even in dreams.
Then one has to say that all who are in search of something are making

---

[5]Lit., "discoveries."

(81) [1]Aucher omits *nι* (οὐ) and the question it prompts.

[2]Aucher has "aut continuum studium virtutis?" See § 25, n. 1.

(83) [1]Lit., "obscure *and* uncertain." Aucher renders loosely, "nam
et in illis per aliquod tempus tantum memoria rerum servatur."

(84) [1]Aucher renders, "collectoribus conchyliorum." See § 80, n. 4.

use of the fifth mode of syllogism! These and other similar *assertions* are delusive fallacies of those more accustomed to the plausibility and sophistry of matters than to the discipline of examining the truth.

(85) We agree that there are some decent and good qualities which are applicable to animals[1] and many other functions which help preserve and maintain their courage; these are observed by sight.[2] There is certainty in everything perceived *or* discerned in all the various species. But surely *animals* have no share of reasoning ability, for reasoning ability extends itself to a multiplicity of abstract concepts[3] in the mind's *perception* of God, the universe, laws, provincial practices, the state, state affairs,[4] and numerous other things, none of which animals understand.

(86) The horse lifts up *its* head and neighs, the deer and the hare are timid, the fox is crafty, and many *others* are attached and devoted to *their* offspring. *But* certainly keen insight of mind is not for every soul; rather, what has been designed *and* fashioned by nature was appropriately given to each as parts of body and soul, to each constituent[1] its proper function toward its own perfection and toward that of the whole *being*.

(87) When lions begin to see clearly their defeat, they flee, running without even looking back; they no longer feel the slightest ambition for power.[1] But they have learned to be debauched like those who crave after power without having education. *Real* power does not subject itself to ridicule. It seems to me that horses, deer, and other animals that long for power[2] should do the same. However, they do not hold a first place in leadership; but rather they are tamed by means of food

---

(85) [1]Lit., "to them."

[2]Aucher renders loosely, "appetitionem (aut imaginationem)."

[3]Or, "immaterial things." Aucher amplifies in rendering, "Rationalis autem habitus est syllogismus ex apprehensione entium, quae minime adsunt."

[4]Or, "state management."

(86) [1]Lit., "to each of them."

(87) [1]Lit., "ambition for glory" or "love of honor."

[2]Lit., "longing for glory" or "craving for honor."

and follow[3] whoever trains them.

(88) One must not refrain from ascribing to these *animals* a basic concern for advancement toward the virtues proper to them, which they never fail to attain[1] according to *their* capability. Consider now the peacocks who display their bright, variegated, and fabric-like feathers and offer a most beautiful sight to beholders; such is not ostentation, but rather a yielding to the cravings of their natural virtue.

(89) But it is ridiculous to think that the Antiochian elephant named Ajax was enraged when about to be deprived of the highest honor. *That rage shows* the licentiousness and ferocity of a wild beast. It may have been that lavishness, excessive attention, and fine food caused it to covet honor.[1]

(90) Never think that dogs, onagers, fawns, *various* species of monkeys, and all that perform marvelously in theatrical shows learn by reasoning. Very often the need for food stirs up suffering[1] in *creatures* such as these. Every *creature* that has partaken of soul is subject to suffering, and it is through suffering hunger that they submit to rules that have been set.

(91) But if you think that ants and bees are household managers incapable of state management, then allow *me* once *more* to refute the thought and to let the matter drop.[1] It is supposed that they are not ignorant of household management; but state management is a child of the same virtue. They are alike, even though they differ in magnitude; one concerns the household and the other the state. But if state man-

---

[3]Aucher has "imitantur."

(88) [1]Or, "allow to fall short of perfection."

(89) [1]Or, "glory." Aucher supplies the first part in rendering, "diligenter alteri comparati invidia ambitiosa accessit."

(90) [1]Aucher has "flagellum" and, twice below, "tormento" and "tormentum."

(91) [1]Aucher renders, "concedas oportet posse falsum esse in specie, quod verum sit in genere sub quo cadit." He supplies "specie," "verum," and "genere," omits *qhшшgbшi* (γνώμων), and disregards the inflection of the rest of the words.

agement cannot be ascribed to them, then it must be said, neither can
household management.[2]

(92) What should one say? That which the ant gathers beforehand it
hoards beforehand in its storehouse. The bee is as dutiful when it goes
about *its* work: it gathers pollen[1] and creates a beautiful honeycomb
which it fills with honey in a most amazing manner. Yet I repeat, because it should be emphasized,[2] that these are not accomplished through
the animals' foresight, but are to be ascribed to nature that manages all.
They do nothing by thought; their work is to attack various created things
and to seize from everywhere whatever is found until their proper *work*
is completed.[3]

(93) With regard to the pinna and the pinna-guard , it was said that
they demonstrate partnership. It is not wrong[1] to say that there are
those who substantiate *their* assertions of truth by displaying fanciful
inventions. But that is nothing; had they known wisdom, they would
have been reprimanded by wisdom.

(94) Be not misled. That these things are altogether doubtful may be
illustrated[1] by the trees as well as the bushes. Even though such have
not partaken of soul, they manifest no less intimacy or indifference;
they move and grow, they kiss *and* embrace each other like lovers, such
as the olive tree *and* the ivy and the elm *and* the vine. And there are
certain things which they reject and turn away from. They not only
raise themselves against *other plants*, openly, face to face, but
also turn away, as if they had feet, and never come closer. Furthermore

---

[2]Aucher has "dispensatio" instead of "aeconomia" as above. Cf.
§ 41, n. 1.

(92) [1]Lit., "flowers."

[2]Lit., "But I say, concerning these I say."

[3]Aucher renders, "Isti vero (Deo) cura est variis de rebus; ut
nimirum sicut Creator impetum singularum creaturarum constringit, ac
corroborat ad id, propter quod factae sunt, sive ad eam quae singulis
convenit perfectionem." He confuses *արարչակաւան* (τῷ ποιητῷ) with
արարիչ (ποιητής or δημιουργός), which he supplies as "(Deo)" and "Creator," and mistranslates most of the rest.

(93) [1]Aucher has "gratis." Gr. retransl., ἄτοπος.

(94) [1]Lit., "it may be learned."

they do not even put forth buds.[2] If they happen to be in bloom, *some* might bear, but the rest drop out of sight or wither away gradually.

(95) Likewise the vine shuns the cabbage and the laurel too.   But I do not think that anyone, however foolish, would dare to say that any of them behaves in a friendly or hostile manner. According to nature's supreme reasoning, certain *plants* settle in certain spots and others move away without becoming attached to one another. With this illustration one can refute the aloofness, partnership, or any other *act* related to the rationality of all mortal creatures[1] except that of man. All such acts, in their natural order, are wrought by reason and understanding. As for the resemblance or likeness of images in animals, these, of necessity, are inferior and, consequently, disdained. There are some demonstrable but very vague images of respect, readiness, gratitude, and like aspects. But these are not genuine. Real attributes and distinct forms belong to human souls.

(96) Furthermore if the stork does not feed its parents in return, it could not be accused of injustice, even though it would appear to be an act of injustice, for it is involuntary. Nor are drones deemed transgressors when they waste the labor of the bees; they do not do *this* voluntarily but rather *are prompted* by the desire for food. Have you not noticed that no one ever blames a little child for anything he does, since he has not yet attained to an accountable[1] age? Although an infant is immature, he is a rational man by nature, having newly received the seeds of wisdom, which, though not yet developed, will soon mature.[2] Throughout the duration of his growth, the seminal powers[3] spread rapidly like sparks in a forest, fanned by a breeze or wind. But the souls of other creatures do not have the fount of reason.   They are destitute of the reasoning faculty.

(97) Researchers into the history of animals make many other state-

---

[2]Lit., "barren of the first growth."

(95) [1]Lit., "animals."

(96) [1]Lit., "thoughtful."

[2]Lit., "it will rise."

[3]Aucher renders, "seminales vires." See § 20, n. 2.

ments which pertain to virtues and vices found in them. *I refer* to those *men* whose custom it is to be verbose in an unprincipled and obnoxious manner. But there is sufficient support for all who undertake to refute their premise. Animals do nothing with foresight as a result of deliberate choice.[1] Although some of their deeds are similar to man's, they are done without thought. According to the primary design of nature, they scatter *their* offspring everywhere after their kind.

(98) Now what has been said about mental reasoning suffices; we must next consider the uttered. Although blackbirds, crows, parrots, and all the like *can* produce different kinds of utterances, they cannot produce an articulated voice in any manner whatever. Furthermore I think that the holes of wind instruments have a marked similarity, for certainly the powerful sounds they release are unintelligible, coarse, and not clearly explained. So also are the meaningless and insignificant sounds produced by the animals *previously* mentioned. These are not so much real expressions of conceptual words as they are chirps.

(99) Take the trumpet, the lyre, and all kinds of musical instruments as common examples. When the air escapes through them, they emit powerful sounds that resemble the human voice. But these are unclear *and* do not convey[1] distinct meaning. Anyone who wishes to consider the numerous and diverse expressions *of sound* will find it easy to form the same opinion. One may also prove positively that listeners differ one from another: it is not to be expected that all shall have the same *impressions*. These defective utterances are even unlike those of stammerers. Surely all who are not endowed with euphonious speech are full of darkness.

(100) Let us now stop criticizing nature and committing sacrilege. To elevate *animals*[1] to *the level of* the human race and to grant equality to unequals is the height of injustice. To ascribe serious self-restraint to indifferent and *almost* invisible *creatures* is to insult those whom nature has endowed with the best part.

---

[1]Aucher amplifies in rendering, "Nos autem jampridem asseruimus ex diametro argumentum adversus praesumentes contradicere, nihil urgere cur dicamus animalia per scientiam deliberativam provido consilio agere quidquid faciunt."

(99) [1]Lit., "to form" or "to represent."

(100) [1]Aucher supplies "immerentia."

III. COMMENTARY

COMMENTARY

1   *Philo . . . Lysimachus . . . Alexander.* On the identity of the
characters, see Intro., pp. 25-28.

   the recent arguments (τοὺς ἐχθὲς λόγους; cf. Provid II 2; Plu.
Mor. 960B). The treatise falls into two parts, each preceded by a
dialogue between Philo and his interlocutor, Lysimachus (§§ 1-9, 72-
76). Alexander's discourse on the rationality of animals comprises
the first part (§§ 10-71) and Philo's refutation of Alexander's premise, the second (§§ 77-100). There can be little or no doubt that
Philo authored the entire work, revealing such arguments as might
have been actually propounded by Alexander himself (see comment on
§§ 8, 74). Placing a contrary argument in the mouth of one's oppon-
ent--whether imaginary or real--is commonplace in classical literature.
Moreover, there is a deliberate literary pattern discernible throughout the treatise. M. Pohlenz observes that the setting of Anim is
somewhat reminiscent of the Platonic *Phaedrus* ("Philon von Alexandreia,"
Nachrichten von der Gesellschaft der Wissenschaften zu Göttingen,
Philologisch-historische Klasse, Neue Folge I, 5 [1942], 412-415). A
comparison of Anim with the *Phaedrus,* however, reveals not only similarities between the setting of Alexander's discourse and that of
Lysias, but also close parallelism throughout the introductory and
transitory dialogues preceding and following the "borrowed" discourses.
Additional similarities are seen between Philo's refutation of Alexander's discourse and Socrates' refutation of Lysias' discourse, leading to the conclusion that the whole structure of Anim is patterned
after the first part of the *Phaedrus* (227A-243E; see App. III, pp. 265-271.

   The treatise is developed with a wealth of stock illustrations,
commonplace in Academic-Stoic arguments regarding the rationality
of animals (for the philosophical background of the treatise, see
Intro., pp. 48-53). Alexander's polemical discourse begins with

111

sweeping denouncements of man's appropriation of reason to himself (§§ 10-11). He attempts to show among the animals instances of the προφορικὸς λόγος, the reason which finds utterance and expression (§§ 12-15). Then follows a lengthy argument for their possession of the ἐνδιάθετος λόγος, the inner reason or thought (§§ 16-71). Some talking and singing birds constitute the examples for the first kind of reason; but the examples for the second kind of reason are far more numerous, including not only such stock examples as spiders, bees, and swallows (§§ 16-22), but also several performing animals (§§ 23-29) and a large number of others in whom demonstrations of virtues and vices are believed to exist (§§ 30-71). Philo, with lack of kindness to his opponent, argues that animals do not possess reason. He briefly refers to several of the animals mentioned in Alexander's discourse and adds others to prove the contrary. Like the Stoics before him, Philo ascribes the seemingly rational acts of animals to the reasoning of nature that fashioned them all (§§ 77-100).

In spite of the title of the treatise and the frequent references to animals, the work as a whole is basically anthropological. By emphasizing the rationality of animals, Alexander seems to make clear a moral and juridical relationship between man and animals (see comment on § 10). This Philo rejects by insisting that there can be no equality between man, who is privileged with reason, and animals devoid of it (see comment on § 100).

2   *three times since then.* Cf. Pl. *Phdr.* 235A (see App. III, p. 265).

engaged, lit., as pointed in the translation notes, "promised by an agreement" (ὁμολογία). Written forms of betrothal agreements (ὁμολογίαι) are cited in Spec Leg III 72 (cf. I 107), where Philo doubtlessly speaks of the practice of his time (see E. R. Goodenough, *The Jurisprudence of the Jewish Courts in Egypt: Legal Administration by the Jews under the Early Roman Empire as Described by Philo Judaeus* [New Haven, 1929], pp. 94-95, 219, where he compares the Philonic passages with the papyri; so also I. Heinemann, *Philons griechische und jüdische Bildung. Kulturvergleichende Untursuchungen zu Philons Darstellung der jüdischen Gesetze* [Breslau, 1932], pp.

293-298; and A. Gulak, *Das Urkundenwesen im Talmud im Lichte der griechische-aegyptischen Papyri und des griechischen und roemischen Rechts* [Jerusalem, 1935], pp. 39-40--cf. his "Deed of Betrothal and Oral Stipulations in Talmudic Law," *Tarbiz*, 3 [1931-1932], 361-376 [Heb.]; for a brief discussion of the subject, see S. Belkin, *Philo and the Oral Law: The Philonic Interpretation of Biblical Law in Relation to the Palestinian Halakah*, Harvard Semitic Series, XI [Cambridge, Mass., 1940], 241-255).

It is reasonable to assume that Alexander's daughter, who was betrothed to her cousin, Lysimachus, was a minor in *patria potestas* (on the implications of Alexander's parenthood on the date of Anim, see Intro., p. 32). In Jewish law--and possibly other national laws of Greco-Roman Egypt--a girl may be given in betrothal from the day she is born (on laws concerning betrothals of minors, marriageable age--usually 12-13 for girls, and the delicate distinctions between betrothal and marriage in ancient Jewish practice--including the Alexandrian, see B. Cohen, "Betrothal in Jewish and Roman Law," *Proceedings of the American Academy for Jewish Research*, 18 [1949], 67-135; reprinted in his *Jewish and Roman Law: A Comparative Study*, I [New York, 1966], 279-347, especially 297-306, 320-323, 336-338; cf. the Egyptian, Greek, and Roman laws on marriage in R. Taubenschlag, *The Law of Greco-Roman Egypt in the Light of the Papyri* [rev. ed.; Warsaw, 1955], pp. 101-130, especially 112-119).

let us resume the discussion of this . . . subject. Cf. Pl. *Phdr.* 238D (see App. III, pp. 265-266).

3   *great assertions* (εὑρέσεις). Philo often talks of εὑρέσεις in connection with the spacious arguments devised by the popular lec- turers or sophists, whom he ridicules and nicknames "Egyptians" (Post 52-53; Agr 143; Conf 39; Migr 171; Congr 29; Quaes Gen III 27; IV 87-88; cf. Cher 8 and Sobr 9, where Ishmael stands for sophistry, as compared with Isaac, who represents wisdom or σοφία: Cher 47; Sacr 43 [*SVF* III 522]; Quod Det 30; Quod Deus 4; Vita Mos I 76; Fuga 200). He stresses the contrast between the plausible arguments devised by the sophists and the true teachings of the philosophers (Agr 159-164;

Heres 302-306; Vita Mos II 212; Quod Omn 1-5; Quaes Gen II 33; IV
92, 104, 107, 221; etc; no doubt he follows the traditional Platonic
detestation of the sophists, such as depicted in *Prt.*, *Grg.*, and
Phdr.; in the latter work, note 236A [see App. III, p. 266]; in considering these and other Philonic passages. É. Bréhier speculates
on Philo's own education: *Les idées philosophiques et religieuses de
Philon d'Alexandrie*, Études de philosophie mediévale, VIII [3d ed.;
Paris, 1950], 287-295; on εὕρεσις as the primary element in rhetoric,
see F. H. Colson's notes to Cher 105 and Migr 35 [LCL II, 485; IV,
561]). Here, as in §§ 84 and 93 (cf. Provid II 103), Philo accuses
Alexander of sophistry.

*listen to them carefully . . . critically examine what the lecturer
is emphasizing.* Philo himself responds in § 75: "I will thoroughly
examine the truth as one accustomed to do so." In the better process
of learning, "we discriminate and distinguish conceptions so that we
can choose what we should choose and avoid the contrary" (Spec Leg
IV 108; cf. Ebr 158; Somn II 263; etc.; see, in particular, Plutarch's
On Listening to Lectures [*De recta ratione audiendi*, *Mor.* 37C-48D]).

why would he leave his other affairs (τὰ πράγματα, in the sense of
state or public affairs)? Alexander is first criticized for discon-
tinuing his philosophical studies to become occupied with other concerns. He is then criticized for neglecting his public affairs to
become involved in such a subject. This conflict between philosophical and political pursuits is somewhat reminiscent of Philo's own
experience as seen in Fuga 36 and Spec Leg III 1-6. In the section
below, Lysimachus, speaking about Alexander, reminds of the social
obligations attached to high rank. Judging from the social standing
of his family, his participation in the Alexandrian Jewish embassy
to Gaius in A.D. 39/40 (§ 54), and his role as *epistrategos* in the
Thebaid in A.D. 42, Alexander must have held some office in the Jewish πολίτευμα as did Philo (on this administrative organization of
the Jews resident in Alexandria, see E. M. Smallwood, *Philonis
Alexandrini Legatio ad Gaium* [Leiden, 1961], pp. 5-14; on his later
career in Roman service, see Intro., p. 32). It is possible that
Lysimachus too found a life career in public service as may be in-

dicated from P. Fouad I 21, 8 (see Intro., p. 28; also E. R. Goodenough, "Philo and Public Life," *The Journal of Egyptian Archaeology*,
12 [1926], 77-79; *The Politics of Philo Judaeus: Practice and Theory*
[New Haven, 1938], pp. 64-66; and his *An Introduction to Philo Judaeus*
[2d ed.; Oxford, 1962], pp. 3-8).

to entertain. Cf. Pl. *Phdr.* 227B, 228E (see App. III, p. 266).

a relative . . . that person (ἀδελφός . . . ἐκεῖνος). Both refer
to Philo. On the speaker's referring to himself by the use of the
demonstr. pron. ἐκεῖνος, see the term (1e) in W. Bauer, *Griechisch-
Deutsches Wörterbuch zu den Schriften des Neuen Testaments und der
übrigen urchristlichen Literatur* (5th ed.; Berlin, 1958), col. 475.

his former courtesies (τὰς πρώτας χάριτας). As observed in the
translation notes, an allusion to Provid I-II is to be suspected.
Both books are in the form of a dialogue between Philo, who maintains
belief in providential sustenance of the world, and Alexander, who
maintains disbelief. The argument on providence is culminated in
this discussion on the rationality of animals. On the latter being
an inseperable part of the question of providence, see Intro., p. 51.
Philo expresses displeasure with Alexander's philosophic position
in both works. Cf. Pl. *Phdr.* 227C (see App. III, p. 266).

You will not get very far with your request. Cf. Pl. *Phdr.* 228C,
236C (see App. III, pp. 266-267).

4   *his want of leisure* (ἀσχολία), i.e. leisure from public affairs
(see comment above; cf. Pl. *Phdr.* 228A [see App. III, p. 267]).

5   *you are interested . . . you are always eager to hear new things.*
Cf. § 76 and Pl. *Phdr.* 228B (see App. III, p. 265). A man with such
eagerness to learn was always held in high esteem. See, e.g., Quod
Deus 107; Ebr 158-159; Heres 18; etc.

you will keep quiet and not always interrupt my speech. Inter-
ruption of lectures is likewise prohibited in Plu. *Mor.* 39B-D; 42F-
43B; 47F-48B; cf. Cic. *Tusc.* i. 16-17; also Pl. *Phdr.* 236C (see App.
III, p. 267).

6   *here I sit . . . here you are seated.* Cf. Pl. *Phdr.* 230E, 236A
(see App. III, p. 267). On the "general and common requirements at
every lecture," with regards to the posture of listeners, see Plu.
Mor. 45C-D.

7   *I am an interpreter and not a teacher* (μὲν γάρ εἰμι ἑρμηνεὺς ἀλλ'
οὐ διδάσκαλος). Philo goes on to define the roles of teachers and
interpreters (fortunately, the original Gr. of these definitions is
extant, see App. II, p. 263; cf. the parallel Platonic passage, *Phdr.*
235C-D, in App. III, p. 267). Philo's calling himself ἑρμηνεὺς here
and in § 74 calls into question the view that places this work
and Provid I-II in that group of his writings said to belong to his
early life, before he undertook the expository works on the Law (on
the date of Anim and the question of the chronological order of
Philo's works, see Intro., pp. 30-34). His role as an interpreter
is all too obvious in those works in which he sets out to find the
spiritual, hidden, or allegorical meanings of Scripture--especially
in Quaes Gen I-IV; Quaes Ex I-II; and Leg All I-III. On Philo as an
exegete of Scripture, see, e.g., C. Siegfried, *Philo von Alexandria
als Ausleger des Alten Testaments* (Jena, 1875); E. Stein, *Die allegorische Exegese des Philo aus Alexandria,* Beihefte zur Zeitschrift für
die alttestamentliche Wissenschaft, LI (Giessen, 1929); I. Christiansen,
*Die Technik der allegorischen Auslegungswissenschaft bei Philon von
Alexandrien,* Beiträge zur Geschichte der biblischen Hermeneutik, VII
(Tübingen, 1969). For a survey of the various understandings of Philo's
exegesis, see V. Nikiprowetzky, "L'exégèse de Philon d'Alexandrie,"
Revue d'histoire et de philosophie religieuses, 53 (1973), 309-329.

By calling himself ἑρμηνεὺς and defining the term as "one who pre-
sents through accurate recall the things heard from others," Philo
seems to allude to his utilization of the thoughts of others. While
most commentators recognize that he used sources and traditions--
whether from Jewish or from Hellenistic realms of thought, there is
no agreement on the extent of his originality (see R. G. Hamerton-
Kelly's review of scholarship on the subject in "Sources and Traditions in Philo Judaeus: Prolegomena to an Analysis of His Writings,"
Studia Philonica, 1 [1972], 3-26). The implication that the inter-

preter's knowledge is not his own is reminiscent of Plato's *Ion*
(530A-542B), where the character, a rhapsode and an interpreter of
Homer, is persuaded to admit that neither his recitation nor his
interpretation is due to his own knowledge but that both are due
to inspiration (see W. J. Verdenius, "L'*Ion* de Platon," *Mnemosyne*
III, 11 [1942], 233-262; J. Behm, "ἑρμηνεύω," *Theologisches Wörterbuch zum Neuen Testament*, II [Stuttgart, 1950], 659-662). It is
interesting to note that the question of interpretation in Philo
is invariably tied to that of inspiration--as seen in his understanding of the prophets as inspired interpreters of divine pronouncements and in his claim to inspiration with regards to his own interpretation (for references, see E. R. Goodenough, *By Light, Light:
The Mystic Gospel of Hellenistic Judaism* [New Haven, 1935], pp. 76,
n. 32; 193, n. 70; see also P. Volz, *Der Geist Gottes und die verwandten Erscheinungen im Alten Testament und im anschliessenden
Judentum* [Tübingen, 1910], pp. 120-130; M. Pulver, "The Experience
of the Πνεῦμα in Philo," *Eranos Jahrbücher*, 13 [1945], 107-121; R. M.
Grant, *The Letter and the Spirit* [London, 1957], pp. 32-38; M. J.
Weaver, Πνεῦμα *in Philo of Alexandria* [unpublished Ph.D. diss.,
University of Notre Dame, Ind., 1973], pp. 115-141).

Philo would thus convey the teachings of others(§§ 75-76; Spec Leg
III 6) without ever calling himself a διδάσκαλος (cf. Socrates'
refusal to call himself a διδάσκαλος and his mode of teaching
διδάσκειν in order not to be identified with the sophists [Pl.
Ap. 33A-B]). In its proper sense, the epithet is reserved for
Moses (Gig 54; Spec Leg I 59) or rather God (Sacr 65; Heres 19, 102;
Congr 114). On other usages of the term in Philo and the Gr. understanding of it as referring to one who contributes his own skill--
especially in conducting a chorus, see K. H. Rengstorf, "διδάσκαλος,"
Theologisches Wörterbuch zum Neuen Testament, II (Stuttgart, 1950),
150-154.

to a few. The passage implies that interpreters do not limit
themselves to a few. Thus, as an interpreter, Philo declares that
he will address himself to everyone (§ 75; Spec Leg III 6). In
like manner he sees the Hebrew prophets as inspired interpreters of

divine pronouncements to the people (see comment above). Likewise, Josephus criticizes the Greek philosophers for speaking only to a few compared with Moses who spoke to the masses (*Ap.* ii. 168-169; but not so with Socrates, who sought to render his services to all his fellow-citizens [Pl. *Ap.* 33A-B; X. *Mem.* i. 10]).

8   *the young man* (νήπιος). In all the MSS this section is ascribed to Lysimachus as part of the introductory dialogue. The remarks, however, belong to Philo's derogatory introduction of his opponent, Alexander, who is likewise called "the young man" in § 75 (see G. Tappe, *De Philonis libro qui inscribitur* 'Αλέξανδρος [Diss. Göttingen, 1912], p. 2). This word, however, does not always refer to age; as a slighting metaphor, it is often used to denote inexperience or ignorance (see § 82; Cher 63; Ebr 146; Sobr 6; etc.; cf. with Philo's calling Alexander "amateur" in Provid I 55; II 40; see also the parallel Platonic passage, *Phdr.* 235A, in App. III, p. 271; and also Ro 2:20; 1 Cor 3:1; 13:11; Gal 4:3; Eph 4:14; Heb 5:13; etc.). To determine Alexander's age at the time when Philo wrote this treatise, the following internal evidences must be taken into consideration: Alexander had a daughter betrothed to Lysimachus (§ 2); he seems to have held some public office (§ 3); and to have participated in the Alexandrian Jewish embassy to Gaius in A.D. 39/40 (§ 54). Moreover, in Provid II 1 he is referred to as "the man" (*ԱՅՐՆ,* ὁ ἀνήρ, rendered as an intensive 3d pers. sing. masc. pron. by Aucher and his Fr. translator, M. Hadas-Lebel, *De Providentia I et II,* Les oeuvres de Philon d'Alexandrie, XXXV [Paris 1973], 212-213). For more on his age and its implications on the date of Anim, see Intro., pp. 30-34.

He sat down. By purporting to have presented Alexander's discourse in his very hearing (see § 9), Philo seems to suggest that the arguments propounded in the discourse were indeed cherished by Alexander himself (cf. Quod Det 134; see § 74 and comment). Moreover, it is reasonable to assume that at the time of writing this treatise Philo could no longer reach Alexander directly; Alexander's departure beyond Philo's reach furnishes the latter with much of the reason for writing this treatise and others as well (see Intro., pp. 29-31). What we have here is a literary actualization of

Philo's desire to have Alexander at hand to listen (cf. Provid II 1).
On obedience as a result of "continuous, insistent urging," see the
parallel Platonic passage, *Phdr.* 228A, in App. III, p. 268.

9   *Philo took them and was about to read.* He must have been inter-
cepted by Lysimachus who reads Alexander's discourse out loud (see
§§ 72-74). The setting here, like the whole arrangement of the
introductory dialogue and the rest of the treatise, owes to Plato's
Phaedrus (see App. III, pp. 265-271), where after an introductory
dialogue between Phaedrus and Socrates, the MS of Lysias' discourse
is brought forth and read (228B, D-E; 230E-234C).

ALEXANDER. Philo ascribes the arguments in §§ 10-71 to his
nephew, Alexander (see comment on § 1). These arguments are refuted
in §§ 77-100.

10   *to fulfill the needs of nature* (τὰς φύσεως ἀνάγκας; cf. Spec Leg
III 97; IV 119). For Philo's association of human beneficence with
nature, see Op 171; Dec 42-43; Spec Leg I 310; II 108-109; IV 178.
For a cognate sense of the entry, note especially Cher 108-112 (on
Lev 25:23), where Philo states that God made none of the created
things complete in itself, so that it should never have no need of
another. "To obtain what it [a created thing] needs, it must per-
force approach that which can supply its need, and this approach
must be mutual and reciprocal." He then proceeds to cite examples
of universal interdependence among created things (cf. Ro 14:7; 15:
1-2).

The Stoics, who held that social obligations are dictated by nature and that they begin with parental affection and extend to
philanthropy, seem to be indirectly addressed here (see Cic. *Fin.*
iii. 62-63 [*SVF* III 340, 369]). They restricted their beneficence
to mankind, beginning, naturally, with relations according to propinquity (Cic. *Off.* i. 58; Epict. *Diss.* i. 16.1-8). In the exchange
of obligations each in turn renders to the other the service that
he needs" (Sen. *Ben.* ii. 18.2 [*SVF* III 573]). But they denied

reciprocal beneficence to animals (see comment on § 63). Since animals have no λόγος, they were deemed unequals; therefore, the Stoics maintained, there is no such thing as a justice which can obtain between animals and humans (see comment on § 100). The charge brought in the section below, however, is that man has failed in his social (or rather natural) obligations as evidenced not only in his treatment of animals but also of women (cf. 1 Pt 3:7; 2 Pt 2: 12-13; Jd 10). Along these lines of argumentation, note the following statements of Plutarch in an anti-Stoic polemic on the rationality of animals (*De sollertia animalium*), a work which--as shall be noted throughout the Commentary--bears the closest similarities to the arguments ascribed to Alexander: "We are to learn to deal innocently and considerately with all creatures, as we are bound to if they possess reason and are of one stock with us," adding that "man is not altogether innocent of injustice when he treats animals as he does" (*Mor.* 964A, D; cf. 999B).

In the light of the preceding, "to fulfill the needs of nature" and its equivalent "complementing one another's weakness" gain additional clarity. The complementing for which the appeal is made is a necessity, a compulsion placed on man by nature. Such beneficence based on the principle of reciprocity is a law of nature, an aspect of universal justice to be rendered not only to one's relations and the rest of mankind but also to animals. Alexander seems to make clear a moral and juridical relationship between man and animals.

selfishness (φιλαυτία), called κακὸν μέγιστον in Congr 130, stands for man's unilateral use of animals and his appropriation of reason to himself (see § 12; note the repeated use of the term in § 15; cf. Praem 12).

the light leading to the knowledge of truth (ἀληθείας. . .φῶς; cf. Leg All III 45; Quod Deus 96; Jos 68; Dec 138; Spec Leg IV 52). The thought owes to the common, metaphorical use of light for knowledge of truth in classical literature--especially since Plato's famous allegory of the cave (*R.* 514A-517A; cf. such New Testament uses as J 1:9; 3:21; 1 J 2:8; Eph 5:9). Truth, said to be "the most precious possession God has ever given to man," is defined in Ebr 6 as "the power which unveils what is wrapped in obscurity." The term occurs fourteen times in Anim (see Ind. IV), where, primarily, it is the truth about the rationality of animals.

the soul . . . its sovereign part (τῆς ψυχῆς . . . τὸ ἡγεμονικόν). Philo sometimes follows the Platonic tripartite division of the soul (λογιστικόν, θυμοειδές, and ἐπιθυμητικόν, located in the head, chest, and around the navel or diaphragm: Leg All I 70-73; III 115-116; Conf 21; Migr 66-67; Spec Leg IV 92-94; Virt 13; cf. *Tim.* 69E-71D; *Phdr.* 246A-256A; *R.* 439D; etc.); but more often he follows the Stoics' elaborations on its rational and irrational parts or faculties (λογικόν and ἄλογον), or masculine and feminine elements (ἄρρην and θῆλυς, Sacr 112; Spec Leg I 200; etc.). The irrational part is divided into seven faculties: the five senses and the organs of speech and reproduction (six in Abr 29, excluding reproduction); the rational or sovereign part (τὸ ἡγεμονικόν) is the ruler of these (Op 117; Leg All I 11; Quod Det 168; Heres 232; Agr 30 [*SVF* II 833]; Mut 111; Quaes Gen I 75 [*SVF* II 832]). Philo sometimes wonders whether the seat of the ἡγεμονικόν is in the head or in the heart (Sacr 136 [*SVF* II 845]; Quod Det 90; Post 137; Somn I 32; Spec Leg I 213-214; the Stoics for the most part decided on the heart [*SVF* II 837, 879-880, 889, 898; III Diog. 29-30], as is also suggested by Philo in Leg All I 59 [*SVF* II 843]; II 6; Spec Leg IV 69). He often follows Plato in locating it in the head (see the passages cited above; also Quaes Gen I 5; II 5; Quaes Ex II 100, 124; cf. *Tim.* 90A). He also follows the Stoic equation of the ἡγεμονικόν with the mind (νοῦς, Op 69; Vita Mos II 82); the faculty of reason (διάνοια, Op 135; *SVF* I 202; III 306, 459), or fount of reason (λόγου πηγή, see comment on § 12).

The irrational part of the soul we and the dumb animals possess in common; the rational or sovereign part was breathed upon the earthborn man (Spec Leg IV 123; cf. Leg All II 23 [*SVF* II 458]; Quod Deus 47). The sovereign part of the soul is the rational spirit (πνεῦμα) in us, patterned after the divine λόγος or image (Spec Leg

I 171; cf. Op 134 - 139; see also Quod Det 80-85; Quod Deus 84; Plant
44; Fuga 133; Somn I 30; Mut 123; Aet 111; Gaium 63; Quaes Gen I 25;
III 8, 22, 27, 49, 59; IV 1, 102; Quaes Ex I 4--on πνεῦμα relating to
mind or soul; cf. also *SVF* I 137-138, 140, 521, 525; II 773-774, 778;
etc.; on man made after the image of God, see comment on § 16). This
element of the soul or mind is called "God's dwelling place" or "temple" (Cher 98-101; Somn I 148-149; II 250-254; Praem 122-123; on the
human body as the temple of God, see comment on § 11). It is also
called "the soul's soul" (Op 66, Heres 55; cf. Migr 5; Quod Det 9), a
phrase of Stoic usage (see S. E. *M.* vii. 233--from F. H. Colson in a
note to Migr 5 [LCL IV, 560]);"it is older than the soul as a whole"
(Leg All II 6). For more on Philo's basically Stoic distinctions
between the human soul and that of animals, see comment on § 78 (*SVF*
II 732). On his understanding of the soul's incorporeality, immortal-
ity, etc., see J. Drummond, *Philo Judaeus: Or, the Jewish-Alexandrian
Philosophy in Its Development and Completion,* I (London, 1888), 326-335;
H. Schmidt, *Die Anthropologie Philons von Alexandreia* (Diss. Leipzig,
1933), pp. 49-67, 138-153; U. Früchtel, *Die kosmologischen Vorstellung-
en bei Philo von Alexandrien. Ein Beitrag zur Geschichte der Genesis-
exegese,* Arbeiten zur Literatur und Geschichte des hellenistischen
Judentums, II (Leiden, 1968), 62-68; H. A. Wolfson, *Philo,* I (rev. ed.;
Cambridge, Mass., 1968), 385-413; R. A. Baer, Jr., *Philo's Use of the
Categories Male and Female,* Arbeiten zur Literatur und Geschichte des
hellenistischen Judentums, III (Leiden, 1970), 84-87; M. J. Weaver,
Πνεῦμα *in Philo of Alexandria* (unpublished Ph.D. diss., University of
Notre Dame, Ind., 1973), pp. 55-114, 169-171.

God (θεός). All five occurrences of the term in this treatise (see
Ind. IV) are less pronounced compared with the predominantly Jewish
conception of God in Philo's works (on the interchangable use of the
term θεός and φύσις, see comment below, on § 11). From among the ex-
tensive discussions on Philo's conception of God and His Powers, see
J. Drummond, *Philo Judaeus: Or, the Jewish-Alexandrian Philosophy in
Its Development and Completion,* II (London, 1888), 1-155; E. Bréhier,
Les idées philosophiques et religieuses de Philon d'Alexandrie, Études

de philosophie mediévale, VIII (3d ed.; Paris, 1950), 69-82, 136-151;
J. Pascher, Ἡ βασιλικὴ Ὁδός. *Der Königsweg zu Wiedergeburt und*
Vergottung bei Philon von Alexandrien (Paderborn, 1931), pp. 160-260;
E. R. Goodenough, *By Light, Light: The Mystic Gospel of Hellenistic*
Judaism (New Haven, 1935), pp. 11-47, 146-147, 166-167, 174-175, 212,
382-383; and H. A. Wolfson, *Philo* (rev. ed.; Cambridge, Mass., 1968),
I, 200-294, 325-359; II, 73-164. Pascher and Goodenough see in Philo's
conception of the knowledge of God strong affinities with the Gnostic
conception.

11   *men ignore the weakness of women* (see comment above). From the
opening lines of this section and those of the preceding, Alexander
seems to have been a defender of women's rights. That he indeed was
a liberator of women has been attested in two papyri dating from the
2d cent. A. D. which quote decisions of his as precedents: one
forbidding women to be made cultivators and the other protecting
them from the performance of certain liturgies (P. Oxy. 899, 28,
dated A.D. 200, B. P. Grenfell and A. S. Hunt, eds., *The Oxyrhynchus*
Papyri, VI [London, 1908]; P. Wien Bosw. 1, 19, E. Boswinckel, ed.,
Einige Wiener Papyri [Leiden, 1942]).

      dumb animals bending downward . . . they themselves stood upright
and erect. Philo comments on the human posture and that of animals
in Quod Det 85 and Plant 16-17, where the thought and much of the
words are borrowed from Pl. *Tim.* 90A, 91E (cf. 69C-71D; 72E-81E,
the lengthy recital on the arrangement of the human body; *Prt.* 321C;
Arist. *PA* 689b12; etc.). The erect posture of the human body, with
the head, the brain, or the eyes in the most elevated position, was
taken as a mark of divine favor allowing man to contemplate the
operations of the heavens (Plant 20; cf. Cic. *ND* ii. 140; Sen. *Ep.*
xciv. 56; *Dial.* viii. 5.4; Pliny *NH* xi. 135; especially because of
the divine element of the human soul, which has its abode in the
heart [see comment on the ἡγεμονικόν, § 10], the body as such becomes
venerable : Sen. *Ep.* xcii. 1-2, 10, 13; cxx. 14; cf. 1 Cor 3:16-17;
6:19, 2 Cor 6:16; on this Stoic view, see E. V. Arnold, *Roman Stoic-*

ism [London, 1911], p. 259). "So far as his [i.e. man's] body
ascends and is raised aloft from the earth, he would justly be said
to be an air-walker" (ἀεροπόρον, Op 147). Man is superior to other
living creatures, though not to stars or heavenly beings (Somn II
132; cf. Op 63, 65, 68, 83-84; Praem 9, 86; Aet 65; Provid I 9 [*SVF*
II 577]; II 105; etc.). His stock is near akin to God (Op 77; Abr
41; Vita Mos I 279; Spec Leg IV 14; Dec 134; etc.; cf. Pl. *R.* 391E;
and E. *Supp.* 201-213, where Theseus praises the deity who raised men
from the brutes and gave them reason and the tongue for speech).
Reason is the bond of this divine relationship, "For everyone who is
left forsaken by reason , the better part of the soul, has been transformed into the nature of a beast, even though the outward characteristics of his body still retain their human form" (Spec Leg III 99;
cf. Abr 33). For more on Philo's treatment of the human body and his
understanding of man's higher nature, see J. Gross, *Philons von
Alexandreia Anschauungen über die Natur des Menschen* (Diss. Tübingen,
1930); H. Schmidt, *Die Anthropologie Philons von Alexandreia* (Diss.
Leipzig, 1933), especially pp. 31-34.

Reason is the best of things that exist (τὸ ἄριστον τῶν ὄντων ὁ
λόγος). Cf. § 100; Op 77; Sacr 83; Quod Det 83; Quod Deus 16; Fuga
94; Somn I 68; etc.; also Chrysippus in Cic. *ND* ii. 16 (*SVF* II 1012);
see also 18, 133 (*SVF* II 1131).

an irreversible reward from nature (ἀπὸ τῆς φύσεως). Elsewhere
Philo argues that reason is man's special prerogative, given to him
by nature, whereby man has been made superior to other animals (Quod
Deus 45 [*SVF* II 458]; Somn I 103, 113; Spec Leg II 173; etc.). Nature
here, as often in this treatise and elsewhere in Philo, may be identified with the divine agency in things (§§ 29, 68-69 [see comment], 71,
80, 86, 92 [see comment], 95, 97, 100 [*SVF* II 730, 732-733]; cf. Sacr
98; Post 5; Heres 114-116; Fuga 170-172; Vita Cont 70; etc.). Philo's
interchangeable use of the terms θεός and φύσις (See H. Leisgang,
"φύσις. 3. θεός est φύσις vel hac voce intellegitur," *Indices*, II
[Berlin, 1930], 838-839), as also of the terms for divine and
natural law (see comment on § 48), does not necessarily imply the

equation of the two--as in Stoicism (*SVF* I 152-177; cf. the personi-
fication of φύσις in 1 Cor 11:14). ALthough Philo's philosophical
system owed much to the Stoics, their conception of God as material
and immanent in the world was unacceptable to Philo, to whom God
was wholly transcendant and immaterial. On special uses of the
term φύσις in Philo, see E. R. Goodenough, *By Light, Light: The
Mystic Gospel of Hellenistic Judaism* (New Haven, 1935), pp. 48-71;
H. A. Wolfson, *Philo*, I (rev. ed.; Cambridge, Mass., 1968), 332-347.

truth (ἀλήθεια). See comment on § 10.

12   *two kinds of reason.* The distinction between uttered reason or
speech (λόγος προφορικός) and mental reasoning or thought (λόγος
ἐνδιάθετος), implied in Plato (*Tht.* 189E; *Sph.* 263E) and Aristotle
(*APo.* 76b24), was emphasized by the Stoics in their debates with the
Academy (*SVF* II 135, 223, etc.; for a fine discussion, see M. Pohlenz,
"Die Begründung der abendländischen Sprachlehre durch die Stoa,"
Nachrichten von der Gesellschaft der Wissenschaften zu Göttingen,
Philologische-historische Klasse, Neue Folge I, 3 [1939], 151-198).
The distinction was subsequently employed by Philo (Quod Det 92;
Mut 69; Abr 83; Vita Mos II 127-130; Spec Leg IV 69) and others
(Plu. *Mor.* 777B-C; 973A; Theophilus Antiochenus *Ad Autolycum* ii.
10, 22; Hippol. *Philosophumena* x. 33; Origenes *Cels.* vi. 65; etc.;
see M. Mühl, *Der Logos endiathetos und prophorikos in der älteren Stoa
bis zur Synode von Sirmium 351,* Archiv für Begriffsgeschichte, VII
[Bonn, 1962]). To Philo, as to the Stoics, the λ. ἐνδιάθετος not only
distinguishes man from animals but also extends itself through the λ.
προφορικός; i.e., the true λ. προφορικός is only a derivative manifestation of the λ. ἐνδιάθετος (see his response, §§ 73, 98-99 [*SVF* II
734], and, in addition to the Philonic passages cited above, Quod Det
40, 129; cf. his allegory of Aaron's relationship to Moses in Migr 78,
84, 169; Quaes Ex II 27; see also G. D. Farandos, *Kosmos und Logos
nach Philo von Alexandria,* Elementa: Schriften zur Philosophie und
ihrer Problem-geschichte, IV [Amsterdam, 1976], 243-252; for a modern-
istic treatment of Philo's theory of language, see K. Otte, *Das
Sprachverständnis bei Philo von Alexandrien: Sprache als Mittel der
Hermeneutik,* Beitrage zur Geschichte der biblischen Exegese, VII

[Tübingen, 1968], 45-77; on the intricacies of language and thought in Stoic logic, see A. A. Long, "Language and Thought in Stoicism," *Problems in Stoicism*, ed. A. A. Long [London, 1971], pp. 75-113).

The metaphor of the spring, stream, or river, as applied to the extension or flow of the λόγος, occurs with almost identical terms not only in the above cited Philonic passages but also in a number of others (Quod Det 40, 83, 126; Post 127 [*SVF* II 862]; Agr 53; Migr 71; Congr 33; Fuga 177-178, 182 [*SVF* II 861]; Somn II 238-239; etc.). Philo follows the Stoics in identifying "the sovereign part of the soul" (τὸ ἡγεμονικὸν μέρος τῆς ψυχῆς) with the faculty of reason (διάνοια, νοῦς, see comment on § 10) or, as the above cited passages suggest, the fount of reason (λόγου πηγή, cf. *SVF* I 148; II 840, 894; III Diog. 29), which, he insists, does not exist in dumb animals (§ 96 [*SVF* II 834]).

both kinds of reason appear to be imperfect in animals (ἀτέλειοι). The same thought is repeated in § 29 and partially endorsed by Philo in § 95 (*SVF* II 730; see comment). Elsewhere he declares: "Perfection depends, as we know, on both kinds of reason: that which suggests the ideas with clearness and that which gives unfailing expression to them" (Migr 73). Similarly, Plutarch states that "the alloyed and imprecise virtue of animals" should not be stigmatized "as absence of reason" but rather "as its imperfection or weakness" (*Mor.* 962B), since "all animals partake in one way or another of reason" (960A).

13   *the abstract concepts which the mind apprehends in connection with sensation are perceived through hearing* (ἣν δὲ τῶν ἀοράτων ἰδεῶν ὁ νοῦς καταλαμβάνει, ἣν τείνας δι' αἰσθήσεως . . .). For the early history of concepts on the interaction of mind and sense-perception in acquiring knowledge of the external world, see F. Solmsen, "Greek Philosophy and the Discovery of the Nerves," *Museum Helveticum*, 18 (1961), 150-163, 169-197. Philo's understanding of the impact of the mind upon the senses and that of the senses upon the mind is made sufficiently clear in his allegories on Adam and Eve, where he contrasts mental apprehension, symbolized by man, with sense-perception, symbolized by woman (e.g., Leg All I 25; II 13, 38, 70; III 50, 200, 246 [*SVF* III 406]).

Allegorizing on the Genesis account of the fall of man, Philo
points out in Op 165-166 (*SVF* II 57) and Quaes Gen I 46 how the
superior mind, which in its rightful place of sovereignty controls
the senses, can be influenced and deluded by them. At such an instance the superior mind becomes one with the senses, "it resolves
itself into the order of the flesh which is inferior, into sense-
perception" (Leg All II 50; cf. III 185). Philo's works abound in
contrasts between the heavenly, rational mind and the earthly, irrational senses (Leg All III 108; Cher 70; Migr 206; Heres 185;
etc.). But he does not overlook the role of the senses in acquiring knowledge; knowledge is dependent on sense-perception (Cher 40-
41; Heres 53; Congr 155). Sense-perception apprehends simultaneously with mind and gives it occasions of apprehending the objects
presented to it (Leg All I 28-30 [*SVF* II 844]; III 56-58; Post 126
[*SVF* II 862]). Mind, helpless in itself, by mating with sense,
comes to comprehend phenomena (Cher 60-64; Leg All III 49; but cf.
Heres 71; Virt 11-12; Quaes Gen III 3). Thus, sense-perception is
the handmaid of reason (Vita Mos II 81). Moreover, the sensible
and the mental must be so blended that "in the world of sense we may
come to find the likeness of the invisible world of mind" (Migr 105),
that it is by experiencing the world of sense that we get our knowledge of the world of mind (Somn I 186-188; cf. S. E. *M.* viii. 56
[*SVF* II 88], that all elements of knowledge come either from sense
and experience solely, or from sense and experience combined with
reasoning; see also *SVF* II 850-860).

Philo seems to have been puzzled between the Stoic dictum that all
knowledge enters the mind through the senses and the idealism of
Plato, for whom the mind alone was the source of knowledge, the
senses being copies or imitations of the Ideas, sources of all illusion
and error. Comparing the generally positive attitude toward the senses
in Leg All II 5-25, 71-108 with Philo's normally negative evaluation
of sense-perception, W. Bousset attributes the incongruity to the use
of different sources: *Jüdisch-christlicher Schulbetrieb in Alexandria
und Rom. Literarische Untersuchungen zu Philo und Clemens von Alexandria, Justin und Irenäus* (Göttingen, 1915), pp. 45-49.

Philo, following Pl. *Tim.* 47A-E, exalts the sense of sight in particular, holding it to be most closely akin to the rational soul and philosophy (Op 53-54, 77; Sacr 78; Post 167-168; Quod Deus 45 [*SVF* II 458]; Conf 140-141; Migr 47-49; Fuga 208; Abr 57; Spec Leg I 339-340; Quaes Gen II 34; etc.). He also praises hearing, often the second sense in importance, and sometimes accords it the first place with sight (Abr 150; Vita Mos II 211; Spec Leg I 29, 193). The importance of hearing is seen in the advantages it affords in instruction (Conf 72; Spec Leg I 342; cf. Ro 10:17; Gal 3:2, 5). Hearing "lays claim to reason" (τὴν λόγου μεταποιουμένην ἀκοήν, Abr 160; cf. with the following statement attributed to Zeno in Galen's *De placitis Hippocratis et Platonis* iii. 5 [Müller, p. 287 = *SVF* II 891]: ἡ κατὰ τὴν ἀκοὴν αἴσθησις καταφερομένη περὶ τὴν διάνοιαν). Unfortunately, none of the above Philonic passages are included in the selection *De visu et auditu* in *SVF* II (863-872).

No doubt the importance the Stoics attached to sense-perception (*SVF* II 105-121), which is also shared by creatures without reason, gave rise to arguments similar to those in Anim and in Plutarch: that animals cannot make use of sensation without some knowledge or understanding; "the impact on eyes and ears brings no perception if the understanding is not involved"; and that "all creatures which have sensation can also understand" (*Mor.* 960D-961B).

listening for themselves . . . inherently self-taught . . . taught (αὐτηκόοι . . . αὐτομαθεῖς . . . διδακτοί). Throughout the writings of Philo three ways of learning emerge: by teaching, nature, and practice (μαθήσει, φύσει, ἀσκήσει--commonplace in classical literature), characterized by Abraham, Isaac, and Jacob respectively (see H. Leisegang, "'Αβραάμ et 'Αβράμ,'" "'Ισαάκ," "'Ιακώβ," *Indices*, I [Berlin, 1926], 1, 11, 13). On the special meaning of the word αὐτοδίδακτος (or αὐτομαθής), see É. Bréhier, *Les idées philosophiques et religieuses de Philon d'Alexandrie*, Études de philosophie mediévale, VIII (3d ed.; Paris, 1950), 272-279. Note that after the citation of certain habits of birds and insects as examples of self-taught knowledge (§§ 13-22), there follows a discussion of taught skills practiced by animals in theatrical performances (§§ 23-28). Plu. *Mor.* 973E has that "self-instruction implies more reason in animals than does readiness to learn from others."

crows and Indian parrots. See D'Arcy W. Thompson, "Κορώνη," "Ψιττάκη," *A Glossary of Greek Birds* (rev. ed.; London, 1936), pp. 169-170, 335-337. On the notes of birds, their mimicry, and articulate voice as indications of reason, see Pliny *NH* x. 80-124; Plu. *Mor.* 972F-973A; Ael. *NA* vi. 19; xii. 28; etc. For Philo's response, see §§ 98-99 (*SVF* II 734).

the wealthy homes of Alexandria. Though rich, Philo seems to have preferred simple living. See his diatribes against luxurious living in Somn II 48-62; Spec Leg II 18-23; and Vita Cont 48-63.

parrots . . . flatter kings, emperors, etc. Similar records on ravens and parrots greeting celebrities abound in antiquity. According to Macrobius, when Octavian returned to Rome after the defeat of Antony, he bought a raven, a parrot, and a pie which had been taught to greet him (*Sat.* ii. 4.29-30). Pliny tells of a raven taught to salute by name first Tiberius, then Germanicus and Drusus, the sons of Tiberius--the first adopted--and that a talking thrush, a starling, and nightingales were owned by Agrippina, Claudius' fourth wife and niece, and the young princes: Brittanicus, his son from previous marriage, and Nero, her son from previous marriage (*NH* x. 120-121). J. Schwartz sees literary parallelism between Anim and Pliny and goes on to suggest "un nouveau *terminus post quem*, de toutes façons encore fort étoigné de la vraie date." "Note sur la famille de Philon d'Alexandrie," *Mélanges Isidore Lévy,* Annuaire de l'institut de philologie et d'histoire orientales et slaves, XIII (Bruxelles, 1955), 595, n. 1. Schwartz's observation suggests a post A.D. 48 date for the composition of Anim--after Claudius took Agrippina as his fourth wife. A due consideration of § 58 (see comment) leads to the same conclusion regarding the date of Anim.

14   *voice always leads to belief in hidden powers* (εἰς τὰς ἀφανεῖς; i.e. of the mind or of thought; the Arm. word *Ժայն* is used for both φωνή and ψόφος). The concept of voice or mere utterences as

expressions of thought or mental states goes back to Aristotle
(*Int.* 16a3-18; for his definition of voice [φωνή], see *de An.* 420b6,
29-31). Philo in his response distinguishes between the articulate
human voice and the utterances of animals which he equates with the
sound of musical instruments (§§ 98-99 [*SVF* II 734]; see comment).

there are two uses of the voice. Cf. Post 106; Plant 131: ἑκατέραν
φωνῆς ἰδέαν, ᾗ τὸ λέγειν καὶ τὸ ᾅδειν.

15   *Blackbirds, turtledoves, and swallows.* See D'Arcy W. Thompson,
"Κόσσυφος," "Τρυγών," "Χελιδών," *A Glossary of Greek Birds* (rev. ed.;
London, 1936), pp. 174, 290-291, 320-321--on the barbarous twitter
of swallows; cf. Is 38:14). Philo talks of the pleasure derived
from hearing "tuneful voices of creatures without reason: swallows,
nightingales, and other birds which nature has made musical" (Leg All
II 75; cf. Ael *NA* ii.11). Although song-birds are superior to musical instruments, the music they produce is incomparable with the
human voice (Post 105-106). Elsewhere he has that there may easily
be unison in the worst disharmony (Conf 150).

selfishness (φιλαυτία). See comment on § 10.

16   *uttered reason and . . . that which is in the mind* (λόγος
προφορικός . . . ἐνδιάθετος). These terms are defined in § 12
(see comment), with discussion of the first kind of reason following
in §§ 13-15. The second kind of reason is discussed in §§ 17-70,
beginning with such *loci communes* as the spider, the bee, the swallow (§§ 17-22), performing animals (§§ 23-28), and, after a restate-
ment of the thesis in § 29, culminating with the participation of
animals in virtues and vices (§§ 30-70).

foolish . . . hunters. Philo has nothing to commend hunters for
in his treatment of Nimrod and Esau (Quaes Gen II 82; Heres 252; on
Gen 10:8-9; 25:27-28). He considers Nimrod (whom he takes to be a
"mighty hunter" not "before the Lord" but "against the Lord") as a
type of all great sinners, being utterly removed from rational
existence; "for he who lives among wild beasts chooses to be on a
level with animals in bestial habits caused by evil passions."
Hunting, however, was generally recognized as useful training for
war. On this both Alexander and Philo agree in Provid II 91-92,

103 (cf. Vita Mos I 60-61; Spec Leg IV 119-121). On how lions
prevent hunters from following their tracks, see Plu. *Mor.* 966C-D;
Ael. *NA* ix. 30. Three accounts on lion hunting are given in Opp.
C. iv. 77-211.

some men . . . observe the different species of animals. In § 61
these men are referred to as "witnesses" whose accounts should be
believed. In § 97 they are the "researchers into the history of
animals" whom Philo ridicules.

whether only the human mind was made after the divine image
(πότερον μόνος ὁ ἀνθρώπινος νοῦς κατ᾽ εἰκόνα γεγενῆσθαι). The
divine image is the archetype of the human νοῦς. It is the universal νοῦς after which the νοῦς in each of us is patterned. (Op 24-25,
69, 134-139; Leg All I 31, 42; III 96; Quod Det 80-85; Plant 18-19;
Heres 56-57; 230-231; Fuga 71; Somn I 74; Vita Mos II 65; Dec 134;
Spec Leg I 81, 171; III 83, 207; Virt 203-205; Quaes Gen I 4-5--
on Gen 1:27 and 2:7; cf. with the Stoic notion that the human νοῦς
is by descent akin to the universal [S. E. *M.* ix. 93]). It is
worth noting that in these passages Philo uses both Gen 1:27 and 2:7
to establish the God-like nature of the human νοῦς, and not, as J.
Jervell supposes, to distinguish between a "heavenly," "image-like,"
or "ideal" man in Gen 1:27 and an "earthly" man in Gen 2:7 (*Imago
Dei. Gen 1:26f. in Spätjudentum, in der Gnosis und in den paulinischen Briefen* [Göttingen, 1960], pp. 52-70; for a careful analysis
of these Philonic passages, see C. Kannengiesser, "Philon et les
Pères sur la double creation de l'homme," *Philon d'Alexandrie,*
Colloques Nationaux du Centre National de la Recherche Scientifique,
à Lyon du 11 au 15 septembre 1966 [Paris, 1967], pp. 277-296, on the
passages in Op primarily; R. A. Baer, Jr., *Philo's Use of the Categories Male and Female,* Arbeiten zur Literatur und Geschichte des
hellenistischen Judentums, III [Leiden, 1970], 20-35; and L. K. K.
Dey, *The Intermediary World and Patterns of Perfection in Philo and
Hebrews,* SBL Dissertation Series, XXV [Missoula, Montana, 1975],
20-30).

The second of the two alternatives here propounded, "whether
only the human mind was made after the divine image" or "whether

God gave a common advantage [i.e. reason] to all creatures," is
contrary to Philo's thought. He responds: "One can refute . . .
the rationality of all mortal creatures except that of man" (§ 95
[*SVF* II 730]; on the natural constitution and characteristics of
plants, animals, and man, see comment on § 78 [*SVF* II 732]). The
human mind is called by him "the man within the man, the better
part within the worse, the immortal within the mortal" (Congr 97),
the divine (Op 146; Quod Det 29) or best element in us (Fuga 148;
Jos 71; cf. the references to the ἡγεμονικόν in the comment on §
10). The creation of man in the image of God is often seen by him
not only in the God-like nature of man's rational soul or mind
but also in the human longing for the divine model (see J. Giblet,
"L'homme image de Dieu dans les commentaires littéraux de Philon
d'Alexandrie," *Studia Hellenistica*, 5 [1948], 93-118).

17   *the friends of truth* (οἱ ἀληθείας ἑταῖροι). The polemical, opening lines of this passage seem to be directed against the Stoics,
who upheld either "recognition" (κατάληψις) or the "recognizable
presentation" (τὴν καταληπτικὴν φαντασίαν) of sense-impressions upon
the mind as the criterion of truth for the acquisition of knowledge
(see *SVF* II 105-121; cf. S.E. *M.* vii. 151, 227; viii. 397 [*SVF* II
90, 56, 91]; Cic. *Ac.* i. 42 [*SVF* I 60, 69]; *Fin.* iii. 17 [*SVF* III
189]). On the Stoics' defense of this ambiguous and much debated
criterion, which came under the heavy attack of the New Academy
ever since Arcesilaus and was later contradicted by the Stoics' own
tenets of accepting scientific certainties as the standard of truth
for the acquisition of knowledge, see M. Pohlenz, *Die Stoa. Geschichte
einer geistigen Bewegung* (Göttingen, 1964), I, 59-63; II, 35-36; J. M.
Rist, *Stoic Philosophy* (London, 1969), pp. 133-151; F. H. Sandbach,
"Phantasia Kataleptikē," *Problems in Stoicism*, ed. A. A. Long (London,
1971), pp. 9-21. Moreover, in the example of the spider and its web
there seems to be a manipulation of Chrysippus' use of the spider model
as an illustration of the way in which the soul functions in getting
knowledge (see comment below). "The friends of truth" are called
"enemies of truth" at the conclusion of these arguments ascribed to
Alexander (§ 71).

the beautiful (τὸ καλόν). Diogenes Laertius records this Stoic
doctrine concerning the beautiful: "The reason why they characterize

the perfect good (τὸ τέλειον ἀγαθόν) as beautiful is that it has. . .
perfect proportion"(τὸ τελέως σύμμετρον, vii. 100 [*SVF* III 83]).
Moreover, it has four aspects, which prove to be the cardinal virtues, except that self-restraint (σωφροσύνη) is replaced by order-
liness (κοσμιότης, *ibid.*). The equation of the term with the cardinal virtues is derived from equating the beautiful (τὸ καλόν)
with the morally good (τὸ ἀγαθόν) and the goods (τὰ ἀγαθά) with the
virtues (αἱ ἀρεταί, *ibid.* 101-102 [*SVF* III 30, 92]). Diogenes
then cites the various definitions of the term: "By the beautiful
is meant . . . that good which renders its possessors praise-
worthy . . . good which is worthy of praise . . . a good aptitude
for one's proper function . . . that which lends new grace to anything" (*ibid.* 100). All of these meanings are implied in the example of the spider's web.

*the faculty of reason is implanted in every creature endowed with
a soul* (διάνοια μὲν γὰρ ἐγγίνεται πᾶσι τοῖς ἐμψύχοις). Among the
numerous examples given in support of this statement note especially
§§ 41, 43-44, where the Arm. words equivalent to διάνοια are encountered again and rendered "reasoning mind(s)"; moreover, the statement
is repeated in the concluding paragraph of Alexander's discourse
(§ 71) and refuted by Philo (§ 96 [*SVF* II 732]). Comparable statements are found in the anti-Stoic speeches in Plu. *Mor.* 960C-963A
(*SVF* III Ant. 47). W. C. Helmbold rightly observes in a note to
960C (LCL XII, 326-327), "There seems to be a great deal more anti-
Stoic polemic in the following speeches than von Arnim has admitted
into his compilation." The Stoics identified διάνοια with the sovereign or ruling part of the soul (τὸ ἡγεμονικὸν μέρος τῆς ψυχῆς,
see comment on § 10), the fount of reason (λόγου πηγή, see comment
on § 12), of which animals have no share. Philo likewise states
that the faculty of reason (διάνοια) has no place in creatures
other than man (§ 96 [*SVF* II 834]; Quod Deus 47; cf. Op. 146; Leg
All II 23 [*SVF* II 458]; Quod Det 29; Fuga 148; Jos 71). Aristotle,
however, states that "many animals have no faculty of reason" (τῶν
ζῴων πολλὰ διάνοιαν οὐκ ἔχειν, *de An.* 410b25), thereby allowing the
possibility that some animals have such a faculty. On the distinctions between ἔμψυχα and ἄψυχα, see comment on § 78 (*SVF* II 732).

the spider. §§ 17-19 are on the spider, its mode of weaving a web and dealing with a catch (cf. Arist. *HA* 623 a 7-23; Pliny *NH* xi. 79-84; Plu. *Mor.* 966 E-967 A; Ael. *NA* i. 21; vi. 57; Philostr. Jun. *Im.* ii. 28; etc.). The webs of spiders, the honeycombs of bees, and the nests of swallows (§§ 17-22) are commonly cited examples of mental reasoning (λόγος ἐνδίαθετος) in discussions on animal intelligence. For a collection of such passages, see S.O. Dickerman, "Some Stock Illustrations of Animal Intelligence in Greek Psychology," *Transactions and Proceedings of the American Philological Association,* 42 (1911), 123-130. Chrysippus uses the spider model as an illustration of the way in which the soul functions in getting knowledge (*SVF* II 879; cf. with Heraclitus' soul/spider analogy, Diels *VS* 22B67[a]). See comment on Philo's response to these arguments in §§ 77-81 (*SVF* II 731-732).

Who rates second in art (τέχνη)? Contextually, the emphasis is on the aesthetic aspect of art, the naturally beautiful. Such art, however, is said to be incomparably inferior to that which has to do with knowledge (ἐπιστήμη, see comment on § 77 [*SVF* II 731]). Note also the comparison with human achievement in this and the following examples: spiders work better than weavers and are as careful as artisans; bees cooperate very much like men, they build their hives like walled palaces; "and the structure of the [swallow's] nest is more perfect than any of the most artful structures made by man" (§§ 18-22; cf. with the Democritean essentials in which animals surpass man, Diels *VS* 68B154).

a lyre. For the figure, which connotes harmony, see Cher 110; Sacr 73; Quod Deus 24 (cf. Heraclitus, Diels *VS* 22B51).

the circular is always more durable than the straight. Similar remarks are made on the swallow's nest in Plu. *Mor.* 966E. The notion is derived from the Platonic doctrine of the spheres (*Tim.* 33B), adopted by the Stoics (Cic. *ND.* ii. 47; D. L. vii. 140 [*SVF* I 95, 404; II 520, 543; III Ant. 43, Apollod. 5]) and Philo (Provid II 56 [*SVF* II 1143]), which exalts the circular motion as the motion of reason and the circle as the perfect figure.

18   *the spider contains within itself all that it needs.* So also
      Pliny *NH* xi. 80; Ael. *NA* i.21. What is claimed for the spider is
      said of the cosmos: "Of the cosmos alone it can be said that it is
      self-supporting, because it alone contains within itself all that it
      needs" (Plu. *Mor.* 1052D [*SVF* II 604]).

19   *sailors. . . physicians. . . artisans.* These *loci communes* abound
      not only in Philo (e.g., Op 10; Leg All III 223, 226; Post 141-142;
      Spec Leg IV 186; Praem 33 [*SVF* II 1171]; etc.) but throughout the
      Greek *paideia.*

20   *the bee.* §§ 20-21 are on bees, their intelligence, industry, and
      social organization. For many comparable passages, see S. O. Dicker-
      man, "Some Stock Illustrations of Animal Intelligence in Greek Psy-
      chology," *Transactions and Proceedings of the American Philological
      Association,* 42 (1911), 123-125; note also Arist. *HA* 623b4-627b22
      and Pliny *NH* xi. 11-70.

      Its intelligence. Cf. Provid I 25, (*SVF* II 1111), where bees are
      said to be intelligent (φρόνιμος); Pliny *NH* xi.12, "What men, I pro-
      test, can we rank in rationality with these insects, which unquestion-
      ably excel mankind in this?"; Ael. *NA* i.59, "There never was any
      creature more gracious (εὐχαριτώτερον) than the bee, just as there is
      none cleverer (σοφώτερον)"; etc.

      the bee receives the dew as if it were a seminal substance (τὴν
      δρόσον ὡς τὸν λόγον σπερματικόν). The last terms which are of Stoic
      usage, occur also in § 96 (*SVF* II 834) and elsewhere in Philo, e.g.,
      Leg All III 150; Heres 119; Aet 85, 93; and in its plural form in
      Gaium 55. Note also Philo's association of δρόσος with λόγος in his
      allegorical interpretation of Gen 27:28 (Quaes Gen IV 215) and Ex 16:
      14-15 (Leg All III 169).

      The Stoic λόγος σπερματικός may be compared with the Platonic Idea,
      the difference being that Plato's ideal world is transcendent, while
      the Stoic "seminal reason" is immanent. This principle or primal
      substance (or cosmic sperm) manifests itself in innumerable forms
      (λόγοι σπερματικοί), which determine all development by permeating
      and activating the whole of matter, the cosmos and everything in it.

Thus the λόγος σπερματικός becomes synonymous with θεός or φύσις, the giver of all potential abilities to various beings (D. L. vii. 135-136 [SVF I 102]; Sen. Ben. iv. 7.1-2 [SVF II 1024]; Aëtius Plac. i. 7.33 [SVF II 1027]; Origenes Cels. iv. 48 [SVF II 1074]; see also comment on § 96 [SVF II 834]). For fuller discussion of the subject see H. Meyer, Geschichte der Lehre von den Keimkräften von der Stoa bis zum Ausgang der Patristik (Bonn, 1914), pp. 7-75, especially 26-46 on Philo; H. A. Wolfson, Philo, I (rev. ed.; Cambridge, Mass., 1966), 342-343.

[humans, for whose consumption God has given . . . plants and animals alike.] The interpolation owes to Gen 1:26-29 and 9:2-3. The thought runs counter to Alexander's thesis and interrupts the description of the bee's honeycomb-making.

inherently malicious animals. See comment on vices, §§ 66-70, and on the nature of animals, §§ 78-80 (SVF II 732).

The hive is like a walled palace. Aelian, in his customary elaborations, compares the hive with the palace of Cyrus in Persepolis and that of Darius in Susa (NA i. 59).

21 The bee takes charge as a captain of the guard and a keeper of the wall (φυλακάρχης καὶ τειχομάχος). Cf. Pliny NH xi. 20, "a guard is posted at the gates, after the manner of a camp." For the wars of the bees, cf. ibid. 58; Verg. G. iv. 67-87; Var. R. iii. 16.9; Ael. NA v. 11; etc.

22 the swallow or house-martin. Numerous passages of this stock illustration are cited in D'Arcy W. Thompson, "Χελιδών," A Glossary of Greek Birds (rev. ed.; London, 1936), pp. 314-325, especially 316-317 on nesting, reproduction, care, and training of the chicks. Thompson points to the epitome of Alexander of Mindos, "that best of all ancient ornithologists," as the source of numerous writers on birds, ibid., p. vi.; so also M. Wellmann, "Alexander von Myndos," Hermes, 26 (1891), 481-566, especially 533, where the parallel passages on the swallow in Plu. Mor. 966D-E and Ael. NA iii.24-25 are traced to Arist. HA 612b21-31; likewise, G. Tappe, De Philonis libro qui inscribitur 'Αλέξανδρος (Diss. Göttingen, 1912), pp. 68-70,

gives the parallel passages in Philo, Plutarch, and Aelian and
shows their dependence on Aristotle, the ultimate source.

prudent in exercising foresight (προμηθείᾳ . . . ἐπιστήμων). This
epithet does not occur in the parallel passages cited above. On
foresight in conjunction with wisdom (σοφία or φρόνησις), or as an
aspect of that virtue, see comment on §§ 30, 34 (on foresight or
forethought prior to parturition), 42 (on the ant's forethought).
See Philo's response in §§ 80-81, 92, and 97 (*SVF* II 732-733).

it appeals to man. . .like those who take refuge in temples. Cf.
Provid II 92, 106; Apol Jud 7.9; J. *Ap.* ii. 213; Porph. *Abst.* iii.
9; iv. 14 (following Josephus, see iv. 11). The passages in Apol
Jud and *Contra Apionem* need special consideration in the light of
Anim and Provid II. In the first instance Philo, alluding to the
law, has, "Do not render desolate the nesting home of birds [Dt 22:6]
or make the appeals of animals of none effect when they seem to fly to
you for help as they sometimes do." In the second instance Josephus,
likewise alluding to the law, has, "Creatures which take refuge in
our houses like suppliants we are forbidden to kill." F. H. Colson
(LCL IX, 430, 540) and H. St. J. Thackeray (LCL I, 379) fail to
explain these passages.

The unnamed birds in Apol Jud 7.9 and J. *Ap.* ii.213 could well be
swallows, as they are in Anim and Provid II 92, 106. Alexander has
argued in Provid II 92 that Providence, if it existed, would not
have allowed edible birds to flee to deserts and "swallows and
crows which are of no profit to build their nests in human dwell-
ings and cities." To this Philo responds:

> If swallows live with us there is nothing to be wondered at
> for we do not attempt to catch them, and the instinct of
> self-preservation is implanted in irrational as well as in
> rational souls. But birds which we like to eat will have
> nothing to do with us because they fear our designs against
> them except in cases where the law forbids that their kind
> should be used as food   (*ibid.* 106).

The above passages help identify the law alluded to in Apol Jud
and *Contra Apionem*. The law is that which prohibits killing or
slaughtering unclean animals for food (see the list of unclean fowls

in Lev 11:13-19; cf. Dt 14:11-20 and Hullin 63a-b; the swallow was
considered unclean: 65a). The deduction from the biblical text
seems to belong to a tradition common to Jewish apologists, a
halakah possibly of Alexandrian origin which, although known in
Palestine, does not occur in rabbinic literature. For more on the
subject see S. Belkin, *The Alexandrian Halakah in the Apologetic
Literature of the First Century C.E.* (Philadelphia, 1936); "The
Alexandrian Source of *Contra Apionem* II," *Jewish Quarterly Review*,
27 (1936), 1-32.

In addition to the authorities cited by D'Arcy W. Thomspon on the
swallows' nesting (see above, especially Ps 84:4), note Plu. *Mor.*
984D, on swallows taking to houses for security, and Pl. *Lg.* 814B,
on birds taking refuge in temples. The thought is based on the
universal right of asylum, granted to suppliants in a temple.

[*It is at odds with mice. . . .*] These words are plainly out of context; they interrupt the description of the nest. Moreover, they do not
occur in any of the parallel passages cited above. A medieval interpolation may be suspected. The word ꞔⱳⱳⱬ, here rendered "cat," is a
hapaxlegomenon in the Arm. corpus of Philo's works and, possibly, in the
entire translations of the so-called Hellenizing School (see Intro.,
pp. 7-8). The equivalent of αἴλουρος in Dec 79 is ꞔⱬⱬ, as is also
the nickname of Timothy of Alexandria.

the nest is perfect (τελεία). The remark is doubtless on the shape
of the nest; cf. Plu. *Mor.* 966E.

it manifests the attachment of a mother. Plutarch has several
insights concerning affection for offspring among animals in his un-
finished work, *De amore prolis* (*Mor.* 493A-497E; cf. 962A; 1038B
[*SVF* II 724]). In the last instance he quotes from Chrysippus'
first book *On Justice* that "the beasts are well disposed (ᾠκειῶσθαι)
to their offspring in proportion to the latter's need." See also
Cic. *Off.* i.11; Opp. *C.* iii. 107-135. On such epithets as φιλόπαις
and φιλότεκνος given to swallows, see D'Arcy W. Thompson (cited above).
Cf. § 35, on maternal watchfulness of partridges. Philo's response to
these arguments for prenatal foresight and maternal care shown by animals toward their young is given in § 79 (*SVF* II 732); cf. § 86.

Neither does it alight on another nest. S. G. Pembroke comments
well on this: "In [Stoic] sources which are admittedly not likely
to go back further than Panaetius or Posidonius, it [recognition]
becomes one of the criteria in establishing the animal hierarchy,
starting with such creatures as are unable to tell members of their
own species apart and moving on. . .to man himself." He refers
also to Nemes. *Nat. hom.* c. 42; Basil. *Homil. in hexaem.* viii.1.;
ix.4. "Oikeiōsis," *Problems in Stoicism,* ed. A. A. Long (London,
1971), pp. 139, 148 n. 125.

23   *self-heard and self-taught knowledge* (αὐτηκόου καὶ αὐτομαθοῦς
ἐπιστήμης). See comment on § 13.

whelps. . . attack huntsmen. In an argument against Providence,
Alexander insists that bears, leopards, lions and other wild animals
do not attack those whom they know to be practiced huntsmen, but only
peaceful cultivators of the soil (Provid II 91; cf. 103 for Philo's
response).

animals can master learned skills. This topic is discussed in
§§ 23-28; Philo's brief response is given in § 90. The claim is well
supported by examples of performing monkeys, fawns, elephants, etc.
On the theatrical performances of these and other tamed and trained
animals see, among others, Mart. *Sp.* 30 (antelopes and hounds); i.
104 (leopard, tigers, stags, bears, boar, bisons, and elephant;
Martial then disgresses on how lions catch their prey without being
trained); v. 31 (bull); xiii. 99 (gazelle), 100 (onager); xiv. 202
(monkey); Plu. *Mor.* 992A-B (puppies, colts, crows, dogs, horses,
and steers); etc. For a fair collection of such *loci communes,* see
G. Jennison, *Animals for Show and Pleasure in Ancient Rome* (Manchester, 1937).

A monkey. . . drove a chariot. This event, like those in §§ 52, 58,
is said to have taken place "the day before yesterday." A similar
story in Ael. *NA* v. 26, however, suggests derivation from a literary
source. On similar adaptations from literary sources to contemporary
scenes, see comment on §§ 37, 58. Numerous stories on domesticated
monkeys are told in classical literature. Pliny speaks of such mon-

keys breeding in the houses where they were kept (*NH* viii. 216). He
had even learned from Mucianus (consul, A. D. 65, 70, 72) of monkeys
taught to play at draught (*ibid.* 215). Juvenal mentions an ape seen
on a boulevard at Rome armed with a helmet and target, and taught to
ride a goat and throw javelins (v. 153-155).

Philo might have attended theatrical shows and games, including
those of performing or fighting animals, acrobats, and horse and
chariot races (Provid II 103; Ebr 177; Quod Omn 26, 141). On his
understanding of the esteemed art of charioteering, see Agr 93.

24 *Once I saw this very thing.* The description of the device is
not altogether clear. The performance has no known parallels in
classical literature. On Philo's attendance of theatrical shows,
see comment on § 23.

from the fawn's horns . . . as from one's hands. The analogy,
repeated at the end of the section, is found also in Pliny *NH* xi.
125.

25 *fools whose eyes of the soul are blind* (τοὺς τῆς ψυχῆς ὀφθαλμούς).
In Philo, as in Aristotle, the eye or the sight of the soul is the
mind (νοῦς, Quod Deus 46 [*SVF* II 458]; Deo 1-2, 6 [*SVF* II 422]; etc.
Cf. Arist. *Top.* 108a11; *EN* 1096b28; 1144a29; *Rh.* 1411b73). This is
further shown in the contrast, "the wise who are clear-sighted and
have enlightened minds." The metaphor, however, is common not only
in Philo (e.g., Plant 22; Conf 92; Migr 47-49; Spec Leg III 6, 161;
Vita Cont 10; Quaes Gen IV 2, 138) but also in classical and other
literature and reflects the peculiar belief that souls are anthro-
pomorphic (cf. Mt 6:22; Lk 11:34; Eph 1:18; also Rv 2:7, 11, 17, 29;
3:6, 13, 22; 13:9 with Virt 147).

the wealth of their rational nature (τὸν τῆς λογικῆς φύσεως πλοῦτον).
The extent of the rationality or sagacity of animals is well stated
by Aristotle, the authority of all subsequent naturalists. Of the
many such passages in his zoological works, see especially *HA* 588a
18-b4 and D'Arcy W. Thompson's notes to his translation. Numerous
examples of sagacious animals are provided in Book VIII, and even
more in Book IX which is generally attributed to Aristotle's successors
in the Lyceum. Note the possible allusion to the *Historia animalium*

and Aristotle's successors in Philo's response (§ 97). In Philo's
classification of creatures endowed with soul, animals are ἄλογα
φύσεις, man and heavenly beings are λογικὰ φύσεις (see comment on §
78 [*SVF* II 732]).

these creatures are esteemed by God and respected and commended
by the God-loving race of mankind (θεῷ. . . φιλοθέῳ ἀνθρωπίνῳ γένει).
That beasts are subject of God's care is a common theme (Ps 36:7; 104:
10-30; etc.). Philo finds in the Mosaic legislation special injunction
of kindness to animals (Virt 81, 116, 125-147; cf. Spec Leg II 69;
IV 205-206; Apol Jud 7.9; note the following biblical passages: Ex
20:9; 22:30; 23:19; 34:26; Lev 22:27-28; Dt 14:21; 22:4, 6-7, 10;
25:4, where all references are to domestic animals) and plants (Virt
148-160; cf. Spec Leg IV 226-229, based on Lev 19:23-24; Dt 20:19-20).
He sums up this very involved discussion by stating, "For he who has
first learnt the lessons of fairness in dealing with the unconscious
forms of existence will not offend against any that are endued with
animal life, and he who does not set himself to molest the animal
creation is trained by implication to extend his care to rational
beings" (Virt 160; cf. Vita Mos I 62; Plu. *Mor*. 996A, 998B; Lk 12:24;
see also 1 Cor 9:9-10 and 1 Ti 5:18, where this view is coupled with
contempt of dumb animals; and 2 En 58:6-7, where man is to be judged
for his mistreatment of animals [cf. Test Z 5:1]).

Unlike the Egyptian veneration of animals as incarnate gods, a practice much ridiculed by Philo (Ebr 95, 110; Vita Mos I 23; II 162; Dec
76-79; Vita Cont 8; Gaium 139, 163), the Greeks held various animals
as sacred by associating them with individual gods: the eagle with
Zeus, the cow and the peacock with Hera, the owl with Athena, the bull
with Dionysus, etc. The Pythagorean precepts which stem from belief in
metempsychosis, especially those of kindness to animals and abstinence
from eating flesh (see comment on § 62), are morally explained by
Plutarch (*Mor*. 959A-965B; 993A-999B) and Porphyry (*VP* and *Abst.*; in the
last work note especially iii. 16, the belief that animals are honored
by the gods and divine men). In Stoic philosophy, all animals, including man, are endeared by nature (D. L. vii. 85 [*SVF* III 178]; cf. with
Philo's thought in Somn I 108, that man, because of his λόγος, is the
dearest to God of all living creatures).

26   *that the wild are incorrigible* is refuted here and in § 27. Although the tiger is "the animal least capable of being tamed" (Leg
All I 69), the most savage animals are said to have become most man-
ageable at the first sight of man (Op 83) and tamed through association
with him (Dec 113; cf Js 3:7). In no insignificant terms that reveal expectations of universal pacification, Philo talks about the
taming of wild animals and man (Praem 85-97).

27   *elephants.* This account is certainly drawn from a literary source,
for the same story is referred to in Pliny *NH* viii. 4 and recounted
with customary elaborations in Ael. *NA* ii.11. According to Aelian,
the twelve elephants which performed at the shows of Germanicus (see
comment below) had been bred from the herd then at Rome and had been
bought when young by the trainer who taught them their tricks.

The source may be traced to King Juba II of Mauretania (*ca*. 50 B.C. -
ca. A.D. 23), a voluminous writer and the first foreign authority cited
by Pliny in his list of authorities to Book VIII (*NH* i), which begins
with stories on elephants (see *NH* viii. 2, 7, 14, 35, especially the
last three instances where Juba is mentioned by name; cf. Ael. *NA* ix.
58). The same source must have been used in § 28 (see comment; in a
parallel story Pliny cites Mucianus as the authority) and possibly in
§ 59 (see comment; in a parallel story Pliny cites Antipater of Tarsus
as the authority) and Aet 128-129 (*SVF* I 106a; cf. Pliny *NH* viii. 32;
Ael. *NA* ii.21). M. Wellmann, "Juba, eine Quelle des Aelian," *Hermes*,
27 (1892), 390-391, 397, questions Pliny's last authorities and concludes that Juba must have been the authority behind these stories on
elephants.

Libyan. Allusion, no doubt, to African elephants (cf. Ael. *NA*
ii.11; vi.56). The superlative "wildest" may imply, among other
implications, comparison with Indian elephants.

Germanicus [Julius] Caesar (lived 15 B.C. - A.D. 19) was the
nephew and adopted son of Tiberius (emperor A.D. 14-37) and the
father of Gaius Caligula (emperor A.D. 37-41). The celebration
spoken of here was probably held in A.D. 12, when he entered on
his first term of consulship following the Dalmatian campaigns

(he entered on a second term of consulship in A.D. 18, following two decisive victories across the Rhine). On the *ludi* given by Germanicus in A.D. 12, see D.C. lvi. 27. 4.

Baebius. The story suggests that he resided in Libya (Roman Africa) during the consulship(s) of Germanicus. This unidentified name is common to the House of the Tamphili, noted since the Roman wars with Hannibal. See [E.] Klebs, "Baebius, 41-46," *PRE*, 4 (1896), cols. 2731-2734; cf. "Baebius, 38," *ibid*.

28   *an elephant learned lettering*. This story too has parallels in Pliny (*NH* viii. 6) and Aelian (*NA* ii.11). Pliny's authority here is Mucianus, who was consul under Nero and Vespasian (A.D. 65, 70, 72). He claims to have seen an elephant write in Greek, "Ipse ego haec scripsi et spolia Celtica dicavi." Aelian claims to have seen an elephant, aided by its instructor, write Roman letters on a tablet. Note that for the Latin writer(s) the elephant writes in Greek, for the Greek writer the elephant writes Roman letters.

The authority cited by Pliny cannot be accepted unquestionably, for to do so would be to endorse the use of a late source in Anim. The citation, however, can hardly be an error since Mucianus is also the first authority cited in the list of authorities to Book VIII (*NH* i). But Mucianus' claim can be refuted. Although his lost book of geographical *mirabilia* is amply used by Pliny, nowhere else in Pliny does he appear to be an authority on elephant stories. The most quoted authority on such stories is the once voluminous Juba (see comment on § 27). The writing quoted on the authority of Mucianus, "I myself wrote this and dedicated these spoils won from the Celts," is an expanded version of the shorter sentence in Anim, "I myself wrote this." These passages suggest the use of a common source with which Mucianus must have dealt in the manner of Aelian--he elaborated on it and claimed to have seen the sight himself. Moreover, the order in which the stories appear in §§ 27-28; Mucianus in Pliny viii. (4), 6; and Ael. *NA* ii.11 (last two stories) bespeaks of a common, earlier source used by all three.

the ode. The origin of this verse on the elephant's pretentious affection is unknown. The authority behind a story of an elephant enamored of a girl in Pliny *NH* viii. 14 is Juba (see comment on § 27).

29 *nature has placed a sovereign mind in every soul* (οὕτως ἡ φύσις ἐν πάσαις ψυχαῖς τὸν ἡγεμονικὸν νοῦν ἐστήριξε). See comment on §§ 11, 68, 92 (*SVF* II 733, on the role of φύσις), 10, 17 (on the ἡγεμονικόν).

impressions (τύποι). The metaphor of seal and impression, derived from the Platonic images of the ideas, occurs frequently in Philo. Note the following passages on the mind's receiving impressions: Leg All I 30, 38, 61 (*SVF* II 844, 843); II 22-23 (*SVF* II 458); Agr 16 (*SVF* II 39); Spec Leg I 30; etc. (cf. with the metaphor of the *tabula rasa* as used by Cleanthes in Aetius *Plac.* iv. 11.1 [*SVF* II 83]). The archetypal seal is God's λόγος, "the image of God," after which human reason, the mind of earthly man, is patterned (see comment on § 16, the references to Philo's interpretation of Gen 1:27 and 2:27, etc.). Philo adds that "images do not always correspond to their archetype and pattern, but are in many instances unlike it." The sovereign part of the mind (or of the heart, τὸ ἡγεμονικόν) "is ever assuming different impressions (τύπους): sometimes that of a coin pure and approved by the test, sometimes of one that is base and adulterated" (Sacr 137 [*SVF* II 842]; cf. with the Stoic view that reason [νοῦς] does not penetrate into the same depth in everything [D. L. vii. 138 (*SVF* II 634)]).

The thought that animals seem to exhibit rudimentary forms of mental images, which in a fully developed form are characteristic of man alone, is anticipated in Arist. *de An.* 433b28-31; *HA* 588a18-b14; etc. Similar reflections appear in the anti-Stoic charges in Plu. *Mor.* 962C-963B (cf. Porph. *Abst.* iii. 23).

The concepts of this passage are somewhat paralleled in § 12, "Both kinds of reason (ἐνδιάθετος and προφορικός) appear to be imperfect in animals." In § 95 Philo admits,

As for the resemblance or likeness of images in animals, these, of necessity, are inferior and, consequently, dis-

> dained. There are some demonstrable but very vague images
> of respect, readiness, gratitude, and like aspects. But
> these are not genuine. Real attributes and distinct forms
> belong to human souls.

Note his emphasis on the word "real."

the deep and distinct impressions are borne upon the image of man
(τύπος δ' ἐναργῆς ἐν ἀνθρωπίνῃ εἰκόνι ἐμφέρεται). See comment above
and on § 16, on man made after the divine image.

30 *"virtues of the rational soul"* (τὰς λογικῆς ψυχῆς ἀρετάς). §§ 30-65
emphasize the Platonic cardinal virtues as developed in Book IV of the
Republic (419-445E): wisdom or prudence (φρόνησις, §§ 30-46), self-
restraint (σωφροσύνη, §§ 47-50), courage (ἀνδρεία, §§ 51-59), and
justice (δικαιοσύνη, §§ 60-65). The exact opposites of these virtues
are discussed under the vices (κακίαι, §§ 66-70). Among the numerous
variations of the cardinal virtues in Philo, the canon as found in
Anim is the most prevalent, as is also the order of the tetrad (e.g.,
Op 73 [*SVF* III 372];Leg All I 63 [*SVF* III 263]; Cher 5; Sacr 84 [*SVF*
III 304]; Quod Det 75; Ebr 23; Vita Mos II 185; Spec Leg II 62; Quod Omn
67, 159; Quaes Gen IV 11 [*SVF* III 207]; Quaes Ex II 112 [*SVF* III 277];
etc. It should be observed that the canon and order of the tetrad in
Anim is prevalent also in the Stoic fragments, as seen in the cross-
references above and elsewhere: Plu. *Mor.* 1034C [*SVF* I 200]; D. L.
vii. 92, 103 [*SVF* III 265, 156]; Stob. *Ecl.* ii. 57.18; 60.9 [*SVF* I
190; III 264]). In the allegorical treatment of the rivers of Paradise (Leg All I 66-73 [*SVF* III 263], on Gen 2:10), however, the order
is different, since self-restraint is held as belonging to the third
or least honored part of the soul (cf. Post 128; Agr 18; Abr 219; Vita
Mos II 216; Praem 160; Quod Omn 70). On the Platonic tetrad of the
cardinal virtues and its later versions, see H. F. North, "Canons and
Hierarchies of the Cardinal Virtues in Greek and Latin Literature,"
*The Classical Tradition: Literary and Historical Studies in Honor of
Harry Caplan,* ed. L. Wallach (Ithaca, N.Y., 1966), pp. 165-183.
Various theories about the origin of the cardinal virtues are review-
ed by O. Kunsemueller, *Die Herkunft der platonischen Kardinaltugenden*
(Erlangen, 1935).

The designation of the cardinal virtues as "virtues of the soul"

(ταῖς περὶ ψυχὴν ἀρεταῖς) occurs also in Sobr 61 (cf. Somn II 256),
where, besides "the virtues of the body" (ταῖς σώματος), such as
health and strength, a third category, "external advandages" (τοῖς
ἐκτὸς πλεονεκτήμασιν), such as wealth and reputation, is mentioned.
These are the threefold "goods" (ἀγαθά, cf. Quod Det 7 [SVF III 33];
Ebr 200-201; Heres 285; Abr 219; Quaes Gen III 19; also Pl. Phlb. 48D-E;
Euthd. 279A-B; Lg. 697B; 743E; Arist. EN 1098b12-14; Pol. 1323a24-26;
Stob. Ecl. ii. 80.22 [SVF III 136]. For more on Philo's classification
of virtues, see H. Leisegang, "ἀρετή, II. 5. numerus, ordo, divisio
virtutem," Indices, I [Berlin, 1930], 110; H. A. Wolfson, Philo, II
[rev. ed.; Cambridge, Mass., 1968], 202-208). The designation of the
four virtues as "virtues of the soul" owes to the Platonic view of the
first three virtues as having their realm in each of the three parts
of the soul (λογιστικόν, θυμοειδές, and ἐπιθυμητικόν),and the fourth
virtue as the unity of the other three (see. Tim. 69E-71D and Book IV
of the Republic). The qualification "rational" is due to the promin-
ence Plato gives to the first part of the soul, upon which generic
virtue depends (on the Stoics' elaborations on this view, see SVF
III 255-260--especially the excerpts from Galen; cf. the Philonic
passages cited by H. Leisegang, "ἀρετή, II. 12. ad animam et mentem,"
"II. 13. ad sapientem et insapientem," and "III. 3. anima semen
virtutis concipiens et virtute gravida," Indices, I [Berlin, 1930],
112-113, 116).

The dependence of virtue upon reason is also seen in the ethical
teachings of Socrates in the Platonic dialogues (La., Hp.Mi., Prt.,
Grg., and Men.), where virtue is identified with knowledge (ἐπιστήμη),
the source from which all virtues flow. In the Republic, Plato likewise emphasizes the identity of the virtues as forms of knowledge (the
early Stoic endorsement of this view is sufficiently clear in the
Zenonian fragments [SVF I 199-204]; cf. the Philonic passages in H.
Leisegang, "ἀρετή, II. 16. ἐπιστήμη, σοφία," Indices, I [Berlin, 1930],
114-115). The association of virtue with reason is also seen in
Aristotle's two classifications of virtue: the intellectual and the
ethical (διανοητική and ἠθική, discussed at length in EN ii [1103a14-
1109b26]) and in maintaining the "mean"--his definition of virtue as

the medium between two extremes or vices, its excess or deficit--by
constant choices (*ibid.*). The Stoics differed in not admitting of
any medium between virtue and vice (*SVF* III 536) and in insisting
that the reason of the virtuous or wise man is always inclined toward
the good, that he does not need to choose constantly between good and
evil; all his acts are virtuous (*SVF* I 202; III 512, 548, 557-566; cf.
Plant 43; Provid I 64). Only virtue is to be chosen and followed for
its own sake (*SVF* III 29, 39-48). They asserted that all virtues are
equally good, all vices equally evil, and that all else is indifferent
(*SVF* I 190, 224; III 527; Cic. *Mur.* 61; Sen. *Ep.* lxvi. 12). They
viewed virtue as a whole (*SVF* I 200; III 256); to have one is to have
all (*SVF* III 280, 295, 303, 557; Sen. *Ep.* lxvii. 10; so also with
vice--he who has one vice has all, Sen. *Ben.* v. 15; Stob. *Ecl.* ii.
106.7; from this arose the notion that a man is either wholly
virtuous or wholly vicious, either wise or fool). Virtue, then, is
the life according to nature or reason (*SVF* I 179; III 4, 16, 139,
188, 208, 312, 560), or that reason is the basis of virtue (*SVF* I
202; III 200a). For similar views of Philo, see H. Leisegang,
"ἀρετή, II. 11. ad λόγον. b) ratio," *Indices*, I (Berlin, 1926), 112
and Leg All I 56-57 (*SVF* III 202), 103-104; Sacr 82, 84 (*SVF* III 304);
Quod Det 141; Congr 129; Somn II 241-244; Abr 100-102; Vita Mos II
140, 181 (*SVF* III 392, 227). It was against such Stoic doctrines as
these that Plutarch wrote his essay *How a Man May Become Aware of
His Progress in Virtue* (*Quomodo quis suos in virtute sentiat pro-
fectus, Mor.* 75B-86A). On the one hand, it is concerned with demonstrating progression in the acquisition of wisdom and its attendant
virtue; and, on the other hand, it endeavors to show stages of imperfection, vice, or sin.

   On animals striving for and possessing the cardinal virtues, see
Plu. *Mor.* 961F-962B. Note the proposition set forth in the last instance and carried through the rest of the treatise: "the alloyed
and imprecise virtue of animals" is not to be stigmatized "as absence
of reason," but rather to be accepted "as its imperfection or weakness"
(cf. 963B; 986F-992E). See also Aelian's *Prologue* to Book I of *De
natura animalium*. Although these views were anticipated in Pl. *La.*

196E and Arist. *HA* 488b12-26, 608a13-17, where the virtues and vices
of animals are equally emphasized, they were doubtlessly taken over
from the arguments raised by the opponents of the Stoics in the Academy, chief among whom was Carneades (214/3-129/8 B.C.). This is well
attested by the almost identical thesis in S. E. *P.* i. 62-77 (cf.
Porph. *Abst.* iii. 1-15; see comment on Philo's response in § 97).
G. Tappe observes similarities of structure in these arguments for
the rationality of animals (*De Philonis libro qui inscribitur
'Αλέξανδρος* [Diss. Göttingen, 1912], pp. 25-29).

   wisdom or *prudence* (φρόνησις), the first of the Platonic cardinal
virtues, is discussed in §§ 30-46. According to Plato in Book IV of
the *Republic* (419-445E), wisdom, the excellence of the ruling part of
the soul, is the highest form of virtue when considered absolutely
(cf. *Epin.* 977D; so also Philo in Leg All I 66-67, 70-71; see *ibid.*
78; Quod Det 75; Mut 79, on the two kinds of wisdom: particular and
universal). Wisdom is often taken as the basis of all virtues, as
Socrates does in X. *Mem.* iii. 9 and Zeno in Plu. *Mor.* 441A (*SVF* I
201). Philo's definition of the term is "concern with things to be
done" (Leg All I 65, 70 [*SVF* III 263; cf. 262, 268; I 375; II 1005];
cf. Arist. *EN* 1143a33-34; 1144a24-25). Elsewhere his definition
follows the Aristotelian doctrine of the "mean" (Quod Deus 164; cf.
EN 1106a26-1108b13; 1144a7-8). Wisdom in Anim, however, implies shrewd-
ness in things to be done in the interest of one's own good rather than
implying a kind of intellectual or moral virtue (cf. Epict. *Diss.* ii. 22
1-37). The term here is synonymous with σύνεσις, "which every creature
receives from nature to enable it to acquire what is proper for it
and to evade what is not" (Plu. *Mor.* 997E; cf. with a certain understanding of οἰκείωσις in Stoic thought, i.e., the creature's self-
awareness and concern for its well-being; see S. G. Pembroke, "Oikeiōsis," *Problems in Stoicism*, ed. A. A. Long [London, 1971], pp. 114-
121). Note that the terms following wisdom: "knowledge, excellent
discerning, superior foresight" (ἐπιστήμη καὶ εὐβουλία καὶ πολλὴ
προμηθεία) are aspects of wisdom. These are simultaneously emphasized
in the examples given in §§ 22, 30-46. See Philo's response in §§ 82-
84 (*SVF* II 726); cf. § 93.

polyp . . . crampfish . . . starfish. For similar accounts, see
D'Arcy W. Thompson, "Πολύπους," "Νάρκη," and "'Αστήρ," *A Glossary of
Greek Fishes* (London, 1947), pp. 206-207, 169-170, 19. In addition
to the authorities cited by Thompson, see Philo's account on the polyp
in Ebr 171-174. For an analysis of the authorities, see M. Wellmann,
"Pamphilos," *Hermes,* 51 (1916), 1-64, especially 40.

31 *Creatures of water, air, and earth alike are endowed with abundant
    wisdom* (περιττῆς φρονήσεως). That all animals of whatever provenance
    are intelligent is emphasized in Plutarch's *Whether Land or Sea Ani-
    mals are Cleverer* (*De sollertia animalium, Mor.* 959A-985C). Note
    that the victory is awarded to neither of the contestants in the dia-
    logue, where it is stated that there are varying degrees of intelligence
    within the groups, just as among men (962D-963A; cf. 992C-E in *Bruta
    animalia ratione uti;* Arist. *PA* 686b22-25). Pliny *NH* viii. 33 has:
    "Every species of animals is clever (sollertia est) for its own in-
    terests." To Philo creatures differ not only in their habitats, sizes,
    and qualities or formations (ταῖς ποιότησιν, Op 63), but also in their
    souls and perceptions, the least elaborate of which has been allotted
    to the race of fish (ψυχῆς, αἰσθήσεως, *ibid.* 65; cf. 68). That the
    first examples of animals possessing the virtue of wisdom are drawn
    from aquatic creatures, "the least elaborately wrought," is note-
    worthy. The "abundant wisdom" of animals is reiterated in §§ 34, 38, 44
    (cf. § 82), and the classes in §§ 43, 60, 78 (*SVF* II 732; cf. Gig 7 and
    F. H. Colson's note [LCL II, 502]; Plant 12; Aet 45; Quaes Ex II 28).
       oyster . . . shellfish, could be any of the subdivisions of τὰ ὀστρακό-
    δερμα: ὄστρεα and κογχύλια (Arist. *HA* 487b9; Heres 211; see the terms
    in D'Arcy W. Thompson, *A Glossary of Greek Fishes* [London, 1947], pp.
    118, 189-191). The ascription to Alexander of the statement that the
    flesh of the oyster is edible is not altogether surprising with regards
    to a renegade Jew. Cf. Provid II 92, where he cites the hare with ani-
    mals fit for food (contrary to Lev 11:6, 10-12; cf. Dt 14:7, 9-10).

32 *They resort to contrived schemes against those animals whose scent
    they eagerly track.* See the example of the hound in § 45 (*SVF* II 726).

33 *concerned about self-preservation* (σωτηρίας). The concern of animals
    for self-preservation is a dominant theme in these passages on wisdom

(§§ 30-46). Philo asserts that the longing for self-preservation (σωτηρίας δὲ πόθος) is found in both rational and irrational souls (Provid II 106). This concern or longing, according to the Stoics, is the primary impulse (τὴν πρώτην ὁρμήν) given to animals by nature (Chrysippus in D. L. vii. 85 [*SVF* III 178]; cf. § 80; Origenes *Cels.* iv. 87 [*SVF* II 725]). It is nature which alienates them from dangers, even from ferocious species never encountered before (Cic. *Fin.* iii. 16 [*SVF* III 182]; *ND* ii. 124 [*SVF* II 729]; Sen. *Ep.* cxxi. 19-21; Plu · *Mor.* 1060C [*SVF* III 146]; Hierocl. i. 38; iii. 19-52; for a discussion of these and other passages, see S. G. Pembroke, "Oikeiōsis," *Problems in Stoicism,* ed. A. A. Long [London, 1971], pp. 116-121); A. A. Long, *Hellenistic Philosophy* (London, 1974), pp. 185-189.

they turn away . . . from the ingenious designs of man. See the stories in §§ 35-36, the one on partridges frustrating fowlers and the other on a sea-monster frustrating fishermen.

deer (ἔλαφος), a vague term for any type of small deer, including gazelles and antelopes (cf. § 68). For comparable accounts on deer after shedding their antlers, see Arist. *HA* 611a25-28; b8-18; Pliny *NH* viii. 115, 117; Ael. *NA* vi. 5. The dependence on Aristotle is demonstrated by G. Tappe, *De Philonis libro qui inscribitur* 'Αλέξανδρος (Diss. Göttingen, 1912), pp. 34-35.

34 *all species of animals have intelligent foresight* (τῆς εἰς τὴν φρόνησιν προμηθείας). That all creatures are endowed with abundant wisdom is stated in § 31. Foresight or forethought is an aspect of wisdom (see, among others, Hdt. iii. 36; Pi. *N.* ii. 46; *I.* i. 40; Th. iv. 62; etc.; according to Ael. *NA* v.11, bees have a share of τῆς ἐς τὸ προμηθὲς σοφίας; for προμήθεια in conjunction with φρόνησις in Philo, see Sacr 27; Ebr 143; Virt 129). The term here is applied to the precautionary care shown by animals prior to and following parturition (see the examples of the swallow in § 22 and of the partridges in § 35; cf. Job 39:1-18). The Stoics, however, ascribed to nature the effort of animals in bringing forth and rearing up their young (Cic. *Fin.* iii. 62 [*SVF* III 340]; cf. Op 171, on the ascription of parental forethought for children to nature, and the illustration in § 79 [*SVF* II 732]).

35   *partridges.* On this common example of affection for offspring, see
the authorities cited in D'Arcy W. Thompson, "Πέρδιξ," *A Glossary* of
Greek Birds (rev. ed.; London, 1936), pp. 235-236. In addition to the
passages cited by Thompson, see Plu. *Mor.* 494E; 971C-D; and Pliny
NH x. 103. These, with Ael. *NA* iii. 16, bear very close similarities
to the account in Anim. The dependence on Arist. *HA* 613b15-21, 31-33
is all too obvious. Philo in his response has that the flight of
partridges from kites involves no thinking, but carrying out nature's
design (§ 80).

36   *the semelē.* The account on this unidentified sea-monster bears re-
miniscences of similar accounts on the fox-shark. See, e.g., Arist.
HA 621a11-14; Ael. *VH* i. 5; Opp. *H.* iii. 144-147 (Plu. *Mor.* 977B con-
fuses this account with that on the so-called *scolopendra;* so also
Ael. *NA* ix. 12, following Plutarch; cf. Arist. *HA* 621a6-11). Cf. G.
Tappe, *De Philonis libro qui inscribitur* 'Αλέξανδρος (Diss. Göttingen,
1912), p. 14, n. 2; D'Arcy W. Thompson, "'Αλώπηξ," *A Glossary of Greek
Fishes* (London, 1947), pp. 12-13.

    they will find their pursuit unrewarding. On fishing as an unre-
warding pursuit, see Pl. *Lg.* 823D-E; Plu. *Mor.* 965F-966B; etc.

37   *the story of the Thracian falcons.* The same story is told less
elaborately in Pliny *NH* x. 23 and, somewhat differently, in Ael. *NA*
ii. 42 (in addition to the comparable accounts on hawking referred to
by D'Arcy W. Thompson, "'Ιέραξ," *A Glossary of Greek Birds* [rev. ed.;
London, 1936], p. 116, see Opp. *C.* i. 62-64). The ingenious adaptation of this literary account (derived from Arist. *HA* 620a33-b5) to a
contemporary and personal scene is noteworthy. For comparable adaptations elsewhere in Anim, see comment on §§ 23, 58).

38   *Animals grow and abound in substantial wisdom* (πρὸς περιττὴν
φρόνησιν). See comment on §§ 30-31.

    when it comes to most essentials, they surpass man in wisdom. In a
parallel passage in Plutarch (*Mor.* 974A-E), the essentials in which
animals surpass man are those identified by Democritus: weaving and
mending, home-building, and singing, in all of which man is said to
have imitated spiders, swallows, swans and nightingales (Diels *VS*

68B154; cf. §§ 15-22; see comment on § 17). To these, however, the three divisions of medicine: the cure by drugs, diet, and surgery are added in Plutarch (*ibid.*; cf. 991E-F). Among the numerous examples cited by him there are parallels to two of the three examples in Anim (§§ 38-39), all of which belong to the cure by drugs. The three instances are stock examples in the traditional arguments for self-healing among animals (see comment below).

the self-taught art of healing (τὴν αὐτομαθῆ ἰατρικὴν [τέχνην]). On self-teaching among animals, see comment on § 13. As for the art of healing, in addition to the above cited passages from Plutarch, see especially Arist. *HA* 611b34-612a8, 24-33; Pliny *NH* viii. 96-101; etc. Among the ancient writers on poisons and their antidotes the leading authority was Apollodorus (3d cent. B.C.), whose works partially survive in the poems of Nicander.

Note that this argument is not dealt with in Philo's response. Perhaps typical of the Stoa's response is that found in Origenes *Cels.* iv. 87 (*SVF* II 725), where it is stated that the inclination of animals to use antidotes is not due to reason but to nature; had it been due to reason, they would choose more than one particular antidote.

deer . . . consider hops. Cf. Pliny *NH* viii. 97, where instead of hops for remedy, stags are said to eat crabs when pricked by the *phalangium* (Pliny's confusion may be explained by a comparison with Plu. *Mor.* 991E). On this venomous spider, see Arist. *HA* 555a27; b12; 622b28.

goat . . . looks for what is called dittany. See Arist. *HA* 612a3 and D'Arcy W. Thompson's note to his edition; cf. Cic. *ND* ii. 126, not cited by Thompson.

39 *tortoises . . . eat marjoram.* See Arist. *HA* 612a24-28 and D'Arcy W. Thompson's note to his edition. Pliny *NH* viii. 98 has that "the tortoise eats *cunila,* called ox-grass, to restore its strength against the effect of snake-bites."

40 *animals . . . feign sickness.* This statement with its accompanying example has no known parallels in classical literature. However, in

the light of the large number of stock examples in Anim, it can hardly be original with Philo. Bordering on this citation is the following reprovement of the Stoics in Plu. *Mor*. 961D: "They themselves punish dogs and horses that make mistakes, not idly but to discipline them; they are creating in them through pain a feeling of sorrow, which we call repentence" (λύπην . . . ἣν μετάνοιαν ὀνομάζομεν). On the cleverness and subtlety of horses, see, e.g., Pliny *NH* viii. 159; Ael. *NA* xvi. 25; Opp. *C*. i. 221-222; etc. For Philo's response, see § 83.

Aristogiton. The "tyrannicides" Aristogiton and Harmodius were friends who plotted to assasinate the tyrant Hippias and his brother Hipparchus, who had insulted the sister of Harmodius. Hipparchus was murdered in 514 B.C. and Hippias ordered the execution of the plotters. After the expulsion of Hippias by the Spartan king Cleomenes in 510 B.C., the deeds of Aristogiton and others received ample recognition. He was later claimed to have given Athens its liberty and numerous songs were dedicated to him (Hdt. v. 55; vi. 123; Th. i. 20; vi. 52-59; etc.). On the later stories about him, see J. Miller, "Aristogeiton, 1," *PRE*, 3 (1895), cols. 930-931.

41 *cattle . . . have reasoning minds* (διάνοια). See comment on § 17.

They have no lack of knowledge (ἐπιστήμη) .   See comment on § 30 .

house management (οἰκονομία . Aucher by mistranslation separates this line from the thought of the next section on the ant). The Stoics maintained that οἰκονομία as a virtue is peculiar only to the wise man, who alone is οἰκονομικός (Stob. *Ecl*. ii. 99.9 [*SVF* III 567]). Moreover, house management and its cognate virtue, state management (πολιτεία, see comment on § 65), were deemed inseparable: to have one implied having the other (see comment on § 91 [*SVF* II 733]). For the broader meanings of the term, as it relates to human beings and property, see H. A. Wolfson, *Philo*, II (rev. ed.; Cambridge, Mass., 1968), 322-323.

42 *the ant*. Of the more common examples of intelligent animals, bees, spiders, and swallows are discussed in §§ 17-22. To such examples belong also the ants (see S. O. Dickerman, "Some Stock Illustrations

of Animal Intelligence in Greek Psychology," *Translations and Proceedings of the American Philological Association*, 42 [1911], 123-130).
In Provid I 25 (*SVF* II 1111) the ant is called "forethinking" or "provident" (προμηθής), an epithet rightly implied here though not stated. The epithet here is "household manager" (οἰκονομικός, see comment above). The thought owes to Aristotle (*HA* 488a10), who lists as social animals (πολιτικά): man, bees, wasps, ants, and cranes (on the social life of ants and other animals as contrasted with that of humans, see D.C. xl. 32. 40-41). In another list of the most industrious animals the ant is mentioned first and its industry briefly discussed (*ibid.* 622b20-27). The closest similarities to the account in Anim, however, occur in Pliny, Plutarch, and Aelian, the latter adding: "It is to Nature then that ants too owe these and other fortunate gifts" (*NH* xi. 108-110; *Mor.* 967D-968A; *NA* ii. 25; cf. iv. 43; vi. 43, 50). Among other important accounts, see especially Pr 6:6-8; 30:24-25; For Philo's response, see §§ 91-92 (*SVF* II 733).

43 *animals . . . use foresight* (προμηθεία). See comment on §§ 22, 30, 34, and above; for Philo's response, see §§ 80 (*SVF* II 732)-81, 92 (*SVF* II 733), 97. That they lack no necessary provisions is stated also in Provid I 70.

   they possess a reasoning mind (διάνοια). See comment on § 17.

   rational soul (λογικῇ . . . ψυχῇ). See comment on §§ 30, 78 (*SVF* II 732).

44 *coping with opposites* (πρὸς τὰς ἐναντιότητας). On the Heraclitean doctrine of the cosmos as a harmony of opposites, see comment on § 61. In Heres 133 Philo declares: "The subject of division into equal parts and of opposites is a wide one. . . . We will . . . content ourselves with the vital points only." He then treats the subject in much of the rest of the treatise, allegorizing on Gen 15:10-18, etc. §§ 207-214 contain a list of opposites much lengthier than the one given here (cf. Abr 239). The thought here presented is apparently that where any harmony of such opposites exists, there must necessarily be reason (based on the notion of the λόγος as harmonizing factor between opposites and bond of the universe; see comment on § 61).

   reasoning mind (διάνοια). See comment on § 17.

They long for that which produces pleasure and flee from that which is loathsome and painful. Pleasure and freedom from pain were long held as natural inclinations. Plato defined pain as departure from the natural condition and pleasure as the return back to it (*Tim.* 64D). In much the same terms, Aristotle defined pleasure as a motion of the soul and a settlement into its own proper nature (*Rh.* 1369b33-35), an activity of the natural state (*EN* 1153a14), having mentioned earlier that pleasure is felt not only by man but also by lower animals (*ibid.* 1104b34-35). The Stoics, beginning with Zeno, held that pleasure is a natural impulse (ὁρμή) of living creatures, though not the primary impulse, which they held to be self-preservation (D.L. vii. 85-86, 149 [*SVF* III 178; II 1132; in the last instance note the reaction against the Epicurean teaching that pleasure is the *summum bonum*]; cf. Cic. *Fin.* iii. 17, 62 [*SVF* III 154, 340], adding in the last instance that shrinking from pain is due to natural impulse, and v. 17-18, on the views of other schools relative to the primary impulse; also the elaborations on Zeno's definition of virtue as consistency with nature: the addition of the words by Diogenes of Babylon, "to take a reasonable course in choosing or refusing things in accordance with nature", and the redefinition of virtue by Antipater of Tarsus as "life with preference for what is natural and aversion to what is against nature" [Stob. *Ecl.* ii. 75.11 (*SVF* I 179); III Diog. 44; Ant. 57]; in like terms Cicero ascribes to nature the animals' pursuit of the healthful and avoidance of the injurious [*ND* ii. 34, 122]). Philo likewise held that "a created being cannot but make use of pleasure" and attacked the Epicurean dictum of pleasure as the *Summum bonum* (Leg All II 17; for references to Philo's attack on Epicurean hedonism and to his agreement with the Stoic doctrine of ἡδονή as πάθος, see H. Leisgang, "ἡδονή. 2. ἡδονή=πάθος, secundum Stoicorum doctrinam unus est ex quattor affectibus," *Indices*, II [Berlin, 1930], 343-344). Typical of the charges brought against the Stoics, note these lines from Plu. *Mor.* 960E-F: "The acts of . . . eluding or fleeing from what is destructive or painful, could by no means occur in creatures naturally incapable of some sort of reasoning and judging, remembering and attending."

mental conceptions (διανοήσεις). See comment on § 17 and below.

abundant wisdom (περιττῆς φρονήσεως). See comment on §§ 30-31.

many characteristics of speech . . . something more evident than voice . . . their actions (λόγου . . . φωνῆς . . . ἔργα). Plutarch similarly exclaims: "It is extraordinary that they [the Stoics] obviously fail to note many things that animals do and many of their movements that show anger or fear or, so help me, envy or jealousy" (*Mor.* 961D; cf. 1038F [*SVF* III 211] and the Stoic doctrine of virtue displaying itself in right action [see comment on § 30]). As for Philo, note Mut 243: "Speech is the shadow of action" (λόγος γὰρ ἔργου σκιά), a saying ascribed to Democritus in D.L. ix. 37 (Diels *VS* 68B145); also Leg All I 74: "Prudence is not seen in speech but in action and earnest doings" (μὴ ἐν λόγῳ τὸ φρονεῖν, ἀλλ᾿ ἐν ἔργῳ θεωρεῖσθαι καὶ σπουδαίαις πράξεσι), a statement contained in his discussion of theoretical and practical virtue. The relationship between "speech" and "action" is a logical and ethical postulate, so that contradiction between them necessarily rests on sophistry (Post 86-87). Thus, although these terms at times appear in opposition, they are none the less to be understood as complementary in most instances (see J. Leisegang, "ἔργον, 7. ἔργον coni. vel opp. λόγος," *Indices*, I (Berlin, 1930), 294-295.

45 *A hound was after a beast.* The argument here propounded originally belonged to Chrysippean dialectic, a logical theory and a paramount theme in the philosophy of Chrysippus, who out of fondness of this discipline went so far as to say that even the dog shares in dialectic (S.E. *P.* i. 69; cf. Plu. *Mor.* 969A-B; Ael *NA* vi. 59). Except for this passage in Anim (*SVF* II 726, including Philo's response in § 84), none of the parallel passages are included in von Arnim's *SVF*. It is equally unfortunate that this illustration of the dog's use of "complex indemonstrable syllogism" (see comment below) is not taken into consideration in the rather profound discussion of Chrysippean dialectic by J. B. Gould, *The Philosophy of Chrysippus* (Albany, N.Y., 1970), pp. 66-88. See especially pp. 83-88 for outlines of the simple and complex syllogistic forms in *SVF* II 241, 242, 245, 1192.

46   *"the fifth complex indemonstrable syllogism"* (τῷ πέμπτῳ διὰ πλειόνων [ἀν]αποδείκτῳ; cf. S.E. *P.* i. 69). See comment above (Aucher unduly separates this from the preceding), especially the last reference, where Chrysippus makes use of another complex syllogism. The number five refers to the propositions of the argument; thus, as R. G. Bury explains in a note to S.E. *P.* i. 69 (LCL I, 42): "Either A or B or C exists; but neither A nor B exists; therefore, C exists." Note Philo's use of the term τρόπος (mode of syllogism) twice in his response (§ 84 [*SVF* II 726]) and Plutarch's substitution of διεζευγμένου (disjunctive) for ἀναποδείκτος (indemonstrable syllogism) in *Mor.* 969A. The meaning of the last term lies in that the certainty of the syllogism needs no demonstration, since it is self-evident (Plant 115; cf. J. B. Gould, *The Philosophy of Chrysippus* [Albany, N.Y., 1970], p. 83, on *SVF* II 242)

resemblances of speech (see comment on § 44).

Aesop, the famed fable writer (*ca.* 620-560 B.C.). See comment on § 73.

foxes. Perhaps the most celebrated story on the astute hearing of foxes is that of their ability to perceive the thickness of the ice on frozen rivers by the sound of the current underneath (Pliny *NH* viii. 103; Plu. *Mor.* 949D; 968F-969A; Ael. *NA* vi. 24; xiv. 26; etc.). The cunning of the fox is of course proverbial (e.g., Arist. *HA* 488b20; Opp. *C.* iii. 45; Lk 13:32; etc.;cf. §§ 66, 86).

monkey. The imitative faculty of monkeys and their mimicry are likewise proverbial (see, e.g., Ael. *NA* v. 26; vii. 21; xvii. 25; Opp. *C.* ii. 605; etc.). It is observed in Athen. 613d; "a monkey is funny by nature."

47   *Self-restraint* (σωφροσύνη) is discussed in §§ 47-50 and its opposite vice (ἀκολασία) in §§ 66-67. Philo takes the etymology of the word to have been derived from σωτηρία and φρόνησις (Virt 14), evidently following Pl. *Cra.* 411E, σωφροσύνη δὲ σωτηρία . . . φρονήσεως. His Stoic definition of the term is that it centers round things to be chosen (αἱρετέοις, Leg All I 65 [*SVF* III 263; cf. 262, 266, 274; I 201, 374, 563]). Self-restraint, which like all other virtues is dependent on reason, takes its stand against pleasure, passion, and desire (ἡδονή, πάθος, and ἐπιθυμία, Leg All I 69-71, 86-87; II 78-108;

etc.; cf. Pl. *Phdr.* 246A-256A; *R.*430E; and 4 Macc, the work of an Alexandrian Jew with an intimate knowledge of the Stoic Posidonius, that
reason can control passion). It is often used by him in the sense of
"continence" (ἀρκεῖσθαι or ἐγκράτεια, as in §§ 48, 49), possibly after
Aristotle's equation of "self-restraint" with "continence" (*EN* 1151b32-
1152a3; this virtue is treated more fully in 1117b20-1118b1, where its
realm is limited to the pleasures of touch and taste, but especially
to those of touch). It may be noted here that after Plato philosophers
of every school, including Plato's, customarily assign self-restraint
only to the appetitive part of the soul, ignoring its function in the
Republic as the excellence of the whole or as harmony operating in all
three parts of the soul (the intellectual, the realm of wisdom; the
spirited, the realm of courage; and the appetitive, the realm of self-
restraint in the stricter sense [432A]). Thus Philo in his allegorical
interpretation of the rivers of Paradise holds that self-restraint is
the least honored part of the soul (Leg All I 70-71).

In §§ 47-49 Alexander advocates simplicity in food and drink and
deplores sexual license with its resultant crimes of adultery. These
passages may be added to the known examples of the diatribe in Philo
(see P. Wendland, "Philo und die kynisch-stoische Diatribe," *Beiträge
zur Geschichte der griechischen Philosophie und Religion*, Festschrift
Diels, ed. P. Wendland and O. Kern [Berlin, 1895], pp. 1-75).

On the self-restraint of animals see comment on § 48. Philo's response to this argument is given in the closing lines of the treatise
(§ 100): "To ascribe serious self-restraint to indifferent and almost
invisible creatures is to insult those whom nature has endowed with
the best part," i.e. reason (see comment on § 11).

feasts, winebibbing, drinking parties are repeatedly condemned by
Philo, especially in Ebr, Sobr, and Plant. Philosophical views on
drunkenness, particularly those of the Stoics, are amply discussed
at the end of the last treatise (140-177; H. von Arnim identified the
source of this passage with a disquisition by a minor Stoic writer
whose name Philo had suppressed: "Quellenstudien zu Philo von
Alexandria," *Philologische Untersuchungen*, ed. A. Kiessling and U.
von Wilamowitz-Moellendorff, XI [Berlin, 1888], 102-140).

superfluous works of cooks and bakers are likewise condemned in de-
nunciations of gluttony, refinement, and luxury as compared to simple
living (see § 48 below; Agr 66, Ebr 214-222, Jos 61, Leg All III 220,
Somn II 16, etc.). Cf. Philo's allegory on the dreams of the chief cup-
bearer and chief baker (Gen 40:9-19) in Somn II 155-214. Such condemna-
tions seem to be directly aimed at the doctrines of Epicurus (see
Athen. 278f; 280a). Chrysippus is said to have remarked "that the
very center of the Epicurean philosophy is the *Gastrology* of Arch-
estratus" (Athen. 104b [*SVF* III 709]).

We who swell with pride, etc. For a similar description of the arro-
gant man see Virt 172-173.

As the poets say. The first two lines of this verse are likewise
quoted in Quaes Ex II 6 (unnoticed by R. Marcus, LCL), where the
variance in the word order betrays some inconsistency on the part
of the Arm. translator (see Intro., p. 12). The verse, with its
ascription to the "poets" (as also in Quaes Ex II 6) and subject
matter as well, is reminiscent of the so-called Orphic poems. The
concept of human devolution from the divine has its origin in Orphic
theogony and cosmogony, where Zeus is said to have created man from
the ashes of the Titans whom he had stricken with his thunderbolt
because they have killed and eaten the divine infant Dionysus. Thus
man came to occupy a middle position between gods and beasts, due to
the mixture of the material from which he was created: the burnt
remains of the Titans and the fragments of the divine child whom they
had devoured (see K. Freeman, *The Pre-Socratic Philosophers: A Companion to Diels, Fragmente der Vorsokratiker* [Oxford, 1959], pp. 9-
18). The Orphics are alluded to in Pl. *Tim.* 40D as "descendents of
gods" (cf. Diels *VS* 1B8; also Ps 82:6-7; Ac 17:28-29).

48   *simple foods.* Cf. Agr 41, where Philo commends shepherds "who prefer what is tasteless but beneficial to what is delicious but harmful,"
and the exaltation of the simple life in Sen. *Ep.* xviii. 5-7; xx. 13;
cviii. 14-16.

air . . . water are called natural riches in which everyone ought to
find contentment (Virt 6 [*SVF* III 707]). Cf. Praem 99 and 1 Ti 6:8,

where the riches are "food and shelter," and Somn II 126, where they are "bread and raiment"; also Mt 6:25-34; Lk 3:14; Hb 13:5. In Quod Omn 8 and Vita Cont 35, following Pl. *Phdr.* 259C, grasshoppers are said to live on air (see F. H. Colson's note to Vita Cont 35 [LCL IX, 132-133]).

the pleasures below the belly. The sequence of discussing self-restraint in sexual intercourse after self-restraint in eating and drinking comes from the understanding that gluttony and drunkenness stir up sexual lust (Op 158; Agr 37-38; Vita Mos II 23; Spec Leg I 92; III 43; etc.). On the self-restraint of animals in this respect, see below.

the law of nature or *natural law* (νόμος φύσεως) is suggestive of *divine law* (νόμος θεῖος), as implied by Philo's interchangeable use of the terms θεός and φύσις (see comment on § 11). Note also his modification of the Stoic concept of "nature's right reasoning" (ὁ τῆς φύσεως ὀρθὸς λόγος) to mean "divine law" (νόμος θεῖος) or "ordinance" (θεσμός, Op 143 [*SVF* III 337]; cf. Jos 28-31 [*SVF* III 323]). This law is understood to have been introduced by God into the world: "that the Father and Maker of the world was in the truest sense also its Lawgiver" (Vita Mos II 48; cf. Op. 171; Spec Leg III 189; Praem 42; Quaes Ex II 42). In the same instance he adds "that he who would observe the laws (νόμοι) will gladly accept the duty of following nature and live in accordance with the ordering of the universe." The νόμοι here stand for the laws of Moses or תורה, as in Vita Mos II 13-14, where the laws (νόμιμα) of Moses stand "firm, unshaken, immovable, stamped, as it were, with the seals of nature herself" (cf. Spec Leg I 31). These stand in opposition to human laws that are liable to change (Ebr 37, 47; Quaes Gen IV 184). The patriarchs, "who lived according to the law of nature's right reasoning" (κατὰ νόμον, τὸν ὀρθὸν φύσεως λόγον) before the written legislation of Moses, are regarded as an exemplification of that law (Abr 3-6; Quod Omn 62). Not only was Abraham led "by a law which nature has laid down, a law unwritten" (ἀγράφῳ μὲν νόμῳ . . . ὃν ἡ φύσις ἔθηκε, Abr 16; cf. 60. 275), but "he was a law and an unwritten statute" (νόμος αὐτὸς ὢν καὶ θεσμὸς ἄγραφος, Abr 276).

There are numerous applications of natural law to sexual morality in Philo. The begetting of children and the perpetuity of the race is fulfilment of the law of nature (Praem 108; Abr 248-249; Jos 43; Vita Mos I 28; Spec Leg II 233; Quaes Gen IV 86; Quaes Ex II 19); however, the bodily pleasure engendered by the union of man and woman is "the beginning of wrongs and violation of law" (Op 152; cf. Leg All II 74 and 1 Cor 7:1-7, 28). Even beholding the nakedness of the opposite sex is "disregarding the statutes of nature" (ἀλογοῦσαι φύσεως θεσμῶν, Spec Leg III 176; cf. 1 Cor 11:6, 10, 13-15). As for adultery, pederasty, and like violations of the law of nature, see comment on § 49. On sexual intercourse and procreation in Philo's writings, see R. A. Baer, Jr., *Philo's Use of the Categories Male and Female*, Arbeiten zur Literatur und Geschichte des hellenistischen Judentums, III (Leiden, 1970), 94-95. For a comparison of these extreme views of Philo with Greek and rabbinical "Sexualethik," see I. Heinemann, *Philons griechische und jüdische Bildung. Kulturvergleichende Untersuchungen zu Philons Darstellung der jüdischen Gesetze* (Breslau, 1932), pp. 261-292.

In the case of animals, see Plu. *Mor.* 988F-991D, a passage showing the inferiority of men to animals in self-restraint. Note especially 990D-F: "Neither does the female continue to receive the male after she has conceived, nor does the male attempt her," adding that beasts have a better claim to "non-violation of nature"; whereas "men do such deeds as wantonly outrage nature, upset her order, and confuse her distinctions" (cf. 493E; Porph. *Abst.* iii. 10). Plutarch adds to the better claim of animals in this respect by citing a Chrysippian observation that among animals mating is free and more in keeping with nature (*Mor.* 1044F [*SVF* III 753]; cf. D. L. vii. 188 [*SVF* III 744] and Zeno's views on marriage in D. L. vii. 131 [*SVF* III 728]). Note also Opp. *C.* i. 236-238: "horses honor nature, and it is utterly unheard of that they should indulge unlawful passion." Cf. Pliny *NH* viii. 112 (on hinds). On exceptional cases of lustfulness and bestiality among men and animals, see §§ 66-67.

The idea of natural law occurs in many other formulations and

specific applications in Philo (see J. Leisegang, "φύσις. 4. naturae
lex et ratio," *Indices*, II [Berlin, 1930], 839) and in Greek thought
(see F. Heinimann, *Nomos und Physis* [Basel, 1945]; H. Koester, "Νόμος
φύσεως: The Concept of Natural Law in Greek Thought," *Religions in
Antiquity: Essays in Memory of E. R. Goodenough*, ed. J. Neusner,
Studies in the History of Religions, Supplement to *Numen*, XIV [Leiden,
1968], 521-541; G. Watson, "The Natural Law and Stoicism," *Problems
in Stoicism*, ed. A. A. Long [London, 1971], pp. 216-238).

49   *sexual indulgence*. On the year round sexual intercourse of humans
see Pliny (*NH* vii. 38,48), who, comparing such indulgence with the
chastity of animals, adds that these are "crimes against nature"
(x. 171-172; cf. Spec Leg III 113).

    They infringe on the marital rights of others. Philo deplores the
various crimes of adultery in two lengthy passages on the sixth commandment, Ex 20:14 (Dec 121-131; Spec Leg III 7-82; cf. Dec 168; Spec
Leg IV 89). On violating the marital rights of others, see Abr 135;
Spec Leg II 50.

    the present laws concerning adultery (μοιχεία). The passage is
clear not only on the fact that adultery was punishable with death
(cf. Jos 44; Spec Leg III 11; J. *Ap.* ii. 201; following, no doubt,
Lev 20:10; Dt 22:22), but also on two aspects of Roman law concerning
adultery: (a) that of "trial before the magistrates (ἔφοροι)" and
(b) that which authorized only the husband to kill the offender
when caught in the act, as a matter of revenge (*sine iudicio*). The
Jewish law differed in that it allowed those who caught the offender
in the act to kill him on the spot. Otherwise, the case had to be
turned over to the courts, which then allowed witnesses to partici-
pate in the execution of the offender. Philo alludes to the Jewish
law in Jos 44. See S. Belkin, *Philo and the Oral Law: The
Philonic Interpretation of Biblical Law in Relation to the Palestinian
Halakah*, Harvard Semitic Series, XI (Cambridge, Mass., 1940), 117-119,
who convincingly argues against E. R. Goodenough's misinterpretation
of Jos 44 as though alluding to Greek and Roman law; *The Jurisprudence
of the Jewish Courts in Egypt: Legal Administration by the Jews under
the Early Roman Empire as Described by Philo Judaeus* (New Haven, 1929),

pp. 77-80; B. Cohen, *Jewish and Roman Law; A Comparative Study*, II
(New York, 1966), 634-635.

sodomy or *pederasty* (παιδεραστεῖν) is among the more deplorable
crimes, characterized ever since Plato (*Lg*. 836C) as against nature
(e.g., Test N 3:4-5; Ro 1:26-27; J. *Ap*. ii. 273; Athen. 605d; etc.).
Philo cites this crime in language very similar to that of Anim (Abr
135-136; Spec Leg II 50; cf. I 325; III 37-39; Vita Cont 59-62).

, *truth herself convicts*. So also in Conf 126; Spec Leg II 49; etc.

unalterable law, i.e. the law of nature. See comment on § 48.

*giving the seed to the immature . . . wasting and destroying the
seed* (τὰ σπέρματα; cf. λόγος σπερματικός, see comment on §§ 20, 96
[*SVF* II 834]). The human semen is conceived of as containing a
potential reasoning principle. Thus, Onan, who wasted his seed on
the ground (Gen 38:9), is said to have not ceased destroying τὸ
λογικόν (Quod Deus 16-19; cf. Post 180-181). The pederast, who is
supposed to become impotent, is likewise accused of wasting or
destroying the seed (Abr 135; Spec Leg III 39; Vita Cont 62; cf. Pl.
Lg. 838E). Among the reasons given for circumcision, it is stated
that "as the seed often flows into the folds of the foreskin, it is
likely that it will be scattered unfruitfully" (Quaes Gen III 48).
The thought is repeated in Philo's endorsement of the law against
touching a woman when the menstrual issue occurs (Lev 18:19; 20:18; see
Spec Leg III 32), and in his disapproval of marriage with a woman
known to be barren (*ibid*. 34-36) or passed the prime of her age
(Quaes Gen I 27).

50 *the Egyptian crocodile*. Crocodiles and hippopotamuses are the
ἀνθρωποβόρα ζῷα of the Nile (Praem 90; Provid II 108).

Although this story cannot be found elsewhere, (except for the obvious use of this passage in the *Commentarius in hexaemeron* attributed
to Eustathius of Antioch, Migne *PG* XVIII, 728), it has some similarities in Plu. *Mor*. 982B-D, an account on the laying of eggs by
tortoises and crocodiles. In 982B, it is said that the male tortoise turns the female on her back; in 982C, it is said of the
crocodile "that this creature's foreknowledge is divine and not
[simply] rational" (ὅθεν οὔ φασι λογικὴν ἀλλὰ μαντικὴν εἶναι τὴν ἐπὶ
τοῦτον τοῦ θηρίου πρόγνωσιν); and in 982D, "the mother [crocodile]

tears to pieces and bites to death" her young, "the one which, upon emerging, does not immediately seize in its mouth anything that comes along." Cf. Ael. *NA* ix. 3.

51   *Courage* (ἀνδρεία) is discussed in §§ 51-59 and its opposite vice (δειλία) in §§ 68-69. Philo treats this virtue in the first part of Virt (1-50), beginning with the thought that true courage is knowledge (ἐπιστήμη), which is the basis of all virtues according to Socrates in the Platonic dialogues (*La.*, *Hp.Mi.*, *Prt.*, *Grg.*, and *Men.*; in the *R.*, to be sure, Plato emphasizes the identity of the virtues when they reach the highest "level," at which stage they are all forms of knowledge.) Following the Stoic definition, courage is the knowledge that centers around things to be endured (ὑπομενετέοις, Leg All I 65, 68 [*SVF* III 286; cf. 262, 264, 280, 285, 295; I 200-201, 563]). Elsewhere he follows Arist. *EN* 1107 b1-3 by stating that "courage is the mean between rashness and cowardice" (μέση δὲ θράσους μὲν καὶ δειλίας ἀνδρεία, Quod Deus 164; cf. Spec Leg IV 146; Ebr 115-116. Aristotle, however, gives a fuller definition of courage in *EN* 1115a6-35).

On the courage of animals, see comment below. Philo's response to this argument is given in §§ 85-89.

animals endowed with strength and vigor (ἰσχὺς καὶ ῥώμη). Philo agrees that animals excell man in these attributes (§ 85 [*SVF* II 726; cf. III 38]; Provid II 20; Post 161; Abr 266; Virt 146; cf. Plu. *Mor.* 963B, etc.), which he classifies among "the bodily good things" (Post 159) or "virtues of the body," distinguishing them from "virtues of the soul," the cardinal virtues (Sobr 61; cf. Leg All II 80; Quaes Gen IV 11 [*SVF* III 207]). Judging from his classification of virtues (for more on the subject see H. A. Wolfson, *Philo*, II [rev. ed.; Cambridge, Mass., 1968], 202-208) and definition of courage (see above), he seems to have subscribed to the Platonic notion that, philosophically speaking, animals are not courageous (*La.* 196C-197C; *R.* 430B).

animals . . . wage quick counterattacks. In Mut 160 Philo vividly describes the various steps bulls take in readiness to attack their antagonists in the arena. These progressive impulses are called ὄρουσις, defined by the Stoics as "mental bearings on what is about

to happen" (φορὰ διανοίας ἐπί τι μέλλον, Stob. *Ecl.* ii. 86.17 [*SVF*
III 169; cf. ὁρμή in IV Index, 105-106]). However, he adds that
the same readiness is also shown by plants when they blossom in
anticipation of bearing fruit (Mut 161-162). On the wrath of bulls,
note these lines of Euripides quoted in Quod Omn 101,

> "Eyes full of fire--you look just like a bull
> Watching a lion's onset."

See *ibid.* 131-135, on cock-fighting; Leg All III 130, on the ram;
and Quaes Ex II 101, on rams, oxen, and goats.

bulls . . . boars. There are numerous parallels in classical literature to the examples given on the courage of bulls and boars (e.g.,
Cic. *ND* ii. 127; Verg. *G.* iii. 234, 241; Pliny *NH* viii. 181; Plu. *Mor.*
966C; Ael. *NA* v. 45; vi. 1; Opp. *C.* 54-58; etc.). A. W. Mair in a
note to the last passage (LCL, p. 59) refers to the ancient practice
of wrestlers who anointed themselves with oil and sprinkled themselves
with dust. On the boars' whetting of their tusks, see the passages
cited from Plutarch and Aelian. In the last author, "whets its tusks
on smooth rocks" is drawn upon Hom. *Il.* xi. 416; xiii. 474-475.

Bears . . . practice all the cunning tricks. The claim is first
encountered in Isoc. *Antid.* 213, "Bears dance about and wrestle
and imitate our skills." For a complete account of a bear hunt,
see Opp. *C.* iv. 353-424.

52 *a python fight with an Egyptian cobra* (δράκοντος πρὸς Αἰγυπτίαν
ἀσπίδα). This incident, like the events in §§ 23 and 58, is said to
have occurred "the day before yesterday." The Egyptian cobra is
mentioned in Vita Mos I 109 and the python, interchangeably with
serpent (ὄφις), in Agr 94-110. Speaking of the python, Philo says
"This is a creature tortuous in its movements, of great intelligence (συνετὸν ἐν τοῖς μάλιστα), ready to shew fight, and most
capable of defending itself against wrongful aggression." This
accurate description, however, is part of an allegory on Gen 49:16-18
(Agr 94-110; cf. Leg All II 94-108), where Dan's serpent represents
endurance and self-restraint (καρτερία καὶ σωφροσύνη), "blocking
the way of familiar pleasure" (τὴν ἐφόδα ἡδονήν, represented by

Eve's serpent), and biting a horse, "the symbol of passion and
wickedness (τὸ πάθους καὶ κακίας σύμβολον). As might be expected,
Gen 3:2--"the serpent is the most subtle (φρονιμώτατος) of all beasts
upon the earth"--is likewise interpreted allegorically in Leg All II
106-107, where Eve's serpent, as always, represents pleasure, "the
most cunning (πανουργότατον) of all things." As for Moses' rod and
brazen serpent (Ex 4; Num 21), see *ibid.* 76-93.

rational beings (λογικαὶ φύσεις). In its singular form the term
usually refers either to the reasoning power of man or to man as a
rational being (see H. Leisegang, "λογικός," *Indices,* II (Berlin,
1930), 485. It is also applied to the human soul and heaven in an
analogy between man and the universe (λογικαὶ δύο φύσεις, Heres 233).
The plural form, however, refers to mankind (Plant 41) and celestial
beings (λογικαὶ καὶ θεῖαι φύσεις, Op 144 [*SVF* III 337]).

53 *centurions and tribunes* (λοχαγοὶ καὶ ταξίαρχοι) are likewise mentioned in § 65 and Agr 87 (cf. Dec 38; Spec Leg I 114, tribunes only).

They . . . press on to valor (πρὸς ἀλκήν). The courage of men
and animals, with emphasis on valor, is well contrasted in Plu. *Mor.*
987C-988E.

54 *When I went on an embassy to Rome* (ὅτε ἐπρέσβευσα εἰς ῾Ρώμην). See
Intro., pp. 30-31.

the contests of wild animals. The description of wild animals in
this section, as in the next, must be of lions (see comment below).
It is remarked in Plu. *Mor.* 963C that "Rome has provided us a reser-
voir from which to draw in pails and buckets, as it were, from the
imperial spectacles." On the various animal shows during the Early
Empire see G. Jennison, *Animals for Show and Pleasure in Ancient
Rome* (Manchester, 1937), pp. 60-82.

the poet has truly said. The quotation (*Il.* xx. 170-171), part
of Homer's remarks on the lion, is alluded to in Ael. *NA* vi. 1 (cf.
Pliny *NH* viii. 49). Philo finds no use for the tail but to swish
off the gnats which settle on it (Praem 125).

55 *shame* (αἰδώς). In Stoic philosophy, shame, the avoidance of de-

served blame, and sanctity (ἀγνεία), the avoidance of offenses against the gods, are subdivisions of caution (εὐλάβεια), the right feeling (εὐπάθεια) of fear (φόβος), which, rightly understood, is consistent with courage (ἀνδρεία, D.L. vii. 116 [*SVF* III 431; cf. 410, 416, 432]). For a good discussion of the Stoic doctrine of εὐπάθειαι, see J. M. Rist, *Stoic Philosophy* (London, 1969), pp. 22-36, 49-53. On Philo's use of the Stoic εὐπάθειαι, see H. Schmidt, *Die Anthropologie Philons von Alexandreia* (Diss. Leipzig, 1933), pp. 98-101; É. Bréhier, *Les idées philosophiques et religieuses de Philon d'Alexandrie*, Études de philosophie mediévale, VIII (3d ed.; Paris, 1950), 254-255, 276; J. Dillon and A. Terian, "Philo and the Stoic Doctrine of Εὐπάθειαι: A Note on Quaes Gen 2.57," *Studia Philonica*, 4 (1976-1977), 17-22.

when lions yield to defeat, etc. Cf. Pliny *NH* viii. 50; Opp. *C.* ii. 72-82 (told of the bull). The statement is contrary to that on lions in Pr 30:30.

This section bears reminiscences of the story of Hagar (Gen 21: 14-19), who flees because of shame (δι' αἰδῶ), "the outward expression of inward modesty" (τοῦ σωφροσύνης ἀπεικονίσματος, αἰδοῦς), and is counseled to be of good courage (εὐτολμίᾳ χρῆσθαι, Fuga 3-6).

56   *The genus of horses is such a lover of honor* (φιλότιμος). Cf. Job 39: 19-25; Ael. *NA* ii. 10. Pliny mentions a number of horses honored with cemetery monuments (*NH* viii. 154-155).

barbarians, or, very often in antiquity, the non-Hellenic races; e.g., Praem 165; Gaium 141, 147; J. *Ap.*i. 58, 116, 161, etc. The term is used with a different sense in § 62.

people from every region gather for various purposes. Philo lists somewhat similar reasons for traveling abroad (Abr 65).

57   *the Olympian, the Nemean, the Pythian, and the Isthmian games.* On these panhellenic festivals, see J. Regner, "Olympioniken," *PRE*, 35 (1939), cols. 232-241; K. Hanell, "Nemea, 4," *PRE*, 32 (1935), cols. 2322-2327; W. Fauth, "Pythia, 2," *PRE*, 47 (1963), cols. 515-547; and [E.] Bischoff, "Isthmia, 1," *PRE*, 18 (1916), cols. 2248-2254.

trainers, who rouse the innate speed of the animals. In spite of

the allegorical overtones in Agr 67-93, Philo's literal description
of race horses cannot be overlooked. Whereas Alexander credits horses here and in § 58, Philo credits the horseman, who brings along
the skill of horsemanship, which the animal is incapable of acquiring
(Agr 71). And although he disapproves of literal race horses breeding, he grants that breeders and trainers have the excuse that spectators of a race, besides being entertained, catch the exemplary
spirit of the horses (*ibid*. 90-91). Note that Homer in *Il*. xxiii.
276 and 374 uses the word ἀρετή to describe the spirit and speed of
noble horses (cf. Pl. *R*. 335B).

58 *four-horse-drawn chariot race*. This event, like the events in §§
23 and 52, is said to have occurred "the day before yesterday." A
similar account in Pliny *NH* viii. 160-161 makes this account appear
to have been taken from a literary source and adapted to a contemporary scene (on similar adaptations in Anim, see comment on §§ 23, 37).
The event in Pliny's account, however, is said to have occurred during the secular games of Claudius Caesar, i.e., in A. D. 47 (C. H.
Rackham's note, LCL III, 112).

This story furnishes a different answer to a rhetorical question
put forth in Quod Det 141, "If a charioteer quit a horse-chariot
during a race, does it not necessarily follow that the chariot's
course will lose all order and direction?" But in Conf 114-115 he
grants the opposite to happen: "ships and chariots. . .the one on the
water and the other on land often goes straight without helmsman or
charioteer." The story also brings reminiscences of Plato's famous
myth of the soul's chariot (*Phdr*. 246A-256A), enunciated in Agr 67-93,
107-109; Leg All I 72-73; II 99-104; III 128; Migr 62.

Philo might have attended Greek games and shows (on Provid II 103;
Ebr 77; Quod Omn 26, 141, see Intro., p. 56; in the first passage
he speaks of being at the hippodrome and observing the excitement of
the spectators).

urged by the eagerness of the spirit . . . if not by the impulses of the
body (ταῖς προθυμίας . . . τοῖς σώμασι). The statement is based
on the traditional contrast between spirit and body, the two extremes of the human nature (cf. Mt 26:41; Mk 14:38; etc.). On the

Platonic background of such a dualism in Philo, which in turn was influenced by Orphic belief (Diels *VS* 1B3; 44B14; etc.), see U. Früchtel, *Die kosmologischen Vorstellungen bei Philo von Alexandrien. Ein Beitrag zur Geschichte der Genesisexegese*, Arbeiten zur Literatur und Geschichte des hellenistischen Judentums, II (Leiden, 1968), 157-160.

59 *the elephant. . .tales*. A brief version of this story is found in Pliny *NH* viii, 11-12. There, on the authority of Antipater of Tarsus, a Stoic of the 2d cent. B.C., the elephants Ajax and Patroclus are said to have belonged to Antiochus; in Anim, the herd belonged to "the king" and was bred by "an Asiarch" (see comment below). The source of another Syrian story on elephants in Plu. *Mor.* 968D is said to be Hagnon of Tarsus, pupil of Carneades. On the questionable sources of these and other stories on elephants, see comment on § 27.

Antioch, possibly in Syria (in the light of the comment above).

A chief of Asia or *Asiarch* ('Ασιάρχης). Although this official title is repeatedly encountered in inscriptions and coins from the western region of Asia Minor, the real function of its bearers is not sufficiently clear. The prevailing view is that this honorary title was temporarily conferred upon municipal benefactors in the Province of Asia. Several of these influential men held local offices besides the Asiarchate, as the city coins and the list of the known Chief Priests of Asia seem to show. Their primary service was to sponsor local contests or spectacles over which they presided (see M Pol 12:2). Thus there could have been several Asiarchs, as in Ac 19:31. Unfortunately, the context of the term in Anim is historically unreliable (the term could not have been used in pre-Roman times and certainly never outside the Province of Asia) and has nothing to offer to our understanding of the office. For the occurences of the term and a fine discussion of the office, see D. Magie, *Roman Rule in Asia Minor to the End of the Third Century after Christ* (Princeton, 1950), I, 449-450; II, 1298-1301, 1526.

Ajax, the son of Telamon. . . Patroclus . . . the arms of Achilles.
Ajax, Patroclus, and Odysseus were co-warriors with Achilles, Homer's
hero of the Trojan War in the *Iliad*. Following the death of Patroclus
and Achilles, Ajax succumbed to wrath when losing to Odysseus in a
contest over the arms of Achilles (*Od.* xi. 541-562). Later stories,
e.g. Sophocles' tragedy *Ajax*, tell that Ajax went so mad with anger
and disappointment that he finally committed suicide. The valor of
these Homeric figures is here claimed for the elephants. Philo responds
to this argument by pointing out that the real motivation of the elephant Ajax was greed (§ 89). Ajax and Achilles are mentioned in Provid I 65.

true happiness (εὐδαιμονία) is here presented as an end to virtue,
under the token of courage (cf. Virt 119-120; Quod Omn 117; Arist.
EN 1098b30). Aristotle had observed that verbally most men agree that
happiness is the final end of life, but that they are by no means in
accord about what it really is (*ibid.* 1095a17-18). He then gives a
catalogue of the different views on happiness, with the conclusion
that all should be right in at least one respect or in most respects
(*ibid.* 1098b23-29). Philo endorses certain of Aristotle's own definitions of happiness: the exercise of perfect virtue in a perfect life
(*ibid.* 1098a16; Quod Det 60; cf. Leg All III 52; the same was adopted
by the Stoics in their doctrine of "living according to nature," to
which Philo alludes in Plant 49 [*SVF* III 7; see especially Seneca's
De vita beata]); that which embraces the external, bodily, and
spiritual good things, all of which the happy man must possess (Arist.
Pol. 1323a24-26; Quod Det 7-8 [*SVF* III 33; cf. 136]). Following
the third and most important category, happiness consists of setting
the mind on intellectual and divine things (Arist. *EN* 1177a11-1178a8;
Op 144; Quod Det 86; Heres 111; Mut 216; Praem 11, 81; cf. with the
Peripatetic view in Cic. *Fin.* v. 11). But Philo goes on to state
that the comsummation of virtues is to be well pleasing to God, "the
definition of supreme happiness" (ὅρος τῆς ἄκρας εὐδαιμονίας, Quod
Deus 118), to know Him (Spec Leg I 345), to serve Him (*ibid.* II 38;
Post 185), to become like Him (Dec 73), since He is the consummate
happiness (Leg All III 205; Cher 86; Plant 66; Abr 202; Spec Leg I 209).
The crowning of virtues with happiness is a gift of God (Cher 49; Quod
Deus 26, 92; Plant 37; Ebr 224). The elaboration of this maxim in the

allegories on the birth of Isaac, who represents "happiness" (Leg All
II 82; III 86-87, 218; Cher 8; Post 134; Heres 86; Abr 32-35), is amply discussed by E. R. Goodenough, *By Light, Light: The Mystic Gospel
of Hellenistic Judaism* (New Haven, 1935), pp. 131, 141-142, 153-155,
192, 232-234.

60   *justice* (δικαιοσύνη). This virtue, the last of the four cardinal
virtues, is discussed in §§ 60 (*SVF* II 728)-65 (60-63 on commutative
justice and 64-65 on political justice). Philo gives the conven-
tional definitions of this virtue as found in the main schools. In
Plato's *Republic* justice, the virtue of the entire soul when each
part does its own work, may be considered superior to other virtues
on the ground that it is the excellence of the whole rather than
merely of a part (this is implied rather than stated). Philo in
his allegorical interpretation of the rivers of Paradise holds that
justice, the fourth virtue, appears when the three parts of the soul
(the intellectual, the realm of wisdom; the spirited, the realm of
courage; and the appetitive, the realm of self-restraint) are in
harmony (Leg All I 72; cf. Abr 27, "justice, the chief among the
virtues, who like the fairest of the dance holds the highest place").
But in the *Republic*, justice proper is a political virtue centering
around the rights and property of individuals and emerges as a kind
of cooperative disposition to do one's own work (433E; 443B).
Aristotle underlines its dual application as distributive and corrective: distributive, i.e., dividing of goods, honors, etc., and
corrective, i.e., regulatory of the inequities in either transactions
or crimes (*EN* 1130b30-1133b28). In both cases justice is a kind of
proportion, also under the doctrine of the "mean" (*ibid.* 1133b29 -
1134a16). The distributive sense is used also by the Stoics in their
definition of justice as "the science of distributing what is proper
to each" (Stob. *Ecl.* ii. 59.4 [*SVF* III 262]) and by Philo in Leg All
I 65 [*SVF* III 263]). Justice for Chrysippus, however, is based on
nature, not convention (D. L. vii. 128 [*SVF* III 308]).

Philo's understanding of justice as conformity to divine law on
the part of the individual and the state is seen in a number of his
works: Jos, Dec, Spec Leg I-IV, Virt, and Praem. In Spec Leg IV
136-238 (subtitled Περὶ δικαιοσύνης) he discusses justice as seen in

the Mosaic Law and concludes with the praises of the virtue whose
mother (μήτηρ) is equality (ἰσότης, 231; Quaes Ex I 6). Elsewhere
equality is said to be the bearer (ἔτεκεν) of justice (Plant 122;
cf. Vita Cont 17), its essence (τροφόν, Heres 163), and source (ἀρχὴ
καὶ πηγή, Spec Leg II 204; cf. Gaium 85), just as inequality is the
source of injustice (ibid.; Vita Mos I 328). On Philo's understanding of political justice, see H. A. Wolfson, Philo, II (rev ed.; Cambridge, Mass., 1968), 374-395; E. R. Goodenough, The Jurisprudence
of the Jewish Courts in Egypt: Legal Administration by the Jews under
the Early Roman Empire as Described by Philo Judaeus (New Haven, 1929);
and his The Politics of Philo Judaeus: Practice and Theory (New Haven,
1938). On δίκη as executor of justice, see his By Light, Light: The
Mystic Gospel of Hellenistic Judaism (New Haven, 1935), pp. 59-62.

The partnership of the marine pinna and the pinna-guard (πίννης
πρὸς πιννοτήρην κοινωνία). The pinna, a bivalve shellfish, and its
guard or sentinel, a commensal crab that lives in the pinna's shell,
were cited by Chrysippus to illustrate partnership (Plu. Mor. 980
A-B; Athen. 89d; cf. Cic. ND ii 123-124; Fin. iii. 63 [SVF II 729-b;
III 369]). Plutarch complains in the cited passage that a better
example could have been given. Philo, like Cicero, grants that this
"partnership" is brought about by nature, not by mutual agreement
(§§ 93-95 [SVF II 730]).

Note that before citing Chrysippus, Athenaeus alludes to the epitome of Pamphilius of Alexandria, who must have followed Arist. HA
547b15-33. On the various other authorities, see D'Arcy W. Thompson,
"Πίννη," "Πιννοτήρης," A Glossary of Greek Fishes (London, 1947),
pp. 200-202.

The Egyptian plover (τροχίλος). The omission of κροκόδειλος (in
Gr. or in its transliterated Arm. form) due to homoioteleuton may
be suspected here. The commensality of the Egyptian plover and the
crocodile is well attested by ancient naturalists (see D'Arcy W.
Thompson, "Τροχίλος," A Glossary of Greek Birds [rev. ed.; London,
1936], p. 287).

The pilot fish (πομπίλος). What follows seems to answer the
charge that animals do not exercise justice since they devour one

another (see Hes. *Op.* 277 quoted in Ael.*NA* vi.50). Cf. Plu. *Mor.*
981E-F, "The male [fish] do not eat their own young" (see Arist. *HA*
621a21-b2, said of the sheat-fish; contrary to Hdt. ii.93). On the
epithet ἱερὸς ἰχθύς given to the pilot fish, see D'Arcy W. Thompson,
"Πομπίλος," *A Glossary of Greek Fishes* (London, 1947), pp. 208-209.

61   *the stork exhibits supreme justice* (τὴν ἀνωτάτην δικαιοσύνην).
Philo observes that in returning kindness man is worsted by beasts.
He cites examples of watch-dogs and sheep-dogs "trained (πεπαίδευται)
to know how to return benefit for benefit," and of young storks
feeding the aged. These birds are said to be self-taught, by nature
or instinct (αὐτοδιδάκτῳ τῇ φύσει, Dec 113-117; cf. § 96). F. H.
Colson in a note to Dec 116 (LCL VII, 612) points out the currency
of associating reciprocation of kindness with the bird (πελαργός) in
the use of the verb ἀντιπελαργεῖν (to return kindness). The idea
occurs also in Ḥullin 63a: "R. Ḥisda said, The *ḥasidah* is the white
stork. And why is it called *ḥasidah*? Because it shows kindness
[*ḥasiduth*] to its companions." For references to similar accounts
in classical writings, see D'Arcy W. Thompson, "Πελαργός," *A Glossary
of Greek Birds* (rev. ed.; London, 1936), p. 223.

   *the universe should be composed not of some only of its parts but
of them all.* The doctrine of the unity of the universe was taught
by Plato (*Tim.* 32C-33A; cf. 31A-B, based on the one ideal model),
Aristotle (*Cael.* 276a18-277b26; *Metaph.* 1074a31-40, based on the
argument for one heaven), and the Stoics (*SVF* II 530-533, based on
their understanding of the whole, which was not simply the sum of its parts,
S. E. *M.* ix. 338-349, 352 [*SVF* II 80]; *P*. iii. 98-101). Philo's adherence
to this doctrine, modified by his belief in the Creator God, is seen in a
number of passages: Op 171; Cher 108-112; Quod Det 154; Plant 7; Migr 178-
180 (*SVF* II 532); Heres 228; Aet 78; etc. This unity is held by God's λόγος,
which is the glue and bond (κόλλα καὶ δεσμός) of the universe (Plant 9;
Heres 188; Fuga 112 [*SVF* II 179]; Quaes Ex II 68; cf. Cic. *ND* iii. 29-30;
also Hb 1:3; 4:12-13). As an integral part of the universe, Alexander
argues, animals must have reason.

   All created beings and things are parts of the interdependent
cosmic order (Provid I 9 [*SVF* II 577], 45, 51; II 110; Somn II 116;
Spec Leg I 210; III 191; Aet 4 [*SVF* II 621]). The unity of the
universe is also shown in the latter's description as a living being

(Heres 155; Spec Leg I 210; Aet 26, 72 [*SVF* II 636], 95; Quaes Gen
IV 188, 215 [*SVF* II 635, 643]; cf. Cic. *ND* ii. 22 [*SVF* I 112-114],
on Zeno's arguments that the world is a living being since it gives
birth to inanimate things and rational beings; D. L. vii. 138-139,
142-143 [*SVF* II 633-634], where the doctrine of periodical destruction and renewal of the world is also connected with the view of
the cosmos as a living being). This unity is also suggested in
the analogy of the cosmos to man as the macrocosm to a microcosm
(Op 82; Migr 220; Heres 155; cf. Pl. *Tim.* 30D; 44D; Arist. *Ph.* 252b-
26-27). For a comparison of Philo's views on the cosmos with those
of the leading schools of philosophy, see H. A. Wolfson, *Philo*, I
(rev. ed.; Cambridge, Mass., 1968), 295-324, especially 312-314 on
its unity. On the cosmos as a harmony of opposites see comment below.

justice . . .injustice . . .reason. Justice here is the correlate
of injustice, just as bad is the correlate of good, the rational of
the irrational, etc. The view of the cosmos and everything in it
as a harmony of opposites is attributed to Heraclitus (or rather Moses)
in Heres 214 and Quaes Gen III 5 (cf. Diels *VS* 22A1§§7, 8; A8, 10; B8,
10, 23, etc.). In Heres 129-236, a lengthy allegory on Gen 15:10 (cf.
Quaes Gen III 5), Philo has the notion of the "dividing logos" (λόγος
τομεύς, 130-131, 140, 165, 215, 225), which not only divides all
things in the universe--material or immaterial--into exact opposites
but also permeates them and holds them together (188, cf. 233; on
the λόγος as the bond of the universe see comment above). The presence of any part, then, is indicative not only of the presence of
its exact opposite or counterpart (213) but also of reason (cf. § 66).
É. Bréhier observes a Heraclitean-Stoic syncretism in the lengthy passage on opposites in Heres (*Les idées philosophiques et religieuses de
Philon d'Alexandrie*, Études de philosophie mediévale, VIII [3d ed.;
Paris, 1950], 87). E. R. Goodenough suggests a Neo-Pythagorean
source ("A Neo-Pythagorean Source in Philo Judaeus," *Yale Classical
Studies*, 3 [1932], 115-164).

the storks . . . the pinna guards . . . the bees . . . also punish,
i.e. they execute justice. On bees punishing drones see the rest
of the passage in Hesiod (*Th.* 594-599); also, Var. *R.* iii. 16.8;
Pliny *NH* xi. 27-28; Ael. *NA* i.9; etc. Philo responds in § 96 that

such acts are involuntary. Hesiod is also quoted in Provid II 36.

62   *animals which attain to the cardinal virtues abstain from eating*
flesh. Cf. Plu. *Mor.* 974C; Ael. *NA* vi.2. This passage contains major
characteristics of the Synic-Stoic diatribe: simple style, Pythagorean example of virtue, barbaric example of civilized people, call
to control of appetite and physical desire, and to chastity (cf. R.
Bultmann, *Der Stil der paulinischen Predigt und die kynisch-stoische*
Diatribe [Göttingen, 1910], pp. 10-64; A. Oltramare, *Les origines de*
la diatribe romaine [Lausanne, 1926], pp. 46-62; for other examples of
such argumentation in Anim.see §§ 47-49 and comment).

It is important to note that this statement is cited as an argument
for justice. In condemning the opposite vice of this virtue, it is remarked on carnivors that they act unjustly (§ 70). The argument is best treated in Plutarch's *De esu carnium* I-II (*Mor.* 993A-999B; cf. Porph. *Abst.*),
where man's feasting on domestic animals and passing a taboo on the wild
are condemned as unjust: "we really have no compact of justice with animals"
(999B). In another work of Plutarch, *De sollertia animalium* (*Mor.* 959A-
985C), Pythagoras is credited with the formula which states that "there is
no injustice in punishing and slaying animals that are anti-social and
merely injurious, while taming those that are gentle and friendly to man
and making them our helpers in the tasks for which they are severally fitted
by nature" (964E-F). The formula makes no allowance for the use of animals
as a source of food.

Philo in Provid II 110-111, like Plutarch in the *De esu carnium* I-II (995E-
996B, 997B-C; cf. 991B-D), says that we need not eat the flesh of animals;
he too considers meat-eating a luxury and adds that those who eat flesh
acquire the characteristics of wild beasts (cf. Spec Leg II 20; IV 100;
Porph. *Abst.* iii. 26; also Ro 14:1-23, especially 2, 21).

barbarians. See comment on § 56 and above. The "primitive,"
"savage" sense suggests itself here. Because of the hostilities in
his environment, Philo seems to have been fond of citing Greeks with
barbarians (see the passages, most of which are in diatribe style, in
H. Leisegang, "βαρβαρικός," "βάρβαρος," *Indices,* I [Berlin, 1926],
137; cf. the NT examples of gentiles for the emulation of Jews and
the *ecclesia*: Mt 6:7-11; 8:5-13; 10:11-15; 11:20-24; 12:41; 15:21-28;
Mk 7:24-30; Lk 7:1-10; 10:1-12; 11:32; Ac 13:44-52; 16:11-15; 17:1-5,
11-13; 18:5-7; Ro 2:14-15; 9:30-31; 1Cor 5:1; 2Pt 2:4-9; Jd 7).

the Pythagorean philosophy. A school, partly scientific and
partly religious, founded by Pythagoras,*ca.* 530 B.C. The major
tenets of the school were the belief in numbers as the ultimate elements of the universe and the transmigration of the soul. Pythagoras
himself left no writings. Xenophanes, a contemporary of Pythagoras
(Diels *VS* 22B40), knew that he taught transmigration (*ibid.* 21B7),
possibly his chief religious doctrine which came down through a system of taboos (see H. Dörrie, K. von Fritz, and B. L. van der
Waerden, "Pythagoras," *PRE*, 47 [1963], cols. 171-300). K. Freeman,
after considering the conflicting testimonies on the Pythagorean
teaching of abstinence from meat-eating, concludes, "All that can
be safely conjectured is that there was a prohibition of some kind
concerning the killing of animals either for sacrifice or for food,
and that this was closely connected, as always, with the transmi-
gration-theory." *The Pre-Socratic Philosophers: A Companion to
Diels, Fragmente der Vorsokratiker* (Oxford, 1959), p. 80. On the
prevalence of this early theory and the various views expressed in
appeals for vegetarianism in later antiquity, see J. Haussleiter,
Der Vegetarismus in der Antike (Berlin, 1935).

63   *The body naturally serves the purposes of the soul* (ἔχει γὰρ ἐκ
φύσεως ἔργον τι σῶμα ψυχῇ). Cf. Sacr 72 and Quaes Ex I 19, where the
body-soul relationship is likened to that of a servant to a mistress.

a tie of domestic relationships (οἰκειότητα). There is sufficient
reason to assume that the polemic here is aimed at the Stoa's doctrine
of οἰκείωσις, which was held as a principle of justice (Zeno in Porph.
Abst. iii. 19 [*SVF* I 197]). A significant passage in Cicero (partly on
the authority of Chrysippus, *Fin.* iii. 62-68 [*SVF* III 340-342]; cf. v.
65) has that it is natural for humans and lower animals to love their
offspring and that in this is found the basis of association and justice. Although one finds among lower animals association with other
species, with humans this relationship is most intimate and peculiar;
so much so that it will be unjust for humans to extend it to animals.
Man is not charged with injustice when he makes unilateral use of
beasts. Cf. the passages cited under *Iuris communionem non pertinere
ad bruta animalia* in *SVF* III (367-376). A number of other important
passages on the subject are cited in S. G. Pembroke's excellent

discussion: "Oikeiōsis," *Problems in Stoicism*, ed. A. A. Long (London, 1971), pp. 114-149. See also F. Dirlmeier, "Die Oikeiosis-Lehre Theophrasts," *Philologus Supplementband*, 30 (1937), 1-100; C. O. Brink, "Οἰκείωσις and Οἰκειότης: Theophrastus and Zeno on Nature and Moral Theory," *Phronesis*, 1 (1955), 123-145.

Porphyry (in the passage cited above) and Iamblichus (*VP* 168) have it that Pythagoras extended οἰκείωσις to animals. Whereas the appeal for man to extend οἰκείωσις to animals is here based on reciprocity, a similar appeal in a fragment of Theophrastus is based on the identical origin, physiology, and psychology of man and animals (Porph. *Abst.* iii. 25; the implication of animal rationality is probably due to Porphyry's amplification; cf. D. L. v. 43). Philo admits kinship of all flesh in his treatment of Lev 17:11, "the soul of every flesh is the blood," but goes on to extol the distinctions of the human soul or reason, the proprium of man (Heres 55-56; cf. Dec 107). Elsewhere, in his treatment of the fifth commandment (Ex 20:12), he challenges man to learn from the animal world lessons of reciprocating kindness (Dec 111-120). But he neither extends οἰκείωσις to animals nor allows it and its opposite (ἀλλοτρίωσις) to exist among them in the truest sense (§§ 93-96 [*SVF* II 730]). The same is also implied by his strict use of the term in describing human relationships on the one hand and of the human with the divine on the other (see H. Leisegang, "οἰκεῖος" - "οἰκείωσις," *Indices*, II [Berlin, 1930], 567-568; cf. Cic. *Leg.* i. 22 [*SVF* III 339]; *ND* ii. 154 [*SVF* II 1131], the *civitas communis doerum atque hominum* to which lower animals do not belong, for they do not possess reason; also Sen. *Ep.* cxxi. 14 [*SVF* II 184]; D. L. vii. 129 [*SVF* III 367]).

Josephus, commenting on the Mosaic legislation on kindness to aliens (Ex 20:10; 22:21; etc.), makes this striking statement: "It is not family ties alone which constitute relationship, but agreement in the principles of conduct" (οὐ τῷ γένει μόνον, ἀλλὰ καὶ τῇ προαιρέσει τοῦ βίου νομίζων εἶναι τὴν οἰκειότητα, *Ap.* ii. 210).

64   *equality and justice* (ἰσότης καὶ δικαιοσύνη). See comment on § 60.

leaders they appoint. The examples given are common; e.g., Pr 24:66 (30:31); Somn I 198; II 288-289; Ael. *NA* vii. 26; etc. On

Philo's indirect and somewhat ambiguous response, see comment on § 87.

divinely imbued (θεόσυτοι), i.e., the mantic power attributed to the congregating of animals; also to their flight, cries, and calls (Arist. *HA* 608b27-30; Pliny *NH* ix. 55; Plu. *Mor.* 975A-B; 976C; Ael. *NA* i. 48; vii. 5; etc.).

65 *a swarm of bees.* On this stock illustration, see the authorities cited in the comment on § 20. Note especially Pliny *NH* xi. 53-54 and Ael. *NA* v. 10-11, 13.

shaken by anarchy--as much as if by ochlocracy. In Somn II 287-289 Philo deplores ochlocracy, which he equates with anarchy, and commends animals that look for their proper leaders and honor them; adding how in the absence of their leaders they scatter and are destroyed. Ochlocracy is the worst of evil polities (Op 171). Elsewhere he contrasts kingship with oligarchy and ochlocracy, both of which stand for lawless rule (Fuga 10; Dec 135-136, 115; absolute sovereignty and tyranny are equally deplored, Abr 46, 242; Leg All III 79; etc.). Ochlocracy is further characterized as counterfeit democracy, which ultimately leads to disorder and anarchy (Agr 45-46; Conf 108; Virt 180). In these and other passages he goes on to call democracy the best of constitutions (Spec Leg IV 237; Quod Deus 176; etc.). For a comparison of these views with those of Plato and Aristotle, see H. A. Wolfson, *Philo*, II (rev. ed.; Cambridge, Mass., 1968), 386-393, especially 390 and the authorities cited in n. 104, on Philo's understanding and use of the term democracy.

centurions and tribunes. (λοχαγοὶ καὶ ταξίαρχοι). See comment on § 53.

the virtue(s) of statesmanship. . . household management (πολιτικῆς ἀρετῆς. . . οἰκονομικῆς). The two virtues in conjunction, or as one and the same virtue, are dealt with in Philo's response (§ 91 [*SVF* II 733]) and elsewhere in Philo (Ebr 91; Fuga 36; Jos 38-39; Quaes Gen IV 165 [*SVF* III 301, 323, 624]; etc.). On house management, see comment on § 41.

judgment and foresight (γνώμης καὶ προνοίας). The terms occur in conjunction in Abr 6.

66 *vices, no less than virtues, are indications of innate reason.*
(ἀρεταί . . .τε καὶ κακίαι φυσικοῦ λόγου δηλώσεις εἰσι). On the relationship of virtues and vices to reason, see comment on § 30. Note
that the four vices listed here (folly, ἀφροσύνη; lack of self-restraint, ἀκολασία; cowardice, δειλία; and injustice, ἀδικία) are
the exact opposites of the four cardinal virtues discussed in §§ 30-
65 (wisdom or prudence, φρόνησις; self-restraint, σωφροσύνη; courage,
ἀνδρεία; and justice, δικαιοσύνη). This Platonic classification was
adopted by the Stoics and Philo, who, after enumerating the four vices
in Conf 90, adds, "and the other members of that fraternity and family"
(καὶ ὅσα ἄλλα ἀδελφὰ καὶ συγγενῆ τούτοις). This, no doubt, is an amplified equivalent to "all that are related to these" in Anim. F. H.
Colson observes in a note to the cited passage (LCL IV, 554-555), that
the Stoics added "as secondary to the primary four, incontinence
(ἀκρασία), stupidity (βραδύνοια), ill-advisedness (δυσβουλία), Diog.
Laert. vii. 93. It is these last three which presumably are meant
here."

In this and the following sections there is an orderly--yet brief--
treatment of the four vices: lack of self-restraint (§§ 66-67),
cowardice (§§ 68-69), injustice and foolishness (§ 70). Note that
the opposite of the first virtue becomes the last of the vices.

animals have been left destitute of truth (ἀληθείας). To this Philo
adds that animals cannot perceive abstract concepts "in the mind's
perception of God, etc." (§ 85 [*SVF* II 726])--neither can the sense
oriented, irrational part within us (Quod Det 91; cf. Quod Deus 52). The
thought that animals and all created things praise God, derived, no
doubt, from Is 43:20; 44:23; Ps 65:14; 98:8, occurs quite frequently in talmudic and midrashic literature; see L. Ginzberg, *The Legends of the
Jews*, trans. H. Szold and P. Radin (Philadelphia, 1947), I, 42-46; V,
60-61.

he-goats and pigeons are lewd in sexual relations. Philo condemns
pleasure-loving men, who mate with their wives not to beget children "but like pigs or he-goats in quest of enjoyment which such
intercourse gives" (Spec Leg III 113). The lechery of goats is almost proverbial (Arist. *HA* 573b30; Ael. *NA* vii. 19; Pliny *NH* viii.

202). Cf. the lewdness of stags (Opp. *C.* ii. 187-205), bears, lynxes, and hares (iii. 139-158, 515-525).

The remarks on cock pigeons, however, hardly apply. See D'Arcy W. Thompson, "Περιστερά," *A Glossary of Greek Birds* (rev. ed.; London, 1936), especially pp. 240-241, the authorities cited on the nesting, conjugal affection, and chastity of pigeons. Note also Pliny *NH* x. 104-105 and Porph. *Abst.* iii. 23, where it is added that cock pigeons relieve hen pigeons in incubation. Remarks similar to those in Anim are made on partridges; see Thompson, "Πέρδιξ," *ibid.*, especially pp. 235-236. Note, among others, Plu. *Mor.* 962E and Pliny *NH* x. 100, where it is said that cock partridges break the eggs in order to consort with the hens.

that inevitable wantonness. On the thought that gluttony stirs up sexual lust, see comment on § 48.

Pasiphaë, daughter of Minos of Crete. The same story is told in Spec Leg III 43-44, where, more accurately, Pasiphaë is said to be the wife of Minos (so also in Arm.). For a thorough discussion of the myth, the varying backgrounds of the story, and a catalogue of pictures where Pasiphaë appears occasionally either with the bull or with the Minotaur on archaic and classical vases, Etruscan ash-urns of the Hellenistic age, and Roman sarcophagi, see K. Scherling, "Pasiphae," *PRE*, 36:3 (1949), cols. 2069-2082. A later development of the story in Porph. *Abst.* iii. 16 has, "Jupiter, when in love with Pasiphaë, is said to have become a bull," indicating the honor which the ancients paid to animals.

Glauce the harpist. See the citation of references to Γλαύκη ἡ κιθαρῳδός [γυνή] in [P.] Maas, "Glauke, 13," *PRE*, 13 (1910), cols. 1396-1397.

promiscuous intercourse of unlike species is condemned in Spec Leg III 45-50, following the story of Pasiphaë (see comment above). The denouncement is based on Lev 19:19. According to Jubilees 5:2-3, animals too had intermingled themselves with other species. For this reason they perished in the deluge.

67   *a dolphin . . . in love with a boy.* The same story, with differ-

ences in locality and detail, occurs frequently in classical literature. A collection of the authorities is found in D'Arcy W. Thompson, "Δελφύς," *A Glossary of Greek Fishes* (London, 1947), pp. 54-55.

68 *Deer . . . their cowardice* (δειλία). See comment on § 69, on the
cowardice of the deer and the hare. The description here fits the
stag (cf. Pliny *NH* xi. 123).

found out by nature. Nature is characterized as the helper and manager of all creatures (§§ 69, 92 [*SVF* II 733]) and the designer of
their mechanism (§§ 80 [*SVF* II 732], 86, 97). Philo's writings abound
with similar thoughts, derived mostly from the common notion of *natura
creatrix* (see H. Leisegang, "φύσις. 2. natura creatrix," *Indices*,
II [Berlin, 1930], 837-838) and, to some extent, from the Stoic
ascription to nature concern for all creatures (e.g., Cic. *ND* ii.
57-60 [*SVF* I 171-172] Plu. *Mor.* 1060C [*SVF* III 146]; D. L. vii. 85
[*SVF* III 178]; etc.; see comment on § 11). Among the more important passages in Philo, note the following: nature provides the
well-armed animals with the equipment they use in self-defense (Op
85; Mut 159) and makes no idle superfluities, "but aids to the
weakness of those furnished with them"(Post 4). "As nature has
fortified other living creatures each with appropriate means of
guarding themselves . . . so has she given to man a most strong
redoubt and impregnable fort in reason" (Somn I 103; cf. §§ 11, 29,
71, 100, where nature is characterized as the giver of reason to
man). She is the friend of all that lives (φιλόζωος, Spec Leg II
205; cf. III 36).

ugly women. For comparable accounts, see Provid II 27; Leg All
III 62; Sacr 21; Fuga 153.

69 *In the contest of fear, the hare, no doubt is next to the deer*.
"Fawns, roe-deer, gazelles, antelopes, hares, which poets style
'cowerers,' are timorous" (πτῶκας . . . ἄτολμα, Ael. *NA* vii. 19; cf.
Arist. *HA* 488b15, δειλά, οἷον ἔλαφος, δασύπους; Opp. *C.* i. 165,
δειμαλέος πτώξ; ii. 181, δίαυλοι [ἔλαφοι]; 186, δειλοῖς λαγωοῖς; etc.).

the poets thought of the hare as a crouching creature. Aucher notes
a play on the words, "Λαγώς enim *lepus* est, et λαθεῖν, *latere*." But

nowhere does the latter word occur in conjuction with the former. The adjective here is no doubt πτωκάς or πτώξ, as in Hom. *Il*. xxii. 310 (cf. xvii. 676; *Od*. xvii. 295);A. *A*. 137; Ael. *NA* vii. 19; Opp. *C*. i. 165; iii. 504; etc.

nature, the helper of all (πάντων ἐπίκουρος ἡ φύσις). See comment on § 68.

70   *wild boars, leopards, and lions . . . act unjustly* (ἀδικεῖν). Cf. Hes. *Op*. 227,quoted in Ael. *NA* vi. 50. Plutarch sees no injustice in lions and other carnivors tearing their prey for food; that is the right thing for them to do, but not for man (*Mor*. 991B; 994A; 995A; on justice in slaying anti-social animals, see comment on § 62; cf. the NT description of heretics as beasts: Mt 7:15; 10:16; Lk 10:3; J 10:12; Ac 20:29; 1 Cor 15:32; Phil 3:2; Tit 2:12; 2 Pt 2:12-13, 22; Jd 10; Rv 22:15).

the dog . . . is frantic and fierce (λυτταλέος). The raging madness of dogs is proverbial, even in Philo (Dec 115; Vita Mos I 130; Quaes Gen IV 165). For a comparable account with emphasis on the loyalty of dogs instead of their foolishness, see S. E. *P*. i. 67-68 (cf. Cic. *ND* ii. 158).

71   *not only men but also various other animals have inherited the faculty of reason* (διάνοιαν ἐκληρονόμησαν). See comment on § 17.

they possess both virtues and vices (ἀρετῆς τε καὶ κακίας). See comment on §§ 30, on virtues; 66, on vices; and 97, on Philo's response.

ignorance (ἄγνοια). Philo distinguishes two kinds of ignorance: one mere lack of knowledge, the other the belief that we know, when we do not (Ebr 162). "The former is the lesser evil, for it is the cause of less serious and perhaps involuntary errors, and the second is the greater, for it is the parent of great iniquities. . . such as are actually premeditated" (*ibid*. 163). The former condition of ignorance is reiterated in Sacr 48, "Ignorance (ἀμαθία) is an involuntary state, a light matter, and its treatment through teaching is not hopeless." But ignorance due to defiance of, or unsusceptibility to, educating influences is strongly deplored (Ebr 11-12, 154-157; cf. the last part of Sacr 48).

those who have been endowed by God and natural agencies, i.e. mankind.
On the role of nature in giving reason to man, see comment on § 11.

enemies of truth (τοῖς ἐχθροῖς τῆς ἀληθείας). Alexander's discourse
not only begins and ends with emphasis on truth, but is also punctuated
with the notion of truth (§§ 10, 11, 17, 25, 44, 49, 52). At the
beginning of his response Philo declares, "But I will thoroughly examine the truth" (§ 75).

These concluding remarks are reminiscent of Plu. *Mor.* 963F, where,
after talking about the affliction of the faculty of reason in rabid
dogs as indicative of the existence of that faculty in animals (cf.
the last part of § 69 and the first part of § 70), it is concluded,
"A man, I say, either must be disregarding the evidence or, if he
does take note of the conclusion to which it leads, must be quarrel-
ing with truth" (φιλονεικεῖν πρὸς τὴν ἀλήθειαν).

72   *Lysimachus . . . Philo . . . Alexander, our nephew.* On the identity of the characters and their relation, see Intro., pp. 25-28 and
the explanation given in the translation notes.

These are the matters. In this section, which marks the beginning
of the transitory dialogue, and in the next two, Lysimachus is presented as the reader of Alexander's discourse. He must have prevented
Philo from reading it (see comment on § 9 and App. III, p. 268 for
parallel passages in Plato's *Phdr.*).

73   *time is longer than life* (μακρότερον ἡλικίας χρόνον)! Possibly a
proverbial saying; i.e., perhaps, in view of the brevity of life,
there are more important things to busy oneself with.

I was nurtured with such instruction (παιδείας). No doubt
in the schools of Philo's day the Aesopian fables were used for
their educational value in moral instruction. The Homeric epics
were set aside by Demetrius of Phalerum, a pupil of Theophrastus
and head of the famed library at Alexandria, who collected the
Aesopian fables in *ca.* 300 B.C., introduced them into the written
literature of the Greeks, and brought them to be used as a favorite
textbook (see W. Schmid and O. Stählin, *Geschichte der griechischen*

Litteratur, I, pt. 1, Handbuch der klassischen Altertumswissenschaft, VII [Munich, 1929], 663-683). Cf. Plutarch's *Whether Land or Sea Animals are Cleverer* (*De sollertia animalium, Mor.* 959A-985C), much of which is presented as a school exercise (see especially 960B, 963B, 965C-E, 975D, etc.).

bacchanals and corybants (οἱ βακχευόμενοι καὶ κορυβαντιῶντες, cf. Heres 69; Vita Cont 12). Philo continues to follow the pattern of the Platonic *Phaedrus* (cf. 234D in App. III, p. 269). Originally, the bacchanals were drunken revelers, orgies, or votaries of Bacchus or Dionysus who took part in the *Bacchanalia*, a festival in honor of Bacchus. The excessive drinking with the accompanying wild music and frenzied dancing carried away the participants despite and beyond themselves. So also the corybants, who originally were priests of Cybele or Rhea, mother of Zeus and other Olympian gods. She was worshipped in Phrygia with festivities resembling the *Bacchanalia* (see Euripides' *Bacchae*; also [G.] Wissowa, "Bacchanal," *PRE*, 4 [1896], cols. 1396-1397; [F.] Schwenn, "Korybanten," *PRE*, 22 [1922], cols. 1441-1446).

the tongue . . . the sovereign part of the soul (ἐκ τῆς γλώττης . . . ἐκ τοῦ τῆς ψυχῆς ἡγεμονικοῦ). On the sovereign part of the soul, see comment on § 10; see also comment on § 12, on the differences between λόγος προφορικός and λ. ἐνδιάθετος, the one formed in the mouth, the other derived from the sovereign, rational part of the soul or mind. Although the derogatory words on the enthusiastic bacchanals and corybants are addressed to Lysimachus as a remark on his reading of Alexander's discourse, they are also applied to those whose "self-proclaimed revelations are not consistent with the reports of researchers and interpreters." These words may thus be interpreted as being indirectly addressed to Alexander (Philo's unkindness to his opponent is seen also in §§ 3, 84, 93, where he accuses Alexander of sophistry; see also § 82).

The comparison between the unintelligible utterances resulting from the looseness of the tongue and the sensible utterances originating in the mind has much in common with 1 Cor 14:1-19. On the unintelligible sounds produced by musical birds and instruments, see com-

ment on §§ 98-99 (*SVF* II 734). Among other Philonic passages which
have a bearing on the subject, note Cher 42; 145-152; Migr 70-73;
Heres 249, 263-266; Vita Mos II 163-164; Spec Leg I 64-65; IV 48-52;
Vita Cont 40-47; etc. See J. Behm, "γλῶσσα," *Theologisches Wörterbuch zum Neuen Testament*, I (Stuttgart, 1949), 719-726, especially
722-723, on the phenomenon of glossolalia in Greek religion.

74   *The affection of a father or of a mother for their children is un-
      equaled.* Cf. Plutarch's *On Affection for Offspring (De amore prolis,
      Mor.* 493A-497E*)*, the purpose of which is to show that human affection
      for offspring is unequaled by that of beasts for their young. Philo's
      point, however, is that in recounting experiences to their children
      parents add "nouns and verbs" (cf. Migr 49; Congr 53) to the original
      experiences. By appending this remark to Lysimachus' presentation of
      Alexander's discourse and by commending him for not adding or omit-
      ting anything, Philo seems to be saying that Alexander's thoughts
      have been faithfully presented, "much as the author himself would
      have presented them."

      from the interpreter's point of view (κατὰ τὸν ἑρμηνέα), i.e. from
      Philo's point of view, or according to him. On Philo calling himself
      ἑρμηνεύς, see comment on § 7.

      much as the author himself would have presented it. See comment
      above and on §§ 1, 8. Cf. the Platonic parallel in the *Phdr.*, App.
      III, p. 270.

75   *the young man* (ὁ νήπιος), i.e. Alexander. See comment on § 8.

      I am not persuaded. Cf. Heraclitus' statement that it is the mark
      of a fool to be excited at every new theory (Diels *VS* 22B87) and the
      Stoic paradox that the wise man is free, he cannot be compelled to
      err (Quod Omn 58-61 [*SVF* III 362-363]).

      I will thoroughly examine the truth (ἀλήθειαν). See comment on
      §§ 3 and 10.

      I will make it known to everyone (μηνύσω). See comment on § 7.

      what our nephew has already written. The same relation is stated

in § 1 (cf. § 72; see note). On the identity of the characters and
the question of authorship, see Intro., pp. 25-30.

76   *love for learning and hunger for truth* (παιδείας . . . ἀληθείας).
Cf. with the attitude shown by and towards Lysimachus in §§ 2 and 5.

    to teach (μύειν, as in Frag. 3, App. II, p. 263).

77   *that . . . we do not sin against the sacred mind* (ἵνα μὴ ἁμαρτήσωμεν
εἰς ἁγνὴν διάνοιαν, i.e. against nature's right reasoning--ὁ τῆς
φύσεως ὀρθὸς λόγος, see Spec Leg II 29-30; cf. 170; Heres 110; Somn II
251). This section (*SVF* II 731) marks the beginning of Philo's
formal response to Alexander. The clearing of conscience here and
at the end of the response may be compared with the clearing of conscience at the end of Socrates' response to Lysias' discourse in Pl.
Phdr. 242C-243E (see comment on § 100). Primarily, §§ 77-81 (*SVF*
II 731-732) constitute Philo's refutation of §§ 17-22 of Alexander's
discourse.

    the spiders . . . the bees. See §§ 17-19, on the spider's craft,
and § 20, on the industry of bees. These accomplishments are carried
out neither by skill in art nor by innate reason, but, as Philo insists in the next section (*SVF* II 732), by nature. Cic. *ND.* ii. 123
likewise has that the spider's craft or cunning is bestowed by nature.
In talking about the household management of ants and bees in §§ 91-
92 (*SVF* II 733), Philo digresses to talk about the industry of bees,
which he again ascribes to nature (see comment).

    knowledge which is the basis of the arts (ἣν [i.e. ἐπιστήμη]
ἀρχὴν τεχνῶν εἶναι; cf. Congr 140, ἐπιστήμη γὰρ πλέον ἐστὶ τέχνης).
The dependence of art on knowledge is repeatedly emphasized by Plato,
for whom the first characteristic of any art is knowledge based on
understanding of the real nature of the subject or object. His low
estimate of the plastic arts is seen in his theory of art as imitation. Art is the copy of an object of the senses, and this again is
only a copy of an Idea. Hence a work of art is only a copy of a
copy. Plato sums up his concept in *Grg.* 465A, where no ἄλογον
πρᾶγμα deserves to be called a τέχνη. The same view is emphasized
throughout one of his short dialogues, *Ion* (530A-542B).

    Philo follows the conventional Platonic understanding of the rela-

tion of art to knowledge and, like the Stoics, likens this relation
to that of sense and mind (Congr 139-145 [*SVF* II 95; cf. 93-94; I 68,
73; etc.]). Works of art are only copies of the works of nature and
are not φύσεις themselves (Plant 110; Ebr 90 [*SVF* III 301]; Migr 31,
167; Fuga 170; Mut 260; Somn II 60; Spec Leg II 159, 161). Although
he subscribes to the Platonic low estimate of the arts (Gig 58-59),
he generally shows a positive attitude toward them. His keen appre-
ciation of art is aptly discussed by E. R. Goodenough, *By Light, Light:
The Mystic Gospel of Hellenistic Judaism* (New Haven, 1935), pp. 256-257.

78   *bees . . . spiders.* See comment above, § 77 (*SVF* II 731), of which
this section (*SVF* II 732) is the continuation.

Each . . . does it by its nature (ἕκαστον . . . τῆς ἑαυτοῦ φύσεως;
cf. τὴν ἑκάστου φύσιν in Leg All III 157; Sacr 30; Quod Det 177; Abr
189; Heres 312; see H. Leisegang, "φύσις 8. φύσις disting. μάθησις,
διδασκαλία, ἄσκησις, μελέτη, τέχνη," *Indices*, II [Berlin, 1930], 843;
also the idea of being or doing something "by nature" [φύσει] in Ro 2:14; Gal
2:15; 4:8; Eph 2:3; also ἐκ φύσεως in Ro 2:27 and κατὰ φύσεως in Ro 11:21, 24).
Time and again Philo attacks the claim for reason among the brutes
with that time-worn argument that their sagacity is instinctive, not
intelligent (see comment on § 92 [*SVF* II 733]).

look at the trees. Philo's basically Stoic analogy is poor, since
plants are soulless (ἄψυχα) and animals are endowed with soul (ἔμψυχα).
Such commonplace Stoic explanations of animal characteristics (see
comment below, on the vine, § 79 [*SVF* II 732], and on other plants,
§§ 94-95 [*SVF* II 730])seem to have further irritated their opponents.
Plutarch complains that the intelligence of certain animals is to be
compared with that of other animals and not with soulless beings
(*Mor.* 992D; cf. 960C-963A [*SVF* III Ant. 47]). Porphyry in *Abst.*
iii. 19 (*SVF* I 197) likewise objects: "Comparing plants with animals
does violence to the order of things: for the latter are naturally
sensitive . . . but the former are entirely destitute of sensation."

According to Philo, all that distinguishes plants from animals is
their soulless condition; this is their natural constitution of
being ἄψυχα. He does not classify plants with animate beings; he
places them among inanimates, making this distinction: inanimates
have a cohesive principle described as "habit" (ἕξις), in addition

to which, plants have an inner principle of growth described as
"nature" (φύσις). The basic characteristic of the soulless is lack
of sense perception. Animals differ from plants in that they are
endowed with soul (ψυχή); this is their natural constitution of being ἔμψυχα. The basic characteristic of the soul is sense perception
with its dependent mental presentation and resultant impulse (φαντασία
καὶ ὁρμή or, the first movement of the soul, τῆς ψυχῆς πρώτην κίνησιν,
Op 62, 66-67 [SVF II 722, 745], 73 [SVF III 372]; Leg All I 30 [SVF
II 844]; II 22-23; Quod Deus 35-45 [SVF II 458]; Plant 13; Heres 137;
Somn I 136). And although animals and man are endowed with soul, the
former are ἄλογα and the latter λογικά (immortal, heavenly, or
divine beings, such as stars, constitute the better part of the latter group). Man is a rational creature by nature; this is his inherent, natural characteristic even when in the seminal stage (λογικὴ
φύσις, Op 150; λογικὴ ἕξις, Leg All I 10; see comment on § 96 [SVF
II 834]). The souls of animals differ in that they are devoid of
the eighth, rational, or sovereign part of the soul, equated with
the faculty/fount of reason or the divine spirit (πνεῦμα) in man
(cf. the numerous references in J. Leisegang, "φύσις. 1. vis vitalis,
a qua distinguuntur ἕξις, ψυχή, etc.," Indices, II [Berlin, 1930],
836-837; see comment on the ἡγεμονικόν, § 10; on man made after the
divine image, § 16; and on the higher nature of man, § 100). For a
schematic discussion of Philo's hierarchic order in nature, see U.
Früchtel, Die kosmologischen Vorstellungen bei Philo von Alexandrien.
Ein Beitrag zur Geschichte der Genesisexegese, Arbeiten zur Literatur
und Geschichte des hellenistischen Judentums, II (Leiden, 1968), 41-
45.

The natural characteristics of creatures endowed with soul, as cited
by Philo, reflect Aristotelian criteria (cf. de An. 413b2, 427b15-16,
433b28-29--although Aristotle classifies plants with animate beings,
ibid. 410b22-23, 413b7-8). However, Philo's distinctions between
animals and plants on the one hand and animals and humans on the
other are basically Stoic (cf. SVF I 515-526; II 708-772; see E. V.
Arnold, Roman Stoicism [London, 1911], pp. 186-189, 244-247; M.
Pohlenz, Die Stoa. Geschichte einer geistigen Bewegung [Göttingen,
1964], I, 81-93; II, 48-53 and his "Tierische und menschliche

Intellegenz bei Poseidonios," *Hermes*, 76 [1941], 1-13; A. A. Long, *Hellenistic Philosophy* [London, 1974], pp. 170-174; and the particularly interesting study by A. Dyroff, "Zur stoischen Tierpsychologie," *Blätter für das bayerische Gymnasialschulwesen*, 33 [1897], 399-404; 34 [1898], 416-430.

Due to their low esteem for the souls of animals, the Stoics at times seem to have made little or no distinction between animals and plants. In Cic. *ND* ii. 160 (*SVF* II 1154), a passage attributed to Chrysippus, the pig's soul is said to have been given to serve as salt, to keep it from rotting. Philo says the same about fish in Op 66 (*SVF* II 722). Misconceptions that led to such an esteem for the souls of animals arose probably from Aristotle's separate classification of creatures intermediate between animals and plants (*HA* 588b4-23; cf. *PA* 681a10-682a2). Aristotle's observations were reiterated with less certainty by later naturalists, e.g. Pliny: "For my part I hold the view that even those creatures which have not got the nature of either animals or plants, but some third nature derived from both, possess sense-perception--I mean jelly-fish and sponges" (*NH* ix. 146). But elsewhere, perhaps following his sources more readily, he states: "It must be agreed that creatures enclosed in a flinty shell, such as oysters, have no senses. Many have the same nature as a bush, for instance the sea-cucumber, the sea-lung, the starfish" (154; cf. 183).

no skill in the arts (ἀτεχνία). See comment above, § 77 (*SVF* II 731).

79   *Have you not seen the vine . . . ?* Cf. Mut 162 and the resumption of analogy between plants and animals in §§ 94-95 (*SVF* II 730). This elaborate description of the vine bears a resemblance to that of Cicero in *Sen*. 53 (cf. *Leg*. i. 45 [*SVF* III 312]; *ND* ii.35). The resemblance is such as to warrant the conclusion that the two passages are influenced by a common source. See S. O. Dickerman, "Some Stock Illustrations of Animal Intelligence in Greek Psychology," *Transactions and Proceedings of the American Philological Association*, 42 (1911), 125-130.

Philo seems to be addressing himself to the arguments for prenatal foresight and maternal care shown by animals toward their young in § 22 (cf. §§ 34-35). Among other Stoic responses to such Academic arguments, see Cic. *ND* ii. 129 and Sen. *Dial.* vi. 7.2.

through its . . . nature (ἑαυτῆς φύσει). See comment above, on § 78 (*SVF* II 732) and § 92 (*SVF* II 733). It is interesting to note that here Philo talks about the growth of plants, in § 81 he refers to the growth of animals, and in § 96 (*SVF* II 834) he discusses human growth.

80   *one analogy*, i.e. the analogy of the vine in § 79 (*SVF* II 732). Philo, however, returns to the analogy between plants and animals in §§ 94-95 (*SVF* II 730).

invisible natures and all orderly and artful things are wrought by rational beings (τὰς ἀόρατας φύσεις . . . διὰ τῶν λογικῶν γίνον-ται; cf. Plant 41; Heres 75, 233; etc.). On rational beings, see comment on § 52, and above, on § 78 (*SVF* II 732).

those that are soulless (ἄψυχα). See comment on § 78 (*SVF* II 732).

animals . . . cannot do anything through foresighted care and thinking (προμηθείᾳ καὶ λογισμῷ). See comment on §§ 22, 30, 34 (cf. §§ 42-43). Philo insists in §§ 81, 92, and 97 (*SVF* II 732-733) that animals do nothing with foresight.

the peculiarity of their design . . . the peculiar designs of nature (φύσεως μὲν ἰδίᾳ κατασκευῇ; cf. D. L. vii. 107 [*SVF* III 493]: αἱ κατὰ φύσιν κατασκευαί). The sense of these lines and that of the preceding sections may be summed up as follows: Such animals as bees and ants, scorpions and serpents, or birds and bulls appear to act rationally, but their activities are clearly governed by design. And it is not to be supposed that they are reasoning beings. They attain their ends naturally or instinctively. On the role of nature, see comment on §§ 68 and 92 (*SVF* II 733); cf. §§ 86 and 97.

partridges. See § 35.

kites . . . clams . . . the scorpion. These creatures are not cited in Alexander's discourse. On the scorpion, see comment on snakes, below.

bulls. See § 51.

deliberately . . . by its own choice (λογισμοῦ . . . τῆς ἑαυτοῦ
προαιρέσεως). Philo's views on this broad subject may be summed up as
follows: Man is the only being upon this earth who is conscious of his
ends (Conf 178). He alone among mortal creatures has "the ability to
reason of his own motion" (Op 149). Free will, God's "most peculiar
possession," bestowed upon man, makes him "accountable for what he
does wrong with intent and praised when he acts rightly of his own
will," since his actions "stem from a volitional and self-determining
mind, whose activities for the most part rest on deliberate choice "
(προαιρετικαῖς, Quod Deus 47; cf. 49-50; Quod Det 11; Gig 20-21; Somn
II 174). Certainly "involuntary acts done in ignorance do not count
as sin" (Leg All I 35; cf. III 210 [*SVF* III 512]; Quod Deus 90; Agr
179; Fuga 65-76; Spec Leg III 120-123). As for "plants and animals,
neither praise is due if they fare well, nor blame if they fare ill,
since their movements and changes in either direction come to them
from no deliberate choice or volition of their own" (ἀπροαιρέτους,
Quod Deus 48). "Unreasoning natures, because as they have no gift of
understanding, they are also not guilty of wrongdoing willed freely
as a result of deliberate reflection" (λογισμοῦ, Conf 177). And the
enmity of wild beasts, unlike that of men, is actuated by natural
antipathy without deliberation (Praem 85). Note also his remarks
on the unaccountability of children and unreasoning natures in §§ 96
(*SVF* II 834)-97 and the broadly developed theme of free will under
divine providence in Provid I-II. Philo's views on free will, deliberation, voluntary and involuntary acts, and responsibility for virtue
and vice are reminiscent of Arist. *EN* 1109b30-1115a2, which became
central in Stoic ethical thought. See H. A. Wolfson, "Philo on Free
Will and the Historical Influence of His View," *Harvard Theological
Review*, 34 (1942), 131-169, reprinted in his *Philo*, I (rev. ed.;
Cambridge, Mass., 1968), 424-462; D. Winston, "Freedom and Determinism
in Philo of Alexandria," *Studia Philonica*, 3 (1974-1975), 47-70; cf.
J. M. Rist, *Stoic Philosophy* (London, 1969), pp. 219-232; A. A. Long,
"Freedom and Determinism in the Stoic Theory of Human Action," *Prob-
lems in Stoicism*, ed. A. A. Long (London, 1971), pp. 173-199; A. -J.
Voelke, *L'idée de volonté dans le stoicisme* (Paris, 1973).

the genera of snakes. With this example and that of the scorpion, Philo is perhaps alluding to what was said about the stingers of bees in § 21 and the snake fight in § 52. He talks about the natural viciousness of snakes and scorpions in Op 157; Vita Mos I 192; Spec Leg III 103; and Praem 90. Speaking of scorpions, Pliny remarks, that "their tail is always engaged in striking and does not stop practicing at any moment" (*NH* xi. 87; cf. Ael. *NA* vi. 20, 23).

providing of food. See §§ 22, 42-43 and comment on § 92 (*SVF* II 733).

healing of diseases. See §§ 38-40.

self-preservation (σωτηρία). See comment on § 33.

81   *the mouse . . . the brood of swallows.* Philo is responding to the arguments in § 22.

Which of the animals really has a sense of the beneficial (τὰ ὠφέλιμα)? Here he is refuting the claims made in §§ 43-44; cf. § 92 (*SVF* II 733).

foresight (προμηθεία). See comment on §§ 22, 30, 34 (cf. §§ 42-43). Elsewhere in Philo's response, the term occurs in §§ 80, 92, and 97 (*SVF* II 732-733).

the creatures' growth. See comment on the growth and development of plants and children (§§ 79, 96 [*SVF* II 732, 834]).

82   *great fraudulance and substantial wisdom* (. . . περιττῆς φρονήσεως) is repeatedly claimed for animals in §§ 30-46 (especially §§ 31, 34, 38, and 44). In §§ 82-84 Philo addresses himself to these arguments for wisdom shown by animals.

those who attest to such a wisdom do not realize that they themselves are ignorant, i.e. had they truly known what wisdom is, they would not have ascribed it to animals (cf. § 93 and Cic. *ND* ii. 147). Alexander himself is here accused of ignorance (cf. § 71).

childish compared with . . . the mature (νήπιος μέν ἐστι πρὸ . . . τῶν τελείων). See comment on § 8.

83   *the horse of Aristogiton.* See comment on § 40.

cheating is rational (ἡ δ' ἀπάτη λογική ἐστι). In negating the statements made in § 40, Philo makes a distinction between mere appearance, or illusion, and reality (cf. Leg All III 61).

was reminded. According to Aristotle, some animals have memory (μνήμη, *Metaph.* 980b25; cf. Pliny *NH* viii. 146; Plu. *Mor.* 960F, 992B [horses and steers remember]; Ael. *NA* iv. 35, 44; vi. 10(ii) [animals retain the memory of their experiences], 48 [a mare's memory]; vii. 23, 48; viii. 3, 32; x. 48; xi. 14). On memory's intermediate role between sense-perception and mental apprehension in Stoic thought, see E. V. Arnold, *Roman Stoicism* (London, 1911), pp. 134-135.

84   *hounds track by making use of the fifth mode of syllogism* (τῷ πέμπτῳ τρόπῳ). The argument in §§ 45-46 (*SVF* II 726) belonged to Chrysippean dialectic (see comment). It may be inferred from the two passages in Sextus Empericus (*P.* i. 69 and *M.* viii. 270 [*SVF* II 727]) that the Academic opponents of the Stoics used this Chrysippean example against the Stoics themselves to show that the dog (a) makes use of a complex syllogism and (b) recognizes the scent as a sign of the wild animal.

those who have no sense of philosophy are "those more accustomed to the plausibility and sophistry of matters than to the discipline of examining the truth." Moreover, they are "those who are persuaded to prove their knowledge to be true by inventions" and have no knowledge of "wisdom" (§ 93). On Philo's accusing Alexander of sophistry, see comment on § 3; note the references to the contrast between the plausible arguments of the sophists and the true teachings of the philosophers.

not even in dreams. For similar expressions, see Quod Det 35; Agr 43, etc.

85   *there are some decent and good qualities which are applicable to animals . . . their courage* (ἀνδρεία). In §§ 85 (*SVF* II 726)-89 Philo responds to the arguments for the virtue of courage exhibited by certain animals (§§ 51-59). He admits that animals surpass man in bodily endowments or "virtues of the body" (see comment on § 51, also for his definition of courage). He agrees that animals possess strength and vigor (*ibid.*, see comment for references; cf. § 88). In Abr 266, e.g., he grants them a multiplicity of bodily endow-

194

ments, in all of which they surpass man: in strength, such as the bull
and the lion; in sight, such as the hawk and the eagle; in hearing,
such as the ass; in scent, such as the hound; and in health (cf. Agr
115; Somn I 49; etc.). On similar admissions by the Stoics, see G.
Tappe, *De Philonis libro qui inscribitur* 'Αλέξανδρος (Diss. Göttingen,
1912), p. 18, n. 2, where he refers to S. E. *M.* xi. 99 (*SVF* III 38);
cf. Origenes *Cels.* iv. 78 (*SVF* II 1173).

surely animals have no share of reasoning ability (λογικῆς δὲ ἕξεως
ἐξ ἀνάγκης ἀμέτοχον ἐστι). These words constitute the most direct
response to the thesis of Alexander's discourse (§§ 10-71). On the
philosophical background of the treatise and Philo's position, see
Intro., pp. 49-53.

*reasoning ability extends itself to a multiplicity of abstract
concepts or immaterial things* (ἡ λογικὴ ἕξις μέχρι πολλῶν τῶν οὐκ
ὄντων τείνας [ἑαυτήν]). See comment on §§ 13, 66. The examples that
follow are much discussed subjects in Philo. Note especially the
mind's perception of God, as, e.g., in his allegorical interpretations
of Gen 18:1-15 (Abr 107-132; Quaes Gen IV I-19; see comment on § 10).
But in commenting on Balaam's donkey (Num 22:21-35), Philo writes: "It
was evidently a divine vision , whose haunting presence had for a considerable time been seen by the terrified animal . . . . For the unreasoning animal showed a superior power of sight to him who claimed
to see not only the world but the world's Maker" (Vita Mos I 272).
Cf. Plu. *Mor.* 992E; Origenes *Cels.* iv. 81 (*SVF* III 368).

86　*The horse . . . neighs.* Philo makes the point clearer in Somn I
108, where he states: "For as neighing is peculiar to a horse, and
barking to a dog, and lowing to a cow, and roaring to a lion, so is
speech (λέγειν) and reason itself (αὐτὸς ὁ λόγος) to a man." Horses
are called "neighing animals" (χρεμετίζοντες) in Agr 72. In Op 149-
150 Philo asserts that the names the first man gave to animals reveal the traits of the creatures who bore them.

the deer and the hare are timid. See comment on §§ 68-69.

the fox is crafty. See comment on § 46.

others are attached and devoted to their offspring. See comment on
§ 22 (cf. §§ 34-35, 79 [*SVF* II 732]).

insight. The Gr. equivalent here is possibly the word εὔδησις, which
Philo uses in Plant 36 and Migr 42 in connection with knowledge
(γνῶσις) and mental apprehension (κατάληψις or σύνεσις).

designed and fashioned by nature. See comment on §§ 68 and 80.

parts of body and soul. See comment on §§ 10-11.

*to each constituent its proper function toward its own perfection
and toward that of the whole being* (τελειότης). A teleological order
in nature is emphasized--whether directly or indirectly--throughout
Provid I-II. The interdependence of the parts of the body and their
teleological purpose, like those of the cosmos or the universe, receive ample consideration in Philo's works (see comment on § 61; cf.
the notion in Cic. *ND* ii. 35, that all things strive for perfection
and that universal nature alone can attain it; and 1 Cor 12:12-31,
where the parts of the church, symbolized by the parts of the body,
work together for the good of the whole). On the pre-Aristotelian
teleological concepts in Greek philosophy, see W. Theiler, *Zur
Geschichte der teleologischen Naturbetrachtung bis auf Aristoteles*
(Zürich, 1925). Of particular significance to a teleological understanding of the cosmos are Aristotle's own contributions in his
zoological works, especially in the *De partibus animalium.* The
principle of teleology was anthropocentrically interpreted by the
Stoics, especially by Chrysippus--as was also done by Philo (*SVF* II
1152-1167). Chrysippus' teleological explanations were criticized
by Carneades (Porph. *Abst.* iii. 20 [*SVF* II 1152]; see O. Rieth, "Über
das Telos der Stoiker," *Hermes,* 69 [1934], 13-45; A. A. Long, "Carneades and the Stoic Telos," *Phronesis,* 12 [1967], 59-90). For a historical survey of teleology, see A. S. Pease, "Caeli Enarrant," *Harvard Theological Review,* 34 (1941), 163-200.

87   *lions . . . no longer feel the slightest ambition for power,* lit.,
as pointed in the translation notes, *love of honor* or *glory* (φιλοδοξία). "It was said that . . ." may be supplied for better sense,

196

since Philo is refuting the claims made in § 55 (see comment). He
seems to confuse the arguments for the αἰδώς of lions (§ 55) with
those for the φιλοδοξία or φιλοτιμία of horses (§ 56).

real power does not subject itself to ridicule (ἰσχύς opp. ὀλιγωρία).
Philo goes on to equate power with leadership (ἄρχειν), which he compares and contrasts with ridicule or contempt. The flow of his argument here is similar to that in § 91; i.e., if indeed animals possess
power, they should also have the ability to lead without subjecting
themselves to ridicule; but, since they become followers instead of
leaders and are laughed at, it cannot be said that they possess power
in the true sense of the word. With these remarks on power and leadership, Philo seems to be answering the arguments for the exercise
of leadership by animals in §§ 64-65. Likewise, in Agr 47 he argues
that too good and gentle leaders are subject to ridicule (ὀλιγωρία),
and because their kindness is weakness, they are sometimes forced to
abdicate. He implies that powerful leaders do not allow themselves
to be ridiculed and that they are unyielding. The argument then
clearly appears to be that if animals are truly powerful, they will
neither be laughed at nor will they follow or yield to "whoever
trains them."

Though not uncommon in Philo, a discrepancy is found in Quod Omn
40, where lions are said to intimidate their owners, just as slaves
often rule their masters. Elaborating further in this work on the
Stoic paradox that every wise man is free and every fool is a slave--
like the irrational animals created for servitude to man, Philo
adds: "The freedom of the wise like all other human good gifts may
be seen exemplified also in the irrational animals. Thus cocks are
wont to fight with such intrepidity that rather than yield and with-
draw, though outdone in strength yet not outdone in courage, they
continue fighting until they die" (Quod Omn 131; cf. Leg All III 89).

88   *the virtues proper to them* (τὰς ἰδίας ἀρετάς). Much of what Alexander ascribes to animals as virtues, Philo classifies under "the
bodily good things" or "virtues of the body," which he distinguishes
from the rational "virtues of the soul" and does not deny to animals

(see § 85 and comment on §§ 30, 51). In Philo's thought, as in
Stoicism, irrational beings are devoid not only of reason but also
of its dependent virtues (Op 73 [*SVF* III 372; cf. 245-254; Sen. *Dial.*
iii. 3.4]; Sacr 115 [*SVF* III 505]; Conf 177; cf. Plant 43). He
equates the progress in the principles of virtue with the power to
understand and know and adds that this is of no advantage unless
carried to its full stature (Migr 53-55; cf. Heres 310-311).

peacocks . . . *yielding to the cravings of their natural virtue*,
i.e. their bodily or innate virtue of attracting the hens. Philo
is definitely drawing upon Chrysippus' premise and illustration
(see Plu. *Mor.* 1044C-D [*SVF* II 1163]; cf. Cic. *Fin.* iii. 18 [*SVF* II
1165]).

89   *The Antiochian elephant named Ajax.*   See comment on § 59.

90   *all that perform marvelously in theatrical shows.*   Note that of the
four animals mentioned in this response to the arguments set forth
in §§ 23-28, dogs and onagers are not mentioned in the said passages.
The means of taming and training wild animals were certainly known
to Philo (see § 23 and comment on § 26). After all, African animals
of all sorts were tamed and trained at Alexandria, from where they
were dispatched to various parts of the Roman Empire, especially to
Rome (G. Jennison, *Animals for Show and Pleasure in Ancient Rome*
[Manchester, 1937], pp. 167-168).

Every creature that has partaken of soul is subject to suffering
(πάντα γὰρ ὅσα ψυχῆς μεμοίραται εἰς βασάνους/κολάσεις βαλλόμενα).
This Platonic view emphasized in the *Phaedo* occurs quite frequently
in Philo. "Mortal existence is not an existence but a period of
time full of suffering" (κολάσεις, Op 156). In Heres 267-274 he
discusses the the descent of the soul into the body and the passions
which come about as a result of this descent (see also Sacr 111
[*SVF* III 609]; Vita Mos I 11; etc.; on the souls of animals, see
comment on § 78 [*SVF* II 732]). Moreover, the fairly common notion
of learning through suffering is suggested in the passage. On
this subject, see J. Coste, "Notion greque et notion biblique de la
souffrance éducatrice," *Recherches de science religieuse*, 43 (1955),

488-523, where the relationship of Hb 5:8 to Philo and the larger
tradition of Greek *paideia* is stressed. See also L. K. K. Dey, *The
Intermediary World and Patterns of Perfection in Philo and Hebrews*,
SBL Dissertation Series, XXV (Missoula, Montana, 1975), 222-225.

91   *ants and bees*. The claims made in §§ 41-42, on the ant, and in
§§ 61 and 65, on bees (cf. §§ 20-21), are refuted in this section
and in the next (*SVF* II 733). With §§ 93-96 (*SVF* II 730, 834),
these sections constitute Philo's response to the arguments for
justice among animals (§§ 60 [*SVF* II 728]-65).

  household management . . . state management (οἰκονομία καὶ πολιτεία).
The two virtues in conjuction, or as one and the same virtue, are
dealt with elsewhere in Philo (see comment on §§ 41, 65). On the
Platonic and Aristotelian backgrounds to this equation, see F. H.
Colson's note to Jos 38 (*SVF* III 323) in LCL VI, 600; on the views
of Chrysippus, see Plu. *Mor.* 1050A.

92   *the ant . . . the bee . . . the animals' foresight* (προμηθεία).
See comment on §§ 20-21, 42, and above. Philo here digresses from
the political virtue or social justice, a subject already covered
in §§ 77-78 (*SVF* II 731-732; see comment). The industry of ants
and bees is attributed to the universal workings of providence in
Provid I 51-53 (cf. Sen. *Ep.* cxxi. 21-23; *Cl.* i. 19. 3-4; Origenes
Cels. iv. 81-85 [*SVF* III 368]). The industry of the ant is disposed
of in a somewhat different way in Cic. *ND* ii. 156-157 and Sen. *Dial.*
ix. 12.3. In §§ 80-81 (*SVF* II 732) and 97, Philo refutes the claims
made for the foresight of other animals. In all of these instances
he repeats that they do nothing with foresight.

  nature that manages all (παντ' οἰκονομεῖ ἡ φύσις; cf. Jos 192:
ἴσως ἡ φύσις οἰκονομεῖ τι βέλτιον). The Stoic ascription to nature
concern for all creatures has been noted in the comment on § 68 (cf.
§§ 69, 80 [*SVF* II 732], 86, 97), where also reference is made to
Philo's ascription to nature the granting of defensive mechanisms
to animals. Philo goes so far as to say that the whole universe
is administered by the purpose and will of nature (Op 3 [*SVF* III
336]; cf. 67 [*SVF* II 745], 130, 133, etc.; on his interchangeable

use of the terms θεός and φύσις, see comment on § 11). Nature here,
as often in this treatise and elsewhere in Philo, may be identified
with the divine agency in things (see comment on § 11; cf. with the
role of "divine reason"--λόγος θεῖος, Op 36; Cher 36; etc.). Since
the inherent properties of all things in the universe--animate and
inanimate--are maintained by nature, all their peculiar characteristics are termed "natural." The λόγος supervenes not as the opponent
of these "natural" characteristics but as the artificer or craftsman in charge of them (see U. Früchtel, *Die kosmologischen Vorstellungen bei Philo von Alexandrien. Ein Beitrag zur Geschichte der
Genesisexegese*, Arbeiten zur Literatur und Geschichte des hellenis-
tichen Judentums, II [Leiden, 1968], 57-61; cf. *SVF* II 1132-1140).

Other than attributing the animals' exploitation of foodstuff to
the workings of nature, Philo says nothing here about this "natural"
characteristic of animals. Whereas he (or perhaps his source) fails
to elaborate on the discerning faculties of animals, respondents to
the anti-Stoic attacks of the Academy (such as Cicero's main source
for Book III of *Fin.*, which is generally agreed to be an orthodox
Stoic handbook of the time of Chrysippus' successor Diogenes or his
pupil Antipater, as also his main source for Book II of *ND*, possibly
Posidonius) seem to have replied that nature has bestowed upon animals both sensation and desire (*sensum et appetitum*), "the one to
arouse in them the impulse to appropriate their natural foods, the
other to distinguish things harmful from things wholesome" (Cic. *ND*
ii. 122). But "for man she amplified her gift by the addition of
reason, whereby the appetites might be controlled" (34; on earlier
Stoic doctrine extant in D. L. vii. 85-86 [*SVF* III 178; cf. 179-196],
see comment on § 44). Philo restates the "thoughtless," mechanical
workings of animals in § 97 (cf. §§ 77-80 [*SVF* II 731-732]).

93   *the pinna and the pinna guard . . . demonstrate partnership*
(κοινωνία). Here and in the next two sections (*SVF* II 730) Philo
responds to the arguments propounded in § 60 (see comment).

*those who substantiate their assertions of truth by displaying
fanciful inventions* (εὑρέσεις . . . μηχαναί). With these and similar disparaging remarks Philo alludes to the sophists and thus

accuses Alexander of sophistry (see comment on § 3).

had they known wisdom (σοφία). Philo seems to be ridiculing the sophists for their attribution of wisdom to animals (cf. § 84; see comment on § 30). That the sophists themselves lacked wisdom is repeatedly emphasized by Philo (Cher 9-10; Quod Det 38-44; Post 101; Sobr 9; Migr 85, 171; etc.). See also § 82; cf. Cic. *ND* ii. 147.

94 *the trees as well as the bushes . . . have not partaken of soul* (ἄψυχα, see comment on § 78 [*SVF* II 732]). These remarks against friendships and hostilities among animals continue in the next section (*SVF* II 730) as a refutation of the arguments in §§ 60-61 (*SVF* II 728, see comment). Friendship or endearment (οἰκείωσις) is regularly contrasted with hostility or aversion (ἀλλοτρίωσις, e.g. Conf. 82). These, in Stoic philosophy, are the bases of consciousness, a rudimentary form of which is carried out by nature in plants and animals for their self-preservation (see comment on § 33). Of the many examples of hostilities and friendships among animals, see Pliny *NH* x. 203-208 (cf. Ael. *NA* i. 32; iv. 5; v. 48; vi. 32; etc.).

95 *the vine shuns the cabbage.* So also Thphr. *HP* iv. 16.6 (cf. *CP* ii. 18.4); Cic. *ND* ii. 120 (cf. *Sen.* 52); and Pliny *NH* xvii. 239-240 (cf. xiv. 58). Theophrastus, followed by Pliny, attributes the observation to Androcydes, an advisor on temperance to Alexander the Great (see S. O. Dickerman, "Some Stock Illustrations of Animal Intelligence in Greek Psychology," *Transactions and Proceedings of the American Philological Association,* 42 [1911], 127, n. 1; A. Hart's note to Thphr. *HP* iv. 16.6 [LCL I, 413]).

nature's supreme reasoning ([κατὰ]τὸν τῆς φύσεως ἀνωτάτω λόγον; cf. Somn I 128; Spec Leg III 207; and τὸν τῆς φύσεως ὀρθὸν λόγον in Op 143 [*SVF* III 337]; Abr 3-6; Quod Omn 62; Jos 28-31 [*SVF* III 323]; etc.). See comment on §§ 11 and 48.

the resemblance or likeness of images in animals (τὰς δὲ ὁμοιότητας καὶ εἰκόνας ἐν τοῖς ζῴοις). The thought of these lines, that the souls of animals are capable of receiving inferior imprints of images, was expressed in Alexander's discourse (see § 29 and comment).

Philo agrees that the power of receiving impressions and being the
subject of impulses is shared also by creatures without reason (Leg
All II 23 [*SVF* II 458]); but here we find his concession at its best.
Similar concessions are reiterated by Seneca: that animals do possess
properties which resemble human feelings, such as anger, confidence,
hope, fear; but they do not in a strict sense possess the same feelings as men (*Dial*. iii. 3. 5-6). Although similarities between the
behavior of animals and that of rational beings are easily seen, the
behavior of animals is no more than a μίμησις πρὸς τὰ λογικά, as it
is called by Origen   (*Cels*. iv. 81 [*SVF* III 368]). On the gratitude
of animals, see Ael. *NA* iv. 44; vi. 44; viii. 3, 22; x. 48.

human souls (ἀνθρωπίνων ψυχῶν). See comment on §§ 10 and 78 (*SVF*
II 732).

96   *the stork . . . drones*. See comment on § 61. This section (*SVF*
II 834) has much in common with § 80 (*SVF* II 732), especially on
voluntary and involuntary actions and responsibility for right or
wrong, virtue or vice (see comment).

a little child. Philo's argument here is that a child is potentially
rational being (see comment below and on § 49, on the human semen as
containing a potential reasoning principle; cf. Cic. *Leg*. i. 26, 30
[*SVF* III 343]; Sen. *Ep*. cxxiv. 9; *Nat*. iii. 29.3; Epict. *Diss*. ii. 11.3,
on the inborn rudiments of reason in man). In his condemnation of in-
fanticide, he carries a similar argument even to unborn children as
potentially separate, rational creatures (Spec Leg III 8-9, 109-111,
117-118; Virt 137-138 [*SVF* II 759]). Although man is a rational being
by nature (see comment on § 78 [*SVF* II 732]), he becomes fully rational
during the first seven years of his life (Leg All I 10; cf. *SVF* I
149; II 83, 87, 764, 835). Throughout this time of immaturity, the
soul of a child is without part in either vice or virtue, good or
evil (Leg All II 53; Congr 81-82; Praem 62). In these passages, the
rise of reason is conceived of as spontaneous, brought from within
the biological realm at a certain level of development or maturity

(for a broader survey of the philonic passages on human development, see H. Schmidt, *Die Anthropologie Philons von Alexandreia* [Diss. Leipzig, 1933], pp. 38-42).

The association of child psychology with that of animals, in terms of unaccountability for their behavior, is a recurring theme in Aristotle's *Nichomachean Ethics* (see comment on § 80 [*SVF* II 732]). However, serious interest and due consideration of child behavior and human development is to be credited to the Stoics, especially to Chrysippus' successors: Diogenes or his pupil Antipater, who are considered to be the authorities relied on in Book III of Cicero's *De finibus* (Diogenes is referred to by name in iii. 33 and 49; see S. G. Pembroke, "Oikeiōsis," *Problems in Stoicism*, ed. A. A. Long [London, 1971], p. 120 and the studies on the sources of *Fin.* iii. cited on p. 143, n. 31; on later Stoic sources--especially Posidonian psychology--utilized in the 1st-2d cents. A. D., see A. Dyroff, "Zur stoischen Tierpsychologie," *Blätter für das bayerische Gymnasial-schulwesen*, 33 [1897], 399-404; 34 [1898], 416-430; M. Pohlenz, "Tierische und menschliche Intellegenz bei Poseidonius," *Hermes*, 76 [1941], 1-13).

the seeds of wisdom . . . *the seminal powers or reason* (τὰ τῆς σοφίας σπέρματα . . . λόγοι σπερματικοῦ). The last term, which is of Stoic usage, occurs also in § 20 and elsewhere in Philo (see comment on § 20), who also conceives of a potential reasoning principle contained in the human semen, (see comment on § 49 and above). The imagery of fire and wind in conjunction with the λόγος σπερματικός is significantly Stoic (cf. πῦρ τεχνικόν, *SVF* I 98, 120, 124, 126, 171, 504; II, 774, 1027, etc.). Fire in Philo symbolizes θεός or his δυνάμεις, and wind the impregnating πνεῦμα θεοῦ (cf. Gig 22-27; Vita Mos II 104; Quaes Ex II 28, 37, 47; see E. R. Goodenough, *By Light, Light: The Mystic Gospel of Hellenistic Judaism* [New Haven, 1935], pp. 41-42, 98, 214, 232).

the fount of reason . . . *the reasoning faculty* (λόγου πηγή . . . διάνοια). On the equation of the two, see comment on §§ 12 and 17.

97 *researchers into the history of animals* (οἱ τῶν περὶ τὰ ζῶα ἱστορίων ζητοῦντες). Aristotle's celebrated *Historia animalium* is almost certainly suggested by the Gr. equivalent. The "researchers" are "those men whose custom it is to be verbose in an unprincipled and obnoxious manner." They may thus be identified with the sophists whom Philo ridicules (see comment on § 3; cf. §§ 16, 61). On being equipped against the sophists, see Quod Det 38-44; Agr 161-162. In Conf 39 it is stated that one can refute "the plausible inventions of the sophists" only with God's help.

 virtues and vices (ἀρεταὶ καὶ κακίαι). See comment on §§ 30, 66, and 88.

 animals do nothing with foresight as a result of deliberate choice (προμηθείᾳ . . . προαιρέσει). In § 80 (*SVF* II 732) Philo had already emphasized that animals do nothing with foresight or by deliberate choice (see comment). In § 92 (*SVF* II 733) he repeats that their deeds are accomplished neither by foresight nor by thought.

 the primary design of nature (φύσεως μὲν ἀρχὴν κατασκευήν). On the peculiar designs of nature, see comment on § 80 (*SVF* II 732). The primary design here is said to be seed-procreation (cf. Quaes Gen III 48 [*SVF* II 740]; Arist. *HA* 589a3-4). Procreation, according to Philo and the Stoics, is the seventh and last of the subordinate powers of the soul (see comment on § 10).

98 *mental reasoning . . . the uttered* (λόγος ἐνδιάθετος . . . προφορικός). On the distinctions between the two terms and their use in Stoicism and Philo, see comment on § 12. In this section (*SVF* II 734) and the next, Philo addresses himself to the arguments in §§ 13-15.

 different kinds of utterances . . . articulated voice (ἔναρθρος φωνή). Since the Greek word λόγος expresses both reason and speech, it contributes to the belief that dumb animals (τὰ ἄλογα ζῶα) lack the faculty of reason (although the distinction between reason and speech is clearer in the Latin, still the words *ratio* and *oratio* appear to be naturally related). Philo explains the affinity of words or speech to reason after Pl. *Tim.* 75D-E: "Foods and drinks enter it [i.e. the mouth], perishable nourishment of a perishable body, but words [λόγοι] issue from it, undying laws of an undying soul, by means of which the life of reason [ὁ λογικὸς βίος] is guided" (Op 119; cf. Mt 12:34-35; 15:11, 17-18; Mk 7:20-23; Lk 6:45; Eph 4:29; Col 3:8; Jm 3:2-12). Like most Greek writers, Philo goes on to associate φωνή with λόγος, underlining the dependence of the former upon the latter (Quod Det 92; Conf 194; Somn I 29; etc.; in the last passage he emphasizes that only man is capable of producing articulated voice; cf. Leg All II 23 [*SVF* II 458]; Congr 17). Much of the same thought emerges in his differentiation between λόγος προφορικός and λ. ἐνδιάθετος (see comment on § 12, especially on the former's derivation from the latter) and in his understanding of the interrelation between sense-perception and mind (see comment on § 13). More of the same thought is seen in his treatment of sensible and insensible utterances and their loci of origin (see comment on § 73). A number of parallels to the thoughts expressed here and in the following section, especially on the distinctions between song birds and musical instruments, are found in his allegorization of Jubal (Post 103-111, on Gen 4:21; for more, see comment on § 99).

In Conf 6-8 Philo answers the criticism of skeptics who had said that the story of the confusion of tongues (Gen 11:1-9) was much the same as the fable that all animals originally understood each other's language but lost the privilege by presumption (cf. Jub. 3:28; J. *AJ* i. 41; Porph. *Abst.* iii. 3). As might be expected, he allegorizes the Gen 3 account of the serpent's emitting human voice (vss. 2-6) and applies it to the Epicurean advocacy of pleasure (Op 155-164; cf. Agr 96). In recounting Balaam's story, he deliberately omits the speaking of the ass (Num 22:28-30); he refers to the animal's superior power of sight (Vita Mos I 272; elsewhere, alluding to the same story, he refers to the creature's symbolism of the irrational life, Cher 32).

wind instruments. See comment on musical instruments, below.

99   *musical instruments*. In response to the arguments in §§ 13-15, Philo distinguishes between the articulated human voice and the twittering of birds, which he equates with the sounds produced by

musical instruments. But in Post 105-106 he admits that just as the
human voice surpasses the twittering of birds, song birds surpass
musical instruments (cf. Leg All II 75). In Provid I 43 he praises
the musician, not the instrument. His depreciation of musical instruments may be sensed in Fuga 22, where he speaks of drum beats as
"noises inarticulate and meaningless, inflicting blows on the soul
through the ears" (cf. 1 Cor 13:1). In Dec 32-35, commenting on Ex
19:16, he treats favorably the trumpet-like voice which announced the
commandments (cf. Migr 47, 52, on Ex 20:18, 22). He has it that the
voice was not God's actual voice, for He is not a man, but a miracu-
lous kind of speech created for the occasion (cf. Rv 1:10; 4:1).

listeners differ one from another. The point here made is that not
only sounds but also listeners differ one from another. The remark
may be applied derogatorily to those who make much of the twittering
of birds. In Dec 34 Philo explains how listeners receive different
impressions--depending, on the one hand, on their distance from the
source of voice and, on the other hand, on the depression of voice
(cf. Ebr 177, where he cites for example the different impressions
of tunes people receive at the theater). On his understanding of
how sound travels, see Heres 15 (cf. D. L. vii 158 [*SVF* II 872]).

all who are not endowed with euphonious speech are full of darkness
(σκότους δὲ πλήρεις εἰσίν, as in Leg All III 7; cf. λογικῶν εὐφωνία
in II 75). Behind this remark lies the notion that thoughts are in
darkness until speech brings them to light (Quod Det 126-131).
Speech is conveyance of meaning (Leg All III 120, including the
allegorical: Congr 172); it depends on right understanding (Post 87-
88) or divine knowledge (Quod Det 91; Quaes Gen IV 120). Philo
goes on to say that human speech is not an altogether reliable interpreter of the thoughts of the νοῦς (Migr 72, 78; cf. 84, 169, Quaes Ex
II 27, on Aaron's relation to Moses) or even of things (*ibid.* 12; Spec
Leg IV 60; cf. *SVF* II 136-165, especially 143, where Varro has that
young children and animals do not possess speech, but only "a sort of
speech," *L.* vi. 56).

100   *stop criticizing nature and committing sacrilege* (ἀσεβεῖν; cf.

Spec Leg II 170 and the other Philonic passages cited in the comment
on § 77). The clearing of conscience here and at the beginning of
Philo's formal response to Alexander (§ 77) is to be compared with
the clearing of conscience at the end of Socrates' response to
Lysias' discourse in Pl. *Phdr*. 242C-243E. On Philo's dependence on
the Platonic *Phaedrus* in his introductory and transitory dialogues
and in the structuring of Anim, see App. III, pp. 265-271. Note this
comparable statement from Plutarch's *Bruta animalia ratione uti:*
"Surely it is absurd for you to find fault with nature" (*Mor*. 988E);
and these words from the concluding lines of the treatise: "Is it
not a fearful act of violence to grant reason to creatures that have
no inherent knowledge of God?" (992E).

to grant equality to unequals is the height of injustice. This
line is quoted verbatim in the *Sacra parallela* (see App. II, p. 263)
and is somewhat paralleled in Dec 61, where Philo condemns those who
venerate the creatures instead of the Creator, those most unjust men
who give equal honor to those who are not equal (cf. Ro 1:23, 25). In
Post 81 he pronounces a curse on the man who treats as alike things
that are widely different and in Heres 253 the practice of "treating
things indifferent as indeed indifferent" he calls "food fit for eating." On his theory of justice and equality, see comment on § 60.

Philo here follows the Stoic doctrine that reason unites mankind
(see the excerpts under *de coniunctione hominum, SVF* III 340-348).
Since animals have no reason, they do not belong to this *civitas
communis deorum atque hominum;* therefore, there is no such thing as
a justice which can obtain between them and humans (see the Stoic
fragments under *Iuris communionem non pertinere ad bruta animalia,
SVF* III 367-376). The moral and juridical relationship between
man and animals appealed for in Alexander's discourse (see comment
on § 10) is rejected in these concluding remarks of Philo (see also
comment on § 63).

self-restraint (σωφροσύνη). The last sentence of the treatise constitutes Philo's response to Alexander's arguments for the virtue of
self-restraint among animals (§§ 47-50). On the thought that man alone

can restrain impulse with reason, see comment on §§ 47, 92 (*SVF* II 733).

those whom nature has endowed with the best part, i.e. man who is naturally endowed with reason (see comment on § 11).

APPENDICES

APPENDIX I

THE ARMENIAN TEXT

WITH AUCHER'S LATIN TRANSLATION

Reprinted from the *editio princeps*, edited and translated by J.
B. Aucher, *Philonis Judaei sermones tres hactenus inediti: I et II
de providentia et III de animalibus* (Venice, 1822), pp. 123-172. The
added section enumeration follows C. E. Richter's edition, *Philonis
Iudaei opera omnia. Textus editus ad fidem optimarum editionum*, VIII
(Leipzig, 1830), 101-148.

ՓԻԼՈՆԻ

ՅԱՂԱԳՍ ՐԱՆ ՈՒՆԵԼ ԵԻ

ԱՆԱՍՈՒՆ ԿԵՆԴԱՆԵԱՑԴ

(ՀԱՍ ԱՂԵՔՍԱՆԴՐԻ)

PHILONIS

DE RATIONE QUAM HABERE ETIAM
BRUTA ANIMALIA

(DICEBAT ALEXANDER)[1]

(**ՓԻԼՈՆ**ː)

     ՅիշեՍ զերիկեան բանսն, ով լիւսե֊
մաքոս, զորս աղեքսանդրոս եղբօրորդիս
մեր պատմեաց , յաղագս այնորիկ՝ թէ
ոչ միայն մարդիկդ , այլ եւ անասուն կեն֊
դանիք բանի բաժին ունին ։

     ԼԻԻՍԻՄԱՔՈՍ ː

     Քանզի ով պատուականդ փիլոն՝ յե
րից յայսցանէ , որք ՚ի նոսա Հայեցու֊
թիւնք բունաորագունացն աձեալ մի
տեղմուՇեամբ առ ասողն ։ Քանզի քե.

(**P**HILO.)

     Recordare hesterna verba, o Lysi- 1
mache, quae Alexander, noster ex fratre nepos, retulit; quasi non homines soli, verum etiam bruta animantia ratione
sint praedita.

     LYSIMACHUS.

     Equidem, o venerate Philo, unum 2
est istud trium illorum, quorum inspectiones convincentes efficiunt, ut auditor
iterum veniat ad eum, qui locuturus est.

---

1 Citaverat titulum praesentis quoque sermo-
nis inter opera Philonis nostri Eusebius Hist.
Eccl. L. II. C. 18. hoc tenore: ἔτι τε ὁ Ἀλέξανδρος:
ἢ περὶ᾿ τοῦ λόγον ἔχειν τὰ ἄλογα ζῶα . h. e.
Adhuc etiam Alexander, sive de eo quod rationem habeant bruta animalia . Quamvis enim totus sermo est dialogus inter Philonem , et Lysi-
machum, tota tamen quaestio est de sententia

Alexandri volentis animalia bruta etiam ratione
aliqua gaudere; quae sententia fusius exponitur,
tamquam scriptis mandata ab Alexandro , et ul-
timo loco funditus evertitur a Philone succinctis
verbis fortissimis, ut videbis suo loco . Quare et
S. Hier. de viris illustr. c. XI. sic notat inter
Philonis opera : De Alexandro dicente, quod propriam rationem muta animalia habeant .

ղի է միանգամայն և անէր։ Քանզի
ոչ անգէտ ես, զի դուստրն իմ և բան
նորա խոստովանութեան խոսմանն կին է ։
Լ՛ժ ասացելոցն նորոգուիցիճ, քանզի
Հերդ և տարտամ է այս, որ ոչ յոյժ
ընդելն է։ Վարձեալ և յաղագս ամենայ
բն մեկնութեանն, որ յայտնի երևուին
Հաստատութեան լոյս ձագէր, որ ազ
դեան լինէր․ որ օրինակ ինն շրջեալ
տաբբանայր՝ ոչ առ երևաս լսելիք ։

Փ Ի Լ Ո Ն ։

3 Մեծ զինատ խոստովանեալ բազձա
ցեալ լսել նոցա ։ Քանզի առ 'ի բարձու
սունմունւթիւն ոչինչ այլ այպէս օգնա
կան լինէ երևի, որպէս Հաստատունն
զինել զայն՝ զոր ազդեցուցանէ ուսու
ցոն։ Ո ի թէ արդարև ուսուցանել
Հանելրեացէ ոչ յայլոց Հոգոց պատո
դեալ անդրին ձգեցի։ Ո իարդ՝ ասա
ինձ, յայլ երա ձգի, զերբայր ոք ուրախ
առնել Հանգերձեալ․ և զայս առանց
ծախոց զբանիցն իրախանութիւն լսե
լեաց պատրաստատէ։ Լ՛ JL թերև յա
զագս այսորիկ, զի մի զիւրն ումեք և
կարծեցին, և Հաստար այսորիկ՝ որբ
'ի պատրաստէ ընկենուն զշնորՀն առա
քէ․ յայս սակս ընդ կրուկ նաՀանջիս
յորդորուին աղաչանաց մեծագոյն ընդ
ունել անկնալեալ, զի աղաչանօք Հաղիւ
Հազ յանձին կայցիս ։

Լ Ի Ս Ի Մ Ա Ք Ո Ս ։

4 Նպարապութիւնք, ով փիլոն, պատ
ճառք ոչ են։ Քանզի ոչ անգէտ ես, որ
ջափ աղզականացն, և դասուցն, և բա

Avunculus enim est, ac simul socer: quoniam non es nescius, quod filia ejus mihi juxta suam etiam promissionem desponsata uxor est. Dictorum vero recordatio molesta ac incerta erit ei, qui non est omnino consuetus his rebus; maxime circa arduam quaestionem, cujus evidentia, certitudinis lumen oriri fecisset, si agitata fuisset; et quasi circumversata ut imprimeretur, non superficie tenus audita [1].

Magna res certe est magistrum amare, iis qui audire cupiunt. Attamen ad optimam eruditionem nihil aliud ita auxilio esse videtur, quam costanter scire ea, quae magister suggerit [2]. Qui si veraciter edocere cupiat, haud ab alienis curis attrahi se permittet in occupationem. Dic mihi, quaeso ; an in alienas quis distrahitur res , fratrem quemdam laetificare volens; ac sine impendio parans verborum convivium auditoribus? At tu fortasse eam ob causam, ne videare facile opus peragere, similemque esse iis, qui prompte gratiam projiciunt in conspectu : ideo gressu remoto recedis, supplicationes rogantis majores expectans, quasi precibus demum nonnisi aegre annuas .

Occupationes, o Philo, an rationes non sint? Non es enim nescius, quot cognatorum, et classium, civitatisque me

---

1 Totus introitus hujus dialogi obscuritatem nimiam habet: eruditi lectores aliunde possunt declarationem aliquam adinvenire .

2 Si legas գեւել pro գեւել, tunc sensus erit, attendere magistrum ea, quae vult suggerere .

դորբիս զինենեպատեալ են իրաց բազմած
թիւնք 'ի բնակութիւն ։

circumduxerint examina negotiorum in
ipsa mea habitatione.

ՓԻԼՈՆ:

Արդ վասն զի գիտեմ զքեզ ոչ 'ի բաց
կալ՝ միշտ նորոգ ծարաւեալ լսելեաց,
սկսայց ասել՝ եթէ դադարեսցես. եւ ոչ
յար առ ինադքն բարբառեցեալ յամել
'ի նոյն իրս բռնութիւն արասցես ։

Quoniam itaque scio te non recusa- 5
re, ut pote semper sitibundum audiendi,
ego incipiam dicere, si tu taceas; at tu
non assidue meis sermonibus immoratus,
perstare in iis vim tibi facies.

ԼԻՒՍԻՄԱՔՈՍ:

Ծանր է եւ անիմաստասէր ինն հրա
մանդ. քանզի ինդրել եւ հարցանել 'ի
վարդապետութիւն վՀարագոյն ինն է եւ
յօգուտ : Իայց տեղի տալ է սակայն
Հրամանացգ : Արդ՝ ես աստանօր ՀԷ
դուբեամբ խաղաղ եւ Նուաստ, որպէս
օրէն է աշակերտի,Նստեմ. իսկ դու դէր
յանդիման Նստեալ 'ի վերայ բարձու եւ
րիք, եւ Նաղելի եւ պարկեշտ զքեզ արա
բեալ, յիմաստակ կազմեալ սկսիր ու
սուցանել զվարդապետութիւն ։

Grave est, et a philosophia abhorrens 6
mandatum istud tuum; quaerere enim
interrogareque multum favet expeditque
doctrinae; nihilominus cedendum iusso
tuo. Ego itaque hic inferius, tranquil-
liter humiliterque, ut decet discipulum,
sedebo: tu autem ex conspectu residens
in alto, et venerabilem ac severum te
effingens apparatu sophistico, incipe tra-
dere doctrinam.

ՓԻԼՈՆ:

Սկսանիմ, այլ ոչ վարդապետութիւն.
քանզի եմ մեկնիչ, այլ ոչ վարդապետ ։
Սաան զի ուսուցանեն աֆնորիկ որբ զեւ
րեանց գիտութիւն այլոց ուսուցանեն.
իսկ մեկնեն աֆնորիկ որբ զայլոց լուրս
Հանդիպողութիւն յիշատակի պատմեն.
ոչ սակաւ զաղեքսանդրացւոցն եւ զՀռ
ոմայնեցւոցն 'ի մի վայր դումեբեալ,
զբաջս եւ զլաւս եւ զընտրեալսն, եւ սկզ
բանց առաֆնոցն, եւ երաժշտականա
թեամբ եւ այլ իմաստութեամբ զարդա
րեալ ։

Incipio, sed non doctrinam; interpres 7
enim sum, non doctor. Docent enim ii,
qui propriam scientiam edocent alios.
Exponunt autem illi, qui ab aliis auditu percepta exacte memoriae referunt;
non paucos Alexandrinorum Romano-
rumque in unum locum congregantes,
viros optimos et selectos, atque praeci-
puis principiis, musica aliisque erudi-
tionibus ornatos.

ԼԻՒՍԻՄԱՔՈՍ·   LYSIMACHUS.

8 'Ի մէջ անցեալ պատանին ցոյցս առ_
նելով, ոչ Հանդերձ անամօթ քաշալէ_
րութեամբ, որպէս որք այժմ, այլ Հանն
դերձ Համարձակութեամբ ամօթով շիկ
նեալ, որպէս պարտ է ազատին՝ և որ
ազատաց ծնունդ է. 'ի մէջ նստեալ և
դղյն իբն յաղագս խրատուն իւրոյ, և
Հօրն առատութեան, զոր 'ի կիր առեալ
յաղագս այնորիկ՝ զոր յառաջագոյն աս
սացեալ:

In medium ecce veniens adolescens [1]
ostendendo se haud impudenti confi-
dentia, sicut hodierni solent, sed una
cum fiducia pudore rubescens, ut decet nobilem, nobilibusque natum, sedit
in medio, tum ob suam instructionem,
tum ob patris liberalitatem, quam usurpavit propter id, quod prius dictum fuerat [2].

9 Ծառայի միոյ՝ որում դեալ երե անդ
կալ յետոյ կողմանէ ոչ Հեռի ուրեք, գիրս
կարկառեալ, առեալ ընթեռնոյր Փիլ
լոն:

Servo itaque quodam, qui ibidem
stabat a tergo haud procul, librum por-
rigente, acceptum legebat Philo [3].

(ՃԱՌ ԱԼԵՔՍԱՆԴՐԻ ՅԱՂԱԳՍ ԿԱՐ_
ԵՑԵԱԼ ԲԱՆԻ ԱՆԲԱՆԻՑ·)

(SERMO ALEXANDRI DE PROBANDA BRUTORUM
ANIMALIUM RATIONE.)

10 Հարց՝ տղայ մանկանցն, և շահա
պաց՝ որբոց Հոգաբարձութիւն, և ա_
բանց կանանց խնամածութի՝ յանձն է,
զբնութեանն զկարոտութին լցուցա
նել՝ այնք, որ օգնելն կարեն՝ ոչ դանդա
ղել: Ապա ուրեմն զՀետ դայ և մարդ_
կան և ամենայն կենդանեաց 'ի միասին
ասել՝ լցուցանելով ջնջոս տկարութն·
Ոչ անգէտ եմ՝, զի փիլաստիրութին Հա
սարակ չար է, և լուսոյ խափան լինի
առ 'ի Ճշմարտութիւնն գիտութեան,
որջաքար կաղադգեալ 'ի ներքս, և
յոգւոյն 'ի բաց կորԵլով զառաջնորդ_
կան խջանունութիւն: ԱՅԼ որք միանե
գամ խրատուն անՃաշակք են, Հեճան
դութին է աղոթս, և ծանրութին ա_
զոթիք: Իսկ որք միանդամ ուսանին

Patribus puerorum juvenum, et tuto-
ribus orphanorum procuratio, virisque
uxorum cura commendata est ad sup-
plendam naturae indigentiam, ita ut qui
adjuvare possunt, non negligant. Sequens itaque erit, ut tam homines
quam omnia animantia, ut simul dicam,
debeant mutuam reficere debilitatem.
Non ignoro tamen φιλαυτίαν (amorem
proprium) commune esse malum, obsta-
culumque luminis requisiti ad notitiam
veritatis, tamquam in cavea intus insi-
dens, et abscindens ab anima principatum directorium. Verum quicumque
nolunt gustare disciplinam, aegre ferunt
orationem praesentem, gravanturque
hoc sermone. Qui tamen student sa-

---

1 Ut putatur, Alexander ipse constructor se-
quentis apologiae de brutis animalibus.
2 Sive, quod jam scriptis mandaverat.
3 Amanuensis Arm. nomen Philonis, pro ti-

tulo exaraverat, quod tamen abs re est. Ad elu-
cidationem rerum nos adjecimus sequentem inscri-
ptionem.

գայն՝ որ յխնատութենէ է, մարթբեալ
յախտս, մերկք մնոք գլերք Հչմարտու–
թեան գՀոա եւք. քան գոր պատուական
նազոյն մարդկան ազգի ստացուած ոչ_
ինչ ած պարգեւաց։

Քանզի որպէս զկանանց տկարութի
անտես արարեալ արքն՝ որք ամենայն
օրէնք գՀասարակ իրս քաղաքացն, որ_
քան մեանգամ ըստ պատերազմի կամ
ըստ խաղաղութեան իրք են, մեայն ինք_
քեանց կանխեցին՝ զլագեալ գէգ ազգն
'ի քաղաքավարութիւն անկար Համա_
րեալ. նոյնպէս կարծեմ ամենայն մարդ
գԹկ տեսեալը գանասունն կենդանիս ընդ
երկրի արմատացեալը, բայց մեայն գինն
քեանս 'ի վեր կյս ուղղեալը յերկրէ,
Հակառակ եգին իւրեանց գեղեցկասան
կազմածին չնոցա, մեծանգամանին մեծ
մեմՀ գմարՀուրդս վերամեարձեալը,
եւ բամբՀեալը իբրև գգետնասասանս,
պլանին որ յենն է՝ անղ, եւ նոցա ինչ ոչ
ամենեին այսինն մամն. բայց իւրեանց
մեՀչակեցուցանեին, իբրու 'ի բնութին
անփայզ մրցանակ առեալը։ Ա. յլ կան
զե եւ նա կցորդութին ունի բանի, եղ_
Հակառակութեամք որ 'ի բաց բարձեալ
զընդգդիմնարան, քաջուսումնութեամք
գեղրա գՀչմարիտն կամեցեալ քննել_
գխասագս։

Քանզի քանի երկին տեսակ է. մին
'ի նեըրս 'ի խորՀուրդն եղեալ է՝ իբրև
զաղբիւր, վայր ունեյոյ գոգոյն գառաջ_
նորդականն եւ գեշխանականն։ Իսկ գա_
չոյն ճման է Հոսմանն, 'ի ձեռն բերանոյ
եւ լեզուի 'ի բնական գործարարութին
լսելեոքյնեքն պեղեալ։ Բայց երկրան
չիւր ունեք է տեսանեկ 'ի նոսա, գե թէ_
պետ եւ ոչ կատարելութին, այլ սկին
նուազ սկզբունս եւ սերմանս։

Արդ անտեսանելի՝ տեսականս, գոր
ընկալայ մաթ. իսկ որ 'ի գգայութին
Հասանեկ՝ գոբձեալ լսելեոք. նախայառա_
ջադոյն այցել է այս։ Ой բազումք
են՝ որ յորացեալ է 'ի ձեռն նոցա իսկ_

---

pientiae, purgati a vitio clarius cernunt, quod pura mente puram sequimur veritatem, qua praestantiorem humano generi possessionem nullam Deus donavit.

Etenim quemadmodum viri mulierum 11 fragilitate rejecta, omnia reipublicae in civitatibus negotia sive ad bellum, sive ad pacem spectantia, per se ipsos solos exequi voluerunt, perperam genus foemineum ad politicen insufficiens rati; similiter, ut mihi videor, universi homines videntes animantia loquela carentia terrae adfixa, et se solos sursum e terra erectos, oppositam putarunt eorum constructionem pulchriori parti, quam ipsi praeseferunt, parem cum corpore elationem efferentes, contemnentesque (animalia) velut humi repentia; ita ut quod in rebus optimum est, nempe rationem, nullatenus eis concedant, sed sibi ipsis sorte possidendam datam credant, quasi vero ex ipsa natura donum peculiare acceperint. Quod tamen et animans participationem habeat rationis, contentiosam siquis deponat oppugnationem, facile per studium disciplinae verum quaerentis assequetur exactissime.

Verbi enim duplex est species. Una 12 intus in consilio sita (quae dicitur Ratio,) quae habet sicut fons, aut sedes in se animae principatum directivum. Altera pronuntiativa similis fluvio, per os et linguam naturali instrumento ad aures percurrens. Utriusque videre est in eis (animalibus,) sin minus perfectionem, attamen haud contemnenda principia et semina.

Quum enim invisibiles sint ideae men-13 te perceptae; quae autem sensum attingunt, sentiantur ab auribus, posterius istud visitandum erit primo loco. Nam multa sunt eorum voce articulata uten-

Հարլուր և իներեսուսունեՃն վարեցեալ ըՆուլ
թեսմե . և ըազզումp եՆ՝ որp և ուսու
զեալ լիՆիՆ , որպէս զազզուաւզ ասեՆ ,
և որ 'ի Հնդիկս պապկայեՆ եՆ ։ Բայզ
ես պապկայզն՝ ըանզի առ մեծամեծս
ըերեւալ յաղեըասնզըիա զիսեմ՝, այս
որյեզիսլոոսեն է , մեծ ՀայսՆ՝ զի զռ
զեն ըեր որp 'ի զպոզիՆ մանկուՆpն եՆ ։
Ինչ զուզանելով ուսան՝ որ ըեզ միաՆ
զամ սիրոյ Համբուրի ՀայսՆ Համբու
ըիp որ2այն ասեՆ , և աՆուամբ ընզու
ՆելութեՆն որ2աՆս ունելիՆ ։ Ինչ միա
Նասանզ զիշիասնազն ընզելուզանելոյ
զթազաւորս և զիՆpՆակալս , և զպար
սունելիս, և այլpն եՆ՝ որ այսպիսույ Նամանp
եՆիՆ զովլեՆ ։

14 Իսկ Նորա սկզբնասան լլնել սովորու
թիսն ունեիՆ , և զՀիՆ լուըն յիշատակ
կաս արձարձեիՆ ։ Լսեն թէ առ մակե
զոնազոզ թազաւորաՆ , և մանաւանզ
առ լազիսան յեզիսլոոսի ազոււաp այս
պէս Նմանեզուզանեն մարզկաՆ Հայսի ,
մինչ զի 'ի զուրս զալ որ2այն աս զայս
օրիՆակ ։ Ինչ Բաըզաւորp պոողոՆ ։ Ինչ
ըազզանզ և այլոզ զեա է լեզու Հ2զրիսն
և սատոզ յոուանալ , և այսոզիkp որp ո2
յոյզ ընզելpն ։ Ինչանզի միշո ՀայսՆ 'ի
Հասասո անյայոոիՆ է ։ Ս ասն զի Հայսի
կրկիՆ ասեսալ է , կ որ յասելp է , և կ որ
յերզելp ։ Լրզ ըասակաՆ յաղազս առաշ
Նոյն պասմեզաp , եըեkորզեն յայզ լե
զուp ։ Ինչ ընսունեթեսն և ո2 միոյ ո2 եՆ
ալեսp այսոզիkp , որp ո2 յոյզ իսազեալp
եՆ , այլ որ2 եՆ ականշp ։

15 Բասնզի կեռնելոp (զ. սարիկp ,) և
սատրաpp (եըր սորզիկp ,) և ծիծառ
Նուլp (որպէս սոսսկp) ո2 միայն երզել

---

tia ex natura per se edocta auditu proprio; multa autem, quae etiam edocen-
tur (ab hominibus,) sicut dicitur de cor-
vis, et Indicis psittacis. Ego sane psit-
tacos novi delatos ad magnates Alexandriae Aegypti, qui magna voce clami-
tabant adinstar puerorum ex Schola; et
indicio assuefacti didicerunt modum sa-
lutandi per dilectionem osculumque, ita
ut voce tenus cum osculo salutem di-
cant, et nominatim recipiant salutem
sibi datam; praesertim vero ita edocti,
reges, imperatores, augustosque, et similes, laudabant [1].

Isti quidem solebant initium facere
verbi, et antiquitus audita per memo-
riam renovabant. Verum ajunt sub Ma-
cedoniae regibus, potius autem sub La-
gis (Ptolemaeis) in Aegypto corvos adeo
imitatos esse humanam vocem, ut exeunti foras regi salutem dicere solerent
hoc tenore: *Salve Rex Ptolemaee.* Et
multis aliis convenit linguam veram ac
certam exaequare, illis etiam, quae non
sunt omnino inter cicures. Atqui semper vox dirigitur ad fidem faciendam
de inevidentibus. Porro vocis duplex est
species, una in dicendo, altera in can-
tando. Quum ergo satis de priori retu-
lerimus, secundam perlustremus. Pro-
bationis autem nulla est necessitas apud
eos, qui non omnino surdi sunt, sed sa-
ni auribus.

Merulae enim, et turtures (imo, tur-
di,) lusciniaeque non solum canere solent, verum etiam voce articulata ca-

---

1 Plin. hist. nat. x. 58. « Super omnia humanas voces reddunt, psittaci quidem etiam sermo-
cinantes. India hanc avem mittit, sittacen vocat . . . Imperatores salutat; et quae accipit verba pronunciat «. etc. Id. ibid. 60. de Corvo .

« Mature sermoni assuefactus, omnibus matuti-
nis evolans in rostra, forum versus, Tiberium,
dein Germanicum et Drusum Caesares nominatim, mox transeuntem populum Romanum salu-
tabat ».

թնաւորեալ են, այլ և յօդաւոր ձայնիս
երգեն, մինչ զի և ասել և դրել կարել
գնալադացն բառս։ Իսկ եթէ կամեցի
որ Հաստատուն լսելեօք ուսանել, փոքր
մի ՚ի մերձաւոր դրախտիցդ մի դանդա
դեսցի երթալ, յորս Հշմարտութեամն
են երաժշտական ամենագան Հաւուց, որ
՚ի միասին Հնչեն. և բառ խառնման ինն
յարմարութեան Հակառակաբարմա
նուագեն երգս։ Բանզի դդատացէ ՚ի
նմին իսկ յառաջագոյն ուսեալ ՚ի դոր
ծոսն, զի ստեալ է այլոց կենդանեացն
անասունն ՚ի դերոսութէ վերիպակ ա
բանց՝ որք ինքնասիրութեան Հշմարու
թեամբ պատուեցան։

Այլ զի՞նչ պարտ է երկայնաբանել
յաղագս բանին դողողի, զայն որ ՚ի խոր
Հուրդն է ՚ի բաց թողեալ. արդ երր
թալլի է յայս։ Բանզի անտեղի յիմա
րութիւն է, թէ վարագաց և առիւծոյ
որսորդքն խնդրեն դեղաս և դդարար,
իսկ դխորՀուրդս ամենեին ոչ տեսանել
քննութեամբ, յորս ՀանՃարոյ Հոդ մ
տանէ։ Արդ մեկալք սովորեցաք յորսն,
ոչ անտառս և մայրիս և մացառս և շամբս
և խարծս յոլզելով, քանզի ոչ յանշունչս
անասուս բնակեալ է, այլ դկենդանեացն
դտեսակս խոլգելով և յայս լնելով, և
թէ միայն մարդկային միոքս պատեէ
բաղբեամ երգեալ պատուն առեալ ՚ի բր
նութենէ մեծ ինն դատ և որիշ յայլոցն,
եթէ և այլոցն Հասարակ դոգուն ամ
նեցունս աՃ եա։

Իսկ մբինչ վասն լաւին Հոդ տարեալք
թուին ինձ ամենեքին որբ Հշմարու
թեան են ընկերբ, ձինչս երկրպայութեամբ
խոստովմանել միով բարուք՝ թէ խոր
Հուրդ ամենեցուն որք միանգամ շնչ
Հասին՝ ընակեալ է ՚ի նոսա։ Իսկ Հաս

nunt [1]; ita ut et dici et scribi posse vi-
deantur vocabula cantuum. Quod si
quis id velit suis ipse auribus discere,
eat statim in proximum hortum; ubi
vera est musica avium omnis generis
concinentium, atque juxta mixturam
harmoniae certatim cantantium; et au-
diet, ipso experimento operis prius edoctus, mendacii redargui tum se tum
caeteros omnes, qui irrationalia esse
animantia pro lubitu fingunt, nonnisi
φιλαυτίᾳ idipsum suadente.

Sed quid oportet plura addere de 16
verbo sonante, relicta ratione, quae in
consilio sita est? Eo itaque pergamus.
Profecto absurda dementia est, ubi venatores investigant aprorum leonumque
cavernas, et antra, consilia ipsa prorsus
non videre diligenti examine, quae animus rationalis penetrat. Siquidem (nos
philologi) consuevimus venatum ire,
non sylvas, nemora, arbusta, arundines
exploraturi; non enim in sylvis inanima-
tis nostra venatio consistit : sed animalium species inquirendo perlustrandoque:
sola ne videlicet, mens humana induta
est imagine (divina,) honorem sortita
ex natura oppido magnum diversumque
ac differentem ab aliis; vel caeteris quoque aequalem universis utilitatem dedit
Deus?

Neque melius quidquam videntur mi- 17
hi omnes veritatis cupidi pervidisse,
nisi sine ulla dubitatione fateantur uno
consensu, quod consilium insit omnibus
spirantibus. Firmum autem argumentum accipiet quisquam, praetermissis

---

1 Plin. x. 59. « Agrippina Claudii Caesaris
(conjux) turdum habuit, quod nunquam ante,
imitantem sermones hominum, cum haec prode-
rem. Habebant et Caesares juvenes sturnum, item
luscinias, Graeco atque Latino sermone doci-
les « etc.

17

տատուն նշանակ առցէ որ՝ ոչ միայն յայլոցն կազմածոց, որ ըստ առաւելութեան իմն հանճարոյս տարեալ լինի, զսարդ ոչ տեսանես, որչափ և որպիսի և իբր սքանչելի իմն կազմէ։ Քանզի ո՞ որք տաժանագործ նիւթող և կամ ոստայնանկ քաւական է՝ ոչ հակառակելով ասեմ, բայց նմանեցուցանելով, կարել տանել արուեստի զերկրորդութիւնն։ Քանզի ամենեքին, և որք միանգամ ոչ հեղգաբար, և ոչ առանց քննելոյ, անդէն իսկ 'ի մանկութենէ թևակոխեցին ստանալ, 'ի ճըշմարտութենէն յաղթեալք 'ի բաց գնացին։ Քանզի առեալ իբր ասր անգործ նիւթ՝ հանճարաբար իմաստութեամբ յոյժ արուեստաւ կազմէ։ Օ առաջինն մանելով իբր ձեռօք մանոցօք 'ի բարակագոյն, և ապա՝ է որ ուղղաձիգս, և է որ կողմնաւոր պրկեալս, կարէ քաջ 'ի դէպ զթելսն ոստայնագործէ, և մանէ հրաշապէս գեղեցկութեան արուեստ գործելով, օդոյ զանկուածն բարակագործութիւնք նմանեցուցանել։ և ժուժկալութիւն տևող երկուց իրաց գործելով 'ի նոսա յայնմանէ՝ որ խորհրդով իմն իմաստութեամբ արարեալ է քնարն, յաղէպնդութենէն և 'ի ձևոյն բոլորակութենէն. քանզի միշտ բոլորակաձևն քան զուղղագիծսն 'ի տևելն հաստատուն է։ Նաև նշանակ մեծ է՝ յորժամ իւրաքանչիւր ոք 'ի հողմոց, և այն որ ստէպ ստէպ 'ի միմեանց վերայ լինին, հազիւ հազ և կամ բնաւ իսկ ամենևին ոչ լինի պատառեալ։

aliis momentis, quae pro diversa ingenii amplitudine quisque potest excogitare[1], araneam ipsam (vel araneum). Viden, quantum, aut quale, et quomodo mirum ista operatur? [2] Quae enim, licet aut netrix operosa, aut textrix idonea, non jam certando, sed etiam aemulando, possit ferre artis secundum locum? Siquidem omnes quicunque haud incuriose, neque sine diligentia jam a juventute operam huic arti dederunt, atque accurate id imitati sunt, superati abiere. Sumens enim (aranea) velut lanam materiam non laboratam, ingeniosa sapientia artificiosissime operatur: primum nendo instrumentis propriis, quasi manibus quibusdam, subtilissime: deinde fila quaedam directe, alia vero oblique deducens, optime stamina nectit texitque, miro modo confecta, atque arte pulcherrima; ita ut aëris texturae subtilem laborem assimilare videatur: atque firmitatem perdurantem dupliciter operi indit talem, qualem cernimus in lyra, quae duritiam chordarum, et figurae rotunditatem praesefert: eo quod semper (tam) figura rotunda, quam rectae lineae, diuturnae comperiuntur. Magnum autem argumentum erit, quod ab omnibus ventis, iisque assidue impulsa fila, vixdum aut nunquam prorsus lacerantur.

18 Բայց կայ և այլ իմն 'ի սարդդ սքանչելի. քանզի մարդիկ որ 'ի հանդերձոց վաստակեն, բաժանելով զհանճարսն, ոչ զնոյն թևակոխեն գործել. այլ որոց նիւթելն հոգ է՝ ոչ անկանեն, և որք անկանենն՝ ոչ նիւթագործեն։ Իսկ նա որ միանգամ պէտքն են իրացն՝ զամե-

Caeterum et aliud in aranea cernitur mirificum; quoniam homines vestimenta elaborantes, divisis inter se industriis, non idem operari aggrediuntur; sed quibus cura nendi est, non texunt, qui vero texunt, non nent. Illa vero quaecumque operi requiruntur, omnia simul

1 Vel: quae in reliquis animalibus cerni possunt signa ingenii amplissima.

2 Fuse agit de Araneis Plin. XI. 28. et passim.

 Նախ մնացած յինքեան փակեալ ունի, ոչ ումեք այլում գործակցի կարօտեալ, այլ իբր կատարելագոյն զիւրաքանչիւր ինչ յայցացանէ յուցանելով ։ Լ զինչ արանելի, եթէ բաւական է, իբր զի մի է չանին՝ կարիել՝ մարթել, որ ոչ միայն արուեստանականաց, այլ և գործեալց զ բութէ զարդարեալ է. ինչն է և այն, և արուեստ և գործի, որով կատարելագործ ծեզան լինի իւրաքանչիւր ոք յայցցանէ ։

Նաւաստք ոչ առանց քեղկաց, կամ առագաստից, և կամ այլոց՝ որ մնան գամ գործակից լինեն առ յուղղութիւն անսխալ վարելոյ նաւահ ։ Լ սակայն և ոչ բժիշկք, ոչ առանց կերակրոց կամ րմպելեաց և կամ դեղոց այս ինչ՝ որ առ ի բժշկութիւն է յաջողեալ լինի ։ Լ զի՞ պետ 'ի թէ արբանել զարուեստաս գետ, որոց գործծեցք և տնտեսկութիւն, և զորութիւն 'ի նոսա առանց սոց ոչ գոյ Լ սարդ անեբելոյ և անպիտան թուի գոլ կենդանին, կատարելագ ոմն յստասինակութիւն, միջ անսարոտ է գործելոյ. բաւական է ինքն անձամբ առ այնոսիկ զորս կագիկն կամի 'ի կատա բելութիւն ։ Բանդի աշա գիտեմ՝ նկա բիչ և նկարոչ առաքինի գ քորոս, որբ առեծծ եաո եյ և նկարս և պատկերս՝ որ նոցա մանն գ այ էր, փութ աս ան յաւէ տ ագ ունա կեք գ զիրբեանին ան մնա Հացա գա նեկ լոյ վ ատ ունա կ. վ ան որպ և այ նոց իկ որ մ ասե ն ին, իբբ գամ Հա զս թնա մ իս ա տ ւ ն ։ Օ այս առնելոյ և սարդ՝ յայտ նի է. որ իբր կատարեցցծ գան կա ձմ, բ ստ միջ նող ս սւ եղ ն ֆ իբ ր 'ի կ կ ին ֆ գ ա բ ան ա մն ա մ լ ին ն ա մ ն ին ա ու բ ք շ ւ բ Հայ ե լ ոյ, թէ մ ի ին չ գ ոզ դ՝ որ յ ո դ դ թ ա ֆ ս, ի ո դ եո ֆ են գ ո ա ա ֆ ն. և զ գ ո ւ զ ա ֆ ս ա յ՝ գ ե մ ի ի ն չ ա մ են ե ն ֆ ա ս ա մ ա յ գ ե գ ի լ ե բ ե ա ն զ ա ն ե լ ա մ են ե ն ֆ ։ Լ ը ե թ է ե զ ֆ գ ե ի ն չ, գ զ ա ա ա ձ ա ր ա ո մ ֆ ա ո ա ս ն 'ի բ ա ֆ բ ա ո ն ա յ՝ 'ի ծ ե ո ե ն մ ա ն ո ւ ա ծ ո ֆ ն ա ն ֆ լ ն մ ա ֆ ֆ ա ֆ ա ֆ ի ֆ մ ե ֆ ի ն ֆ ր ե ֆ, ի ո ֆ ֆ ա ա ա ո ւ ե ա ն 'ի ֆ լ ե բ ա ո ֆ ն ֆ ֆ գ ֆ է ։

in se inclusa habet, nullo altero cooperante indigens, sed velut perfectior (homine) singula ista per se praestat. Et quid mirum, si ad haec sufficiens comperitur? quum et potentia et facultate par sit: et non tantum artificis, sed etiam instrumenti vi polleat; quippe quae ipsa est tam ars, quam instrumentum, quo peragitur opus universum.

Nautae nihil agunt absque guberna- 19 culo, aut velo, aut aliis adjumentis ad directionem certam navis. Neque medicis sine cibo, aut potu, aut remediis cedit optata sanitas aegroti. Et quid oportet dinumerare artifices, quorum tota ars per instrumenta exercetur, et quibus nulla est vis sine his? Aranea tamen tantilla, ac inutilis putata inter bestiolas, perfectissima in textura nullo eget instrumento, sed sufficiens est per se ipsam ad perfectionem rerum, quas operari vult. Ecce enim scio pictorum fictorumque ingeniosa experimenta, qui figmenta et picturas imaginesque, quantum in ipsis esset, sategerunt perpetuare, ac immortales reddere labores suos: ac proinde laedentes opera sua tamquam inexorabiles inimicos odio habuere. Idem facere et araneam manifestum est, quae nimirum postquam perfecerit texturam, medio loco velut in centro insidiatur, quaquaversus circumspectans, ne forte volantia per aërem lacerent telam, atque cavet, ne quidquam umquam se invita pertranseat. Quod si accidat, causam damni e medio tollit, per texturam ipsam illico vindictas accipiens; scissurae autem rursus medetur.

20 Այլ է իմն՝ զոր ոչ գեղջուկք միայն,
այլ և մեծ թագաւորք իմանեն․ Մեղու
հանճարով ոչ երբեք զանազանեալ է ա
ւասելապէս քան զմարդկան միտս ունել
սանեցով վարեալ ։ Ի սա գարնանային
ժամանակին՝ յորժամ ամենայն դաշտ և
որ միանգամ արգաւանդ ահող ինչ է ՚ի
լերինն նախ ծաղկեալ, մեծամեծ մեղուաց
 որք ՚ի վերայ թռչին բուրաստանաց, և
դրախտից և շառաւեղաց և դալարեաց,
՚ի վերայ անցեալ նստին անուշահոտ
ծաղկոբ ՚ի կոկոնաց և ՚ի պտկոց իբրև
ցողով ՚ի վերայ յնկեանա ծծեն․ և մա
նաւանդ ՚ի թխմայն՝ և յայլոցն եւ ո
մանց գեանանախանձից՝ որ յարասու
նակէ անուանեն մեղեղոպեայ ։ Հրաշա
պէս կազմածով յատկութիւն աւանձ
նասերք՝ փոփոխէ զցողն ՚ի մեղրէ բնու
թիւն․ և գործէ այսպէս․ զցողն իբրև
զսերմնական ինչ քան ընկալեալ՝ մեղուն
լինի յղի․ և փութայ առ ծնունդն յայս
վայր ուր ոչ վնասեցի զոր ծնանելոցն է
երթալ․ Նա են աբարեալ փեթակք բստ
զործակցութեան մարդկան առ ՚ի լու
ծումն նոցա ։ Որոց աշ զորէսն իբն ժա
ռանգութեան որ ՚ի յերկրէ են, ոչ միայն
անկոց, այլ և զկենդանեաց բաշխեաց
եստ պատուոյ ։ Արդ յորժամ ծնանի,
ստեղծանէ զՀասատակեալ ձևն, և զլէ
ծութիւն որ վայելէ նմա՝ բնանի, այն
որ ՚ի տոն յարութեանն մանաւանդ երբ
թայ․ Իսկ կրկնեցելոյն կրկին կենդան
ոյ օրինակ բաղկացութեան է․ կ որ
նիւթաքոյս մարմնի նմին՝ մ. և կ որ
երբ ոյն՝ որ առ այնւ յածէ՝ մեղր ։ Այլ
յորժամ նմին ցող ընկալեալ, զի մի վաս
տակն անկատար իցէ առականեալ, իր
գեաց ծածկեալ պնդապէս առ այնոսիկ
որ ՚ի վերայ յարձակին կենդանեքն զա
ւ ութիւն, որ եղէ ՚ի քսութենէ․ կ ինչ որ
երբ ՚ի քաղաքէ մօտ ապարանք ցանգով

Est et alterum quoddam, cui non rustici soli, verum etiam magni reges curam exhibent. Apis ingenio haud profecto concedit, plusquam homines visa mentem exercere[1]. Verno tempore, quo universus campus, et totus montis locus fertilis primum floret, examina apum supervolantia et perlustrantia hortos ac viridaria, tum super ramusculos virides, fragrantesque flores sedentia, ex calycibus papillisque rore aspersis ipsum exsugunt, praesertim vero ex thymis,[2] caeterisque humi proximis, quae onomastice melilota nominantur. Mirifica itaque compositione sibi propria apis transmutat rorem in naturam mellis; et ita operatur: rorem velut seminale quoddam in se assumens, gravida evadit: et properat ad partum eo in situ, ubi non laedatur, quem procreandum scit. Facta sunt autem alvearia ad habitationis hominum similitudinem pro diversorio earum, quibus Deus haereditatis modo distribuens dedit fructus in terra repertos, non solum ex plantis, sed etiam ex animalibus. Quando itaque apis parit, effingit figuram solidam, magnitudine sibi conveniente, utpote bene nota, et quae magis apta sit ad continuationem propaginis. Duplicati autem animantis duplex est modus constructionis; alter maxime materiali corpori similis, cera scilicet: alter autem tamquam anima, circa quam versatur, nempe mel. Quando vero cera rorem acceperit, ne labor imperfectus devastetur, clauditur obstruiturque firmiter naturali instinctu contra fraudes animalium invadentium; ita ut, tamquam in civitate palatium juxta palatium undique septum, ac densum cernatur vallum. Interiores

---

1 Confer cum traditis a Plinio prolixe. XI. 4=22. et XXI. 41. etc.

2 Nomen Gr. θύμος, thymus, observat et Interp. Arm. sicut etiam μελίλωτον.

շուրջանակի մօտ առ մինեանս իշխա գան
զ է։ Իսկ զետերքին մասունման նեղ և նուրբ
մտիք արգելեալ է , զի մի դիւրաւ իցեն
'ի վերայ յարձակմունքն . քանզի յընդ
արձակութեան 'ի վերայ յարձակել դիւ
րաւ լինի ։

Այլ ոչ միայն պարտ է զշուրջանակի
փակել Հաստատութեամբ , այլ յետ
պարսպելոյն Հզօրագոյն մեղուն պաՀապա
նապետ և պարսպապաՀ կարգեալ լինի ։
 Իսկ առեալ' իսծ Թուի Թէ զերխաւու
Թեանն զպարս , լուսաւորագոյն յայտնե
անցեալ նստի առ մուտան' իբր 'ի դիտա
նոցէ յականերլով , և ամենայն ուրեք
շուրջ Հայելով ։ Իսկ եթէ դադարեալ
են Հակառակամարտըն , Հանգարտէ և
նա . իսկ իսկ իսու Հատեղրոյն նոցա' անդէն
վաղվաղակի առ 'ի վլեզվինդրութեն
յոլդեալ լինի ։ Իսկ եթէ Հզօրագունի
Ատ պիտոյ լինիցին կազմածոյ , պատմէ 'ի
ներքս գորւգն' եթէ աշա յանկարծ
րէն 'ի վերայ Հասեալ են Թշնամիք ։ Իսկ
նորա 'ի դարանէն յոտն կան , դիմեն միան
գամայն շաշմանէ , և փուՓով յարձակ
ման գնայթգն 'ի վեր բառնայ առ 'ի
զարՀուրումն , զի մի ինչ կրեսցէ , և առ
'ի վրէժինդրութեն , եթէ 'ի Հարկէ
պիտոյ լիցին , րնդիսառէսն ։

Բայց զի ծիծառն . ոչ նախախնամու
Թեամք ինչ կենդանիս այս և Հանձ
րեղ է . 'ի մ?ծէ և 'ի կաւոյ , և յայնցանէ
որ ինչ դեպ լիցի պատահէլ՝ նմա' րզիս
գործէ . յաղագս երկուց իրաց վան ալ
տանապոյ ծեով և մեծութեամբ զա
փէ , և զՀատատութիմն ըստ տեղեացն
պատ»Տ առն է ։ Իսկ 'ի Հզօրագունաջ
Փատու» փախուգեալք' նան առաջին
ադաշանոր առ մարդիկ իբր 'ի մեՀեանս
փախուգեալք անկանին 'ի տունս ։ Իսկ ալ
պա յերկրէ քարձր վլերացեալ 'ի քարձ
րագոյն վլերնս շրջին , և 'ի մարդկս
պարապագոյն շուրջանակի արկանէ պա

1 Consule Plinium . x. 49. aliisque locis .

autem partes angusto subtilique aditu
coarctatae sunt, ne facilis sit invasio:
largum enim spatium irruptioni patet.

Nec solummodo circumdatur locus 21
solida munitione, sed etiam post moenia
ducta, fortior robustiorque apis custo-
diae praeses, et custos muri constitui-
tur. Tum illa collatae sibi, ut mihi videtur, facultatis personam sustinens pro-
dit, ac palam venit sedetque prope ad
introitum, sicut ex specula attendens,
et quaquaversus circumspectans. Si ergo cessant adversarii, acquiescit et ipsa;
eis autem agitatis, confestim ad defen-
sionem armatur. Sin autem potioris
opus fuerit apparatûs, intus sedenti exer-
citui nuntiat adventasse hostes repen-
te. Ille itaque assurgens ex insidiis, impetum conjunctim facit strepens, ac studio expeditionis aculeum extollens; ti-
mensque, ne quid mali patiatur, et in
ultionem ardens, qua necessitas instat,
congreditur.

Quid hirundo? nonne providentia qua- 22
dam praeditum hoc animal, atque inge-
niosum est? Ex ramulis et luto obviis-
que rebus nidum parat; quas duas res
secundum commodum per figuram ma-
gnitudinemque metitur, et soliditatem
secundum locorum circumstantiam dat[1].
A damno fortiorum fuga accepta, primum suppliciter ad homines profugum
in domos velut in templa, refugit; postea
altius de terra elevatum superiora loca
percurrit, et inter trabes potius moenia
circumducens, quam murum, interius
nidum fabricat connexum cum aedificio,

րիպզ, 'ի ներքսն զզոյսն շինէ անկանէ_
լով ընդ շինուածսն . որ Հակառակ իրս
Հասուցանէ մկանց , և Հալածէ . քանզի
որպէս ասէն յեգիպտոս' թէ կատուոց
մազդ է , քանզի մկանք կատուոց կերա_
կուր են ։ |ա կատարեալ գործեցեալ'
քան դամենայն արանց շինող արուեա_
տականագոյն է շինուածսն ։ |ա քաջալէ_
րութիւն' պաշարեալ ծննդոցն գզու_
շութիւն ծնանի . և յետ ծնանելանն ոչ
զատարէ , այլ կերակուրք եղելոցն Հայ_
թայթէ . և ամենայն ուտիք գումարէ_
լով' զզգդն իրաքանչիւր ումէք բաշխէ
սալ , զագաՀութիւն առաւելուլեան
և զՀակառակութեան արգելլով . և 'ի
բերանս դնելով' մօր միտերմութիւն , և
աանուտի և դայեկի զպատաւորութին
ցուցանէ ։ |ունէ և դայն նախախնամա_
գոյն . քանզի զթրքացեալ կերակուրն
Հաշկաւորէ 'ի դուրս ընկենուլ , ձագուցն
սովորութեամբ դարձէալք արտաքս կոյս
ծրտել' երկիւղել , զի մի ձանռութեամբ
քոյսն պաստեալ 'ի բաց անկցի ։ |սկ յա_
զագս միամտութեան' յածեալ 'ի Հայ_
թայթանս Հարկաւորացն անդրէն դառ_
նայ , յոտար բոյն ոչ Հայէ , բայց 'ի վար
արկանէ զանմեղութիւնն զլիզն անդ_
րէն դարձ ճանապարՀէ' փափազմամբ
իւրոցն խնամոց ։

|այց ոչ միայն էնքենալուք և էնքնու_
սունն Հանճարոյ' ունէնք 'ի նոցանէ Հա_
սանեն , այլ բազումք և ուսեալք լաւա_
կան մտադիրութեամբ լսել , և Հաւա_
նէլ այնոցիկ' զոր նոցա վարդապետքն
պատուէրէն , 'ի մեղանչելոյ երկուցեալք,
և խնդալով ընդ վճարել իձն ուղիղ ։
|այց զի պէտո զընդելեացն իձն յիշէլ,
կորիւնք ունմզ ոստ բնական քաջալէ_
րութեան և կերակրոցն յատկութեամբ
որսորդք եղեն , և որսորդս արանց պա_
տահէալք 'ի վերայ յարձակին . յորժամ

quo armatur contra mures, et persequi-
tur ipsos; ita ut in Aegypto felis pilus [1]
dicatur, mus enim cibus est feli . Perfectis autem operibus, homines aedificii
peritos elaborato artificio vincit opifi-
cium . Praeparata itaque habitatione
nascituris, diligenter parit; post autem
procreationem haud cessat, sed cibum
adinvenit natis: et undique collectum ae-
qua portione singulis distribuit, aviditatem superfluam atque rixam inhibendo;
in ora autem apponendo (cibum) matris fidelitatem una cum officio lactantis
nutricisque demonstrat . Providissime
agit et illud, quod cibum digestum re-
solutumque in stercus, projicere docet eo
modo, ut pullos assuefaciat se se con-
vertere, ut foras alvum exonerent, prae
timore, ne aggravatus nidus praeceps
cadat . Quum autem simpliciter ober-
rans huc illuc ob procurationem neces-
sariorum, redux fuerit, in alienum nidum minime intrat, sed falli nescium
ad suum venit in reditu ob desiderium
curae propriae .

Caeterum non solum naturalis auditûs
ac instructionis habent ingenium aliqua
animalium, verum etiam complura edo-
centur meliori diligentia auscultare obse-
quique iis, quae ipsis magistri praeci-
piunt; ita ut vereantur peccare, et lae-
tentur de rebus recte peractis. Sed quid
expedit de cicuribus mentionem facere?
Catuli quidam (leonum, vel canum) juxta
naturalem fiduciam una cum cupiditate
ciborum, venatores facti sunt, ita ut homines venatores obviantes invaserunt.

---

1 Aut hirundinem ipsam, aut nidum suum hoc
nomine appellari notat. Apud Plinium x. 85.

haec tantum habentur. « Aegyptiis muribus durus pilus, sicut herinaceis ».

Ի յայսցանէ է որ` թնակակիցն են
մեզ, որք ուսեալք արուեստս ընդունին։
Ւրդ զկառավարութիւն` որ քան զամե-
նայն անասունս անարգ է 'ի տեսլարա-
նին եւանդ զայս եղոյց գործեցեալ կա-
պիկ ։ Ււնզէ 'ի վէրայ կառաց անցեալ
 համէր 'ի տախտակս քառայարոյս այ-
ծից ընդ կաքաւարանն վարէաց, իբր 'ի
ձիրընթացս մարտիկ ոք, յառաջս յայլ-
թութեան Հակառակեալ շարժէլով զե-
րասանակն, եւ մօրակաՆ Հարկանե-
լով, եւ յականջս այծիցն կանչելով, աքան
չելի եւ Հրաշագործ տեսանդայն զուար-
ծութիւն գործեաց ։

Ի,ւ ուլ` Ււււրբ շւււանի լերկար լերկ
րէ արկեգելլա վագեգեալ` 'ի վէրայ եւ
լեալ, Հաստատուն գնայր իբր 'ի վէրայ
յատակի, յառաջ երիթայր ընդ պղոդ
տայն, կարծելով վասն 'ի վայրն կոյ
ձգելով որ լեւոյ կողմանե ընդդեմ էր
մի յայնկոյս քարշեցե ։ Ււ վանն զաս
եւ տեսի լետ ապորիկ 'ի վէրայ 'ի բար
ձունս գնալոյ, 'ի վէրայ կաքաւելով, եւ
երեքին գնդի խաղայր ։ Ււ Հիանալի ինն
թութեաե գոլ. ապս ոչ ունի առ աւն` որ
'ի նման ձգեալն լինէր ընդունելութիւն ։
Ւ Հա եւ 'ի շրշանակ ինն 'ի վէր երեալ,
լերկար մեծամեծ ինն Հրաշ եւ աքանչե
լիս գործեալ աւելի եղոյց ։ Ււանզի քան
արտաքնումե բոլորակին 'ի նեքըս 'ի ծրին
ամենայն բոլորակին յուղեալ շարժմամբ
ստեղ ստեղ, եւ ջաՀք 'ի մի մի իբրա
քանչիւր ոք ձակլեւեալ սերմանեցան ։
Ււկ նա իբր 'ի նեքըս 'ի գործիս 'ի բոցոյն
որպեսեւ 'ի մեջ կայր ։ Ււ կալլիաս
էր` որ 'ի վէրայ անգեալ կայր նորա ։ Ււ
նա տեսեալ զշրշանակն շուրջ Հրիկգ
գայր, եւ կենդանիքն պատրաստեալ էին
աշա առ 'ի գոյս ։ Ււ 'ի նոցանէ որ թէ
բանով զջաՀն 'ի բաց կորզեալ Հանէր,
եւ յաշ կողմ զպարանոցն զեէլ թիւ

Quum autem aliquis eorum cohabitet nobiscum, artem propositam discendo percipit. Ecce enim aurigae artem vilissimus omnium animalium simius in hesterno spectaculo usurpatam ostentavit. Residens enim in tabulis currûs a quatuor capris vecti, per orchestram aurigam egit; ac velut in hippodromo certator quidam de victoria concertans, movens habenas, flagellumque concutiens, et in aures caprarum vociferans, mirifice spectatores oblectavit.

Haedus autem, subtili funiculo alte a 24 terra strictim colligato, accurrens ascendit per eum, et constanter superambulabat tamquam in solo progrediens per plateam, veritus ne per declivitatem deorsum ferretur, cum ab adversario a tergo stante eam in partem detraheretur. Et hunc ipsum ego vidi postea in alto situ ambulantem, tripudiantem, et per armos (pedes anteriores) globulis ludentem; quod mirificum visum est, quum manus non habeat ad prehendendum ea, quae jactata sunt. Ecce et in circulum quemdam ascendis, diu magna prodigia ac miracula potius monstravit. Quoniam clausus sphaera externa, intus in cyclo tota sphaera agitata, movebatur continuo, et faces in singulis foraminibus connectebantur sparsim. Ipse vero intra machinam quasi captatus a flammis stabat in medio. Callias [1] autem praesidebat; et ille viso circulo, versabatur circumquaque: ad haec animalia jam parata erant ad ostentationem; quorum unum ore faces evellens eduxit, et ad dextram versus collum detorsit, et elevavit, ut offerret Calliae. Ille vero (hae-

---

1 Vox ipsa Graeca videtur servata ab interprete; de qua studiosi lectores dijudicabunt. χάλ- λαια, vel χάλλος, praeter alia est etiam barba seu palearia galli gallinacei.

բեր․ 'ի վեր եթարձ դայն՝ զի մատուցէ
կալլեայ ։ Դակ նա յեղշեռացն իբր 'ի ձե_
ռաց դերբաբար մասամ͵ բնկալեալ ձեռն
առեալ․ և շուրջ ածէր 'ի վեր անդր՝
որպէս կանթեղակիրբեբ մանկունբ՝ յօժ
գոյգս առնելով, առ եզերբն արտաքուստ
կողմանն գդդրձ կալով, զի մի գոզ բոչն
առցէ՝ դայնոբիւբ որ վմեբօբ շուրջ կային
Հայեին ։ Դե ապանչելի ինձ թուեցաւ
իսձ և դայս առնել․ բանզի յանեին
զդայս մխանգամանն և 'ի վարդապետան
Հայեր՝ կասկածեալ, որպես կարծեմ,
զի մի ինչ յանցիցէ․ մխանգամանն և առ
տեսողսն կոա Հայեր, եթէ զխանդ ար
դ_եօբ ունին յինբեանս ուրախութիւն և
կամ'դժուարութիւն, բննելով։ Դակ յետ
ցոյցսն առնելոյ արծակեալ յերկիւղէն
զօրէն․ մարտիկի յաղթողի, պանծայբ
իւրովխտմանե․ և յայս էր յստատացույն
և յանդդարն լնելոյ, և յաչախ․ իբր
զի զիննդուխին զուարթութեան ձա
ղու ինն յինբեանս ցուցանէին․ և այոր
բեկ խնդալոյն Հատրակ նշանակ էր, և
ոյնս ունաղ` և իւր առանձին ունս ըզ
պողմանն շարժէր․ իսկ կալլին իբր ձեռբ
թուեին ։

25   Դե են ումանբ՝ որբ և ծառայական
պետս ունանին, խնամաբար սպաս տա
նեղ դորէին պաշտոնեիցս՝ անդադար փու_
թով Հակառակեալբ․ և սբբբ էր տես
լարանն լի․ և րստ միում միում իւրա
բանչիւբ ումեբ` է ասել, ժողովլի ցոյցս
առնեին անսսունբս այս՝ վարդապետացն
իբր իրա մխանգամանն ։ Դայս այս ոչ
խաղոլ է, այլ փունթոյ առաբինութեան
արժանի ։ Դանդի ծաղր է սա յիմմացւաց
որբ դողւոյն այս կուբացուցին․ և փու_
թ_ուլէ է սա իմաստանցն՝ որոց միտբ լու_
սաւորբ սրատեսիլբ են ։ Դբդ զմեծու_
թին բանական բնութեան մանզիւ
րութեամբ առանց ծիծաղեղլ պարկեշ_
տաբար բննեցէբ, զկարծիս 'ի բաց են

dus) cornibus tamquam manibus facile
quasi recipiens, at manu prehendens,
circumducebat sursum, sicut pueri lam-
padas gestantes, nimia ostentatione facta;
circa oram forinsecus cavens, ne flamma
cuiquam noceret eorum, qui circumstan-
tes spectabant . Mirum autem mihi et
illud factu visum est quod cum haec
agebat, simul et in ludimagistrum oculos fixerat, veritus, puto, ne quid de-
linqueret; atque simul spectatores respi-
ciebat, an exultatione afficerentur vel
moleste ferrent, investigaturus. Postquam autem spectaculum perfecit, timore absoluto, sicut bellator victoria po-
titus jactabunde gloriabatur; quod manifestum erat ex saltu , et continua a-
gitatione, oculisque laetitiam ac hila-
ritatem tamquam risûs praeseferentibus;
et praeter ista communis exultationis
indicia caudam quoque sua haedus mo-
vebat, juxta nutus quos Calliae manus
edebat [1].

Sunt etiam aliqua (animantia,) quae
et servile officium ediscunt, diligenter
obsequi ministrorum more, assidue inter
se contendendo; et hisce plenum erat
spectaculum: atque singillatim propriam
sibi ostentationem exhibebant populo
bestiae istae, quasi dixeris ludentes ad
nutum ludimagistri. Quod tamen non
ludi res est, sed studio virtutis dignum.
Siquidem risum id movet insipientibus,
qui oculos animi obcaecarunt, sed studium erit sapientibus, quorum mentes
lucidae sunt et acutae visu. Itaque ma-
gnitudinem naturae rationalis diligenter
sine risu honeste examinantes, opinio-
nem deponite, veritatem discite, ut in-

կեցէք , ուսարմէք զ\XՀﬕﬕﬕﬕ , 'ի
վերստին Հաստատեցէն , որ մեծարդի է
այ, և պատուեալ Մարդկան ﬓﬓﬓﬓﬓﬓ աղ-
դէ , և դովէլէ է ազգս այս ։

Բըդ ասացեալ անասունքս՝ ասէն ա
ﬓﬓﬓ , եթէ ուսանին որչափ այս որ 'ի
Նոսայս է ընտանի . բայց վայրենիքս ու
սէն ﬁ ուղղաբար․ \Բանգի աﬔﬓ ուﬓﬓﬓ
Նոսա և վայրենեացն և անձեռնրնդէ
լացն պատաՀեցին ուﬓﬓ 'ի վարդապե-
տաﬁﬓ , զի դային ընտանացեալըն ։

Բդ վիդաց ոչ վայրենագոյն է , այլ
սակայս լեբիացւոց բաւրբաս դերﬓﬓﬓ
կայ կայսեքս , աﬕﬓﬓﬓ ուստէք գլուբս
խադոց կոչեցելոց առ 'ի գոցս , յայﬓ
ժաﬓﬓﬓﬓ յոբում դպարապս Հ Սպատﬕﬓ
ﬓﬓﬓﬓﬓﬓﬓ յոբում դպարապս Հ Սպատﬕﬓ
դաղոys ﬁﬕﬕ մեծադործեալ , առապեցy
րնⅾay եբաﬓﬓ ﬕ վիդ 'ի տան ﬓﬓﬓ
լեաyե առ կեբակուրս և րﬓﬕﬕﬕ ﬓﬓ
լսաբ\Xﬓﬓﬓ . \ﬕﬕ դ՞ պիտի ասելդայ
Նոսիկ՝ որք յատաել\Xﬕﬓ մեծﬓﬓﬓ
տեբ\Xﬕﬕ էին Հ \Բանգի եբբ անgyeal
կայցin 'ի մէջ ատրապանXﬓ կարդ ըստ
կարգ է յատﬕﬕﬕﬕﬕ սﬓﬓﬓ եբբ 'ի Հրա
ﬓﬓﬓﬓ լռուﬁﬓﬓ եդ\Xﬕﬕ ժոդովﬕﬕﬕﬕﬕ , դﬓ
ա-ր\Xﬕﬕﬕ ասﬕﬕﬕﬕﬕﬕ ﬕﬕﬕﬕﬕﬕﬕﬕﬕ ﬕﬕﬕﬕﬕﬕﬕ ﬕﬕﬕﬕﬕﬕﬕﬕ
յեբins aﬓﬓﬓ 'ի դﬕX իXﬕﬕﬕ ﬓﬓﬓﬓﬓﬓ
կադցﬕﬕ եբկեբպադ ﬁﬕﬕ․ ապa կaﬕﬕﬕﬕﬕﬕﬕ
և դդատ\Xﬕﬕﬕ 'ի վեբ Հﬕﬕﬕﬕﬕ ﬕﬕﬕﬕﬕﬕﬕ (ﬖ·
aﬕﬕﬕﬕﬕﬕﬕﬕﬕﬕﬕﬕﬕﬕﬕ) շարﬕﬕﬕﬕﬕﬕ , aﬕﬕﬕﬕﬕﬕﬕﬕﬕ
եﬕﬕ ժoduﬕﬕﬕﬕﬕﬕﬕﬕﬕﬕﬕﬕ oﬕﬕﬕﬕﬕ ﬕﬕﬕﬕ · և ﬕﬕﬕﬕ
պատﬕﬕﬕﬕﬕ uﬕﬕﬕﬕﬕﬕﬕﬕﬕ եﬕﬕﬕ ﬕﬕﬕﬕﬕﬕﬕﬕﬕ , uﬕﬕﬕﬕ
եﬕﬕﬕ eﬕﬕﬕﬕ ﬕﬕﬕ ﬕﬕﬕﬕﬕﬕﬕﬕﬕ ﬕﬕﬕﬕﬕﬕﬕﬕ ﬕﬕﬕ
կaﬕﬕﬕﬕﬕﬕﬕ , ﬕﬕﬕﬕﬕﬕﬕﬕ ﬕﬕﬕﬕﬕﬕﬕﬕﬕﬕ ﬕﬕﬕ
ﬕﬕﬕﬕ 'ﬕ ﬕﬕﬕﬕ ﬕﬕﬕﬕﬕ ։ \ﬕﬕ ﬕﬕﬕﬕﬕﬕ 'ﬕ ﬕﬕﬕ
ﬕﬕﬕﬕﬕ ﬕﬕﬕﬕﬕﬕﬕ ﬕﬕﬕﬕﬕﬕﬕﬕ . և ﬕﬕﬕ պatﬕ
ﬕﬕﬕﬕﬕﬕﬕﬕ յatﬕﬕﬕﬕﬕﬕﬕﬕ ﬕﬕﬕﬕﬕﬕ ﬕﬕ
Հﬕﬕﬕ ebﬕﬕﬕﬕﬕ յatﬕﬕﬕﬕﬕﬕﬕﬕﬕﬕﬕ ,
ﬕﬕ ﬕﬕ ﬕﬕﬕﬕﬕﬕﬕ ﬕﬕﬕﬕﬕﬕﬕﬕ ﬕﬕﬕﬕﬕ
ﬕﬕﬕﬕﬕ . և ﬕﬕﬕ ﬕﬕﬕﬕﬕ ﬕﬕﬕﬕﬕﬕﬕﬕ ﬕﬕﬕ
ﬕﬕﬕﬕﬕﬕﬕﬕﬕﬕ ﬕﬕﬕﬕ ﬕﬕﬕﬕﬕ , և ﬕﬕﬕ ﬕﬕﬕ

stauretur persuasio, nempe pretiosum esse apud Deum, et honoratum et laude celebratum apud homines Deum diligentes genus hoc (animantium).

At praelaudata animalia, inquiunt quidam, discunt quidem aliquid, sed eo magis, quantum magis cicures sint: agrestia vero non polite. Ecce enim etiam aliqua agrestia, et minime mansuefacta viderunt quidam in scholas venisse, familiaritate adepta.   26

Elephantorum gens nonne ferarum maxime agrestis est? Attamen ex Libya Baobas, Germanico Caesari undique ad pompam ludi apparatum disponenti, eo tempore quo de consulatu certamina proponebat, novam rem magnifice praestans dono misit gregem elephantum domi mansuefactorum ad cibum et potum deliciose [1]. Et quid oportet dicere de illis, qui ampliores erant in vasto dominio? Quum enim in conspectum venerunt theatri ordinatim dispositi, tamquam ex mandato silentium observante turba, primum omnes simul elephantes facie tenus genuflectentes adorarunt agonothetam. Deinde assurgentes, proboscides elevatas movebant, indicantes salutare populum. Postea spectatoribus eos laudantibus, illi quasi mutuam salutationem recipientes, proboscides velut dexteras deorsum demiserunt. Nonnulli vero eorum discumbebant illico; jam enim praeparati fuerant lecti ferrei solidissimi: provide quidem, ne immani pondere elephantum frangerentur; reliqui vero velut famuli pro ministerio stabant ante eos, et multa quidem convivio usuvenientia obsequiose exhibebant incunctanter. Neque post multum temporis surgens unus velut ex dapibus mensae, praeseferens similitudinem ebrii   27

---

1 Similia occurrunt apud Plinium. VII. 2. et Aelianum. II. 11. atque Martialem. L. 1. epigr. 1025.

Է՛ որ առ 'ի զկնոս ուրախութեան եր
թայ՝ պարտեքին անդանդաղ։ Իսկ ոչ յետս
քաղուս ժամանց յարուցեալ մի յետրա
խանագն՝ երբ 'ի սեղանէ զինուց սպասա
կութեն նմանեցուցանելով, զմարդկան
խազ զեղեղկանուաղ՝ 'ի փող և 'ի քնար
կապաւեալ ։ Իսկ այլքն յայտժամ զգլա
տիրձան շարժեքին. և եր երրեք՝ զի Հայումն
քննն առ կապաւս կատարեքին՝ երբ զովկ
լով՛ ։ Իսկ յետոյ երբ լոյս ոմն եկին եքեր,
զոր երբ սոսուն ժամանակ անդրքն զաւ
նացիկ կարծեալք՝ յոտն կացեալք անցա
նեքին զնայքն ։ Ոչ ևս յաշխարՀե որ առ
բեցուեամ սպաւալեաւ է կերպարանա
զուցանեքին, սայեքութեավ և դեղեկեավ,
և Հագիւ Հագ ոսու կարեքին շարժել 'ի
գնացս, բայզ գրդուեքին և ատանդղեքին
զորեն սրբելոց, միեչ 'ի ոտւարանէն 'ի
բաց չնացին ։

28 Ասի և յաղեքսանդրիա՝ որ յեգիպ
տոս է, թէ ուսաւ 'ի Հինումն դպրութիւն,
միեչ զի կարաց նշանակել զայս ինչ, Ես
ինքն դպա կրեկ ։ Իսկ ուսանել անասունն
այսմիկ և զոտփականն կեղծաւորութ.
Վ՛անզի երգս սանն՝ երբ մանկասիրս.
թեամբք վեգ Հարեալ յուղղոյ՝ Հանբուր
բել և գիրկս արկանել և պատառել, և
'ի նորա բերանն զիւր կերակուրն դնել.
ոչ 'ի գիշերի և ոչ 'ի տուել կարել ժամ
ունել 'ի նմանէ 'ի բացեալ, առանց Հրա
մանի վարդապետի գազանին ։

29 Այսպէս բնութիւն յամենայն ոգ
լոյ զիշխանական միտս Հաստատեաց.
այլ յ յուվմք՝ որ երբ Հունաղոյն է ենկա
բազբութեան ընդ աղօտ, և դիւրաշին
ձև կերպարանի. և է յուվմք՝ յորում
որ երբ անշինք յայտնապէս և դժուա
բալուծ կերպարան եկարեալ է։ Ազդ
անյայտ տեսանն յայլմն է. իսկ Հաստա
տուն և յայտնի տիպք 'ի մարդկան կեր
պարանս կրէ ։

30 Բայզ 'ի վերայ ասացելոց ոչ վայրա
պար և զայս ասել է, եթէ իմաստու
թիւն և գիտութիւն և քաջախորՀրդու

humano ludo ad pulchrum cantum tubae citharaeque saltavit. Caeteri vero tunc proboscides movebant, et identidem sibilabant ad tripudium, quasi laude prosequentes. Postquam demum unus lucernam aliquam adtulit, ea visa, tempus redeundi judicantes surrexerunt, et transeuntes discedebant, necdum omnino ebriorum personas mostrantes, sed labascendo ac nutando, ita ut aegre pedes movere possent in gressu; verum tamen vacillabant titubantes vinolentorum more, dum a theatro abierunt.

Dicitur etiam Alexandriae Aegypti unus istorum olim litteras didicisse, ita ut potuerit haec notare: *Ego ipse haec scripsi*. Necnon edocta dicitur bellua illa simulare animi motus sicut amatores; ajunt (vel canunt) enim elephantem tamquam amore pueri captum ferri in camelum, osculari eum, amplecti atque constringere, suum cibum in os ejus mittere; imo neque nocte, neque die pati ab eo distare, nisi jussu magistri ferarum.

Sic ergo natura in universis animis dominatricem mentem condidit; ita tamen ut in uno languida sit delineatio ac subobscura, et facilis ad delendum figura formae; in altero vero velut indelebilis, clara, et vix delenda forma depicta sit. Inevidens itaque forma est in aliis, constans autem evidensque typus in hominis forma geritur.

Porro praeter jam dicta non superfluum erit illud quoque dicere, quod sapientiam, scientiam, et optimam deli-

թիւն, և այն եա որ յառաջատեսակն է
Հանդերձ նախախնամութեամբ, և որ
խնչ միանգամ այլ ևս խորհրդոյ խնամ
առութեան եղլայր է, բազումէ և այլք
'ի կենդանեաց՝ զպյտիք զորս բանական
ասեն ոգւոյ առաքինութիւնք են՝ ունինն
Բայց անստո յաղագս սոյս զոլ բանի՝
յայտնագոյս Հաստատ՝ պահարանք Հայ
թայթամաց Հարկաւորացն են իբբ, և
մանտք, զորավարութիւն, Հաստ ազդի
ազգիբ : Բանդի բազմատմնիդ յորժամ
յորս կերակրոյ է, զիւր մարմինն միմաց
նմանեցուցանեն՝ առ 'ի պատրել զձկունս.
զի Թուեցեալ թէ առ բարամիբ լա
զինն, բուռն Հարեալ ընձութե : Իսկ որ
ասին նարկա, զգոնմանան ապա սյորժամ
կազմեցգէ այնոցիկ որ 'ի նա Հային
ձկունբն, զիւրաե զնոսա կարթաբար
որպայ զգդժս որովբ ընդդեն դառան
առ խնգ ձգելով: Իսկ որբ կոչին աստղ,
կարի բաժ իմաստնապար կանան և ախորժ
ժելի արարեալ զորէն խաՀանրի զձետ
երբեալ : Բանդի թէ զձկնբեանժ զգաա
զե խնչ որպես խմ Թոււբ՝ ընտութեան
տաաւգտն խնչ, բաժութերեալ աՀա և 'ի
վերայ մեծամեծացն ևս յարձակի . բան
զե առ 'ի Հրոյն տապ ընդդիման ոչ
կարացեալբ, խորովեալբ անցաաազին
կերակրոյն վարի:

Ըո ստանան որպես ջրայիբդ, այս
պես և Թոչունք և ցանակապայնբ յաւե
լորդ խմաստութեն. բանդի աման յացեալ
է և ոչ ոբ յխմաստութեն, թեպես և
տեղեոբ և կամ մարմնով ջ զատեալ որո
չեն 'ի միմեան : Ըո նշանակ խեզեսոր
Թայդ կատարելագոյն է, որոց մարմինն
ունելի է . դարանատււո լևն երկուբ

պատեկիւք. մինչև 'ի մին9է կոդմանէ,
և մևւսանին 'ի մին9է՝ որբ վնովաւ անեկեաւ
պնդեն պատեածնբն : Ո_րդ գտոա վասն
զի.բակեւ անՀւաբ էր, որ սասին Թնսձբ,
(ա. պեդւականբ) դիւբաՀսատարար կյա
նեն յառաջադդոսն, զի կոկողդեն ջերսնւ
Թեամե Հայեսգն, որ դդստեանան պնդ.
պաՀէ սնսսն, և սասյաւ Թուչագեաւ_ար.
ձակեսգեն աղեբն. յայմեմ'ի դղւբս բե
կեգեաւ 'ի ձեռն բնդսրձակույԹէ՝ զոււ
ատեյին 'ի նեբբս կորգեաւ_ մասոււդսնեեն
անձսնդ :

32   Լ_ սսայս և ոչ ստարակոււեսււդն,
մինչ խսսւս անսսսյգեւււԹիււն՝ ուննն կբբ
Թուֆեան նս Հսսբից սյնսգեկ, որսդ
բնդ Հսոսն իմնդան անսսսոււնն, 'ի խոր
յորջս կամ'ի մսյբս Թսբոււգեսւդ դա
բանասբար սպասեն : Լ_ են որբ եբր 'ի
Հ֊ն֊ն֊դոդ են ֊սղ֊օ֊դդ ս֊ուտ֊ ֊չ֊նու ֊ձ֊ձ֊ն
ա֊ն֊դ֊ա֊ն֊է֊բ խ֊սն֊ած ֊ս֊ո֊ւ֊եսս֊᾿ ֊ո֊բ֊ն֊զ֊ս֊ս֊վ֊ս֊ս֊ո
լեն֊ի֊, դ֊ո֊ղ֊ե֊ւ֊ո֊ւ֊֊֊ 'ի ֊մ֊ն֊ո֊ւ֊ո֊ջ ֊ս֊ս֊ս֊ս֊ո֊ւ֊է֊᾿ ֊ս֊ս֊պ֊ս֊ս֊
կ֊ս֊ն֊ե֊ւ֊. ֊բ֊ս֊ն֊գ֊ե֊ յ֊ս֊ս֊բ֊ո֊ւ֊գ֊ե֊ս֊ւ֊ 'ի ֊դ֊ս֊ս֊ս֊ա֊ն֊է֊᾿
ս֊ն֊դ֊ե֊ն֊ ֊վ֊ս֊ս֊դ֊ս֊ս֊ս֊ս֊ս֊կ֊ե֊ ֊ս֊ն֊ս֊ս֊դ֊ ֊ս֊ո֊ւ֊խ֊ս֊ն֊ս֊ :

33   Լ_ դդոււշսսկսն սսպս֊ե֊ս ֊ս֊ս֊ս֊մ֊ե֊ ֊ի֊ս֊
բ֊ե֊ս֊ն֊դ֊ ֊վ֊ե֊ն֊կ֊ո֊ւ֊Թ֊ե֊ս֊ս֊ ֊ի֊ե֊ն֊ս֊ս֊ ֊զ֊ի֊ ֊ո֊չ֊ ֊մ֊ի֊ս֊յ֊ս֊
դ֊ս֊ս֊ 'ի ֊յ֊ս֊յ֊ց֊ո֊ջ֊ս֊ ֊դ֊ս֊ս֊ս֊ս֊ս֊ն֊, ֊ս֊յ֊ ֊դ֊'ի֊ ֊մ֊ս֊ս֊դ֊
կ֊ս֊ն֊ ֊Հ֊ո֊բ֊ս֊ս֊բ֊ս֊ս֊դ֊ ֊խ֊ս֊ս֊ս֊Հ֊ո֊ււ֊դ֊ս֊դ֊բ֊ս֊ ֊խ֊ս֊յ֊ս֊ ֊ս֊ս֊ս֊յ֊
\Ո_ս֊ս֊դ֊ս֊ ֊ս֊ս֊ս֊ե֊ս֊, ֊Թ֊է֊ ֊ե֊դ֊ջ֊ե֊դ֊ս֊ւ֊դ֊ ֊Հ֊ս֊ս֊դ֊ս֊ե֊դ֊
ձ֊ե֊ս֊ս֊ ֊բ֊ս֊կ֊ե֊ն֊ո֊ււ֊ ֊դ֊ե֊դ֊ջ֊ե֊ս֊բ֊ս֊ս֊, ֊ի֊ս֊ս֊դ֊ս֊ե֊ն֊ ֊յ֊ս֊
ս֊ս֊ս֊Հ֊ո֊ււ֊Թ֊ս֊ս֊ս֊ ֊ս֊ս֊ս֊ս֊ս֊ս֊ս֊ս֊Թ֊ս֊ս֊ս֊ս֊,'ի֊ ֊բ֊ս֊ս֊
ձ֊ս֊ս֊ս֊ս֊ ֊լ֊ե֊ս֊ս֊ս֊դ֊ ֊ե֊ս֊ե֊ս֊ս֊բ֊, 'ի ֊մ֊ս֊յ֊բ֊ս֊, և
կ֊ս֊ս֊ե֊ս֊ ֊յ֊ս֊ս֊ս֊ս֊ս֊ս֊ո֊դ֊ց֊ս֊ ֊վ֊ս֊ս֊խ֊ս֊ ֊յ֊ս֊ս֊ս֊ս֊ս֊դ֊ ֊ս֊ս֊ս֊ս֊ս֊
ս֊ս֊ս֊ս֊ս֊ո֊ ֊ե֊ս֊բ֊Թ֊ս֊ս֊ս֊ : Ո_ս֊դ֊ ֊վ֊ս֊ս֊ս֊ն֊ ֊վ֊ի֊ ֊դ֊ե֊դ֊զ֊ե֊ն֊
բ֊ս֊յ֊ս֊ ֊դ֊դ֊ե֊ս֊ս֊ս֊կ֊ս֊ս֊ն֊ ֊դ֊ո֊բ֊ս֊ս֊Թ֊ս֊ե֊ն֊ն֊ 'ի ֊բ֊ս֊ս֊ց֊ ֊ս֊ն֊
կ֊ե֊գ֊ս֊ն֊, ֊ե֊ս֊բ֊ 'ի ֊ս֊ս֊ս֊ս֊ս֊ս֊ս֊դ֊ս֊ս֊ս֊ն֊Թ֊ս֊ս֊ն֊է֊ ֊վ֊ս֊ս֊
ձ֊ս֊ս֊կ֊ս֊ս֊ս֊ ֊մ֊ե֊ս֊բ֊ս֊ ֊ե֊ս֊ս֊ս֊ն֊, ֊ե֊ս֊ս֊կ֊ս֊ս֊ս֊բ֊ ֊ե֊ս֊ն֊
մ֊ի֊ս֊յ֊ս֊ ֊յ֊ս֊ս֊ս֊դ֊, ֊դ֊ս֊ ֊վ֊ս֊ս֊ ֊ս֊ն֊ս֊ս֊ս֊ս֊դ֊ս֊ս֊բ֊դ֊ս֊ ֊կ֊ս֊ս֊ս֊

altera ex altera, circumcludunt eam:
quas quoniam dissolvere desperandum
est, Paelotes dicti [1] facili negotio in-
glutiunt eam prius, ut per calorem gulae
liquescat viscus, qui solidas servat testas,
et paulatim remissae dissolvantur com-
pages: tum extrahentes in largitate, e-
scam inclusam sibi administrant.

Nec tamen in solitudine (vel solicitudine) degentes extrema laborant cibi
penuria: habent enim exercitium indu-
striae; ea nempe, quorum odores se de-
lectant, belluae in latebris et cavernis,
aut sylvis latentes sicut insidiis expe-
ctant. Sunt etiam, quae tamquam ex pin-
guedine ab ipso aëre odorem sugunt
naribus: foetorem vero sentientes, nares
dimovent, et latentes in caveis praesto-
lantur venationem; surgentes enim ex insidiis, statim avide saturantur.

Adeo autem nonnullae cautionem usur-
pant de salute propria, ut non solum
dolum inter se mutuum, verum etiam
fortiora hominum consilia eludant. Dicuntur enim cervi, dum projiciunt cornua, quaerere sicuritatem solitudinis,
et ascendentes montes altos, petere
sylvas, inaccessibilesque anfractus, et de-
serta impervia. Quia vero cum cornuum,
armis aequalem vim exuerunt, nudique
instrumentis defensionis evaserunt, ob
metum ne irremediabile aliquid patian-
tur, in solitudine spem ponunt, quasi
restet eis fuga sola: unde et proverbium

---

1 Παιλοτής, ut ἀπαιολητής, fraudator fortasse est. Quod si legas Polypodes, tunc spectat
quodammodo locum Plinii. IX. 46. « Polyporum
multa genera ... vescuntur conchyliorum carne,
quorum conchas complexu crinium (sive bra-
chiorum) frangunt; itaque praejacentibus testis

cubile eorum deprehenditur ». Quae sunt fere
verba Aristotelis. hist. animal. VIII. 4. Οἱ δὲ
πολύποδες μάλιστα κογχύλια συλλέγοντες, καὶ
ἐξαιροῦντες τὰ σαρκία, τρέφονται τούτοις· διὸ
καὶ τοῖς ὀστράκοις οἱ θηρεύοντες γνωρίζουσι τὰς
θαλάμας αὐτῶν.

գէն, իթր թէ կայ ինայ նոցա առաջի մխայն փախումստ. յորմէ և առակ մի յայսմանէ եղև, Ո՛ր եղջերւոյ դղշէ՛րս ընկենու, արդարև՛ կոյր անապատագոյն է : Վ՛որ վասն զիյետ ընկենուլ զեղջերւսն՛ առ արմատան լինին ցաւք վիրին, թուջիս Խանձք 'ի վերայ և առնեն, և ցաւս դժնդակս 'ի վերայ Հասուցանեն : Վսկ նա 'ի տուէ 'ի խարձն և 'ի թալ ան, տատաս մնեւալ դողէ, յստացոյ և 'ի բաղմախխո սառարթիյն ծածկեալ լինի. իսկ 'ի գիշերի յորժամ թռչունք գղուլ առնուն, ելեալ անԵրկիւղ Հաշրակին : Ինչ գարձեալ եղջիւրս բուսեալ՛ ոչ քաշալ ընն աշա 'ի Հասարակ աւուր ելաներ 'ի գուրս, կարծեցեալը թէ յԵ՛ իս Հաս տատեալ եղջերն՛ Քանզի մինչ գեռ փափուկէն են, սպաԵալ լինին իբր Խայ ռոստւթեանէ յարԵղակնստյին տատոյն, որպէս կաս բրոմ 'ի Հրոյ . Քանզի չէ՛ Ես Հաստատեն (բ. Հաստատցեալ) թէ բարիորք է կառուցեալ Հաստատութէ, 'ի ծառս զգլուխն Հարկանեն, և փորձին գեղջերսն՛ չփելով և միելով, եթէ Հաստատութեամէ է կառուցեալ :

Վսկ որ իմստսւթեամէ նախախնամ մութիւնն է՛ ոչ միոյ, այլ ամենայսին, ամենայն անասնոց ազգ՛ ժամանակին, բստ որում՛ մասսաւանդ ծնունդին արդակին յանդիմնեալ լինի : Քանզի է՛ ինչ 'ի սՆ գանԵ, որ առ. բարձրագոյն լեռանց գիմք, և է որ զխորագոյն սիլբաց ցոյն և լերացւցոյն գանապատան յոյ զ գւա սնե՛. և է որ 'ի թանձր մացաս առ գեալ մՆրիս առնու. և բազում իս ցԵրկ երկրի գոցես յստուս բուսեալ, մինչ 'ի ծնա սնել խստ վարխ : Ինչ Ես սրա իստաա թԵան ցոյցք աւելորդի, Համբերձ իսնա տութեանէ զստխախնամական զգուշու թիւն ծնանդոցն յուցանԵլով :

Վայց սբանքելի ինն գործԵն Հալ բաղԵգ, որ ամԵնայն բասի արժանի է :

illud, *Ubi cervi cornua deponunt, profecto locus est desertissimus*. Quoniam ergo post cornua rejecta, juxta radices extant vestigia vulneris, supervolantes muscae pascuntur, et saevissimos dolores affe-runt. Quare illi per diem intrantes dume-ta, nemora, frondosasque sylvas, latent, ramis condensisque frondibus obducti: no-cte autem, quando volatilia acquiescunt, exeuntes tutius pascuntur. Ad haec re-natis cornibus, haud confidunt statim meridie foras exire, veriti necdum sta-bilita fuisse cornua: quum enim adhuc tenera sunt, quasi lutescunt ac discin-duntur ab radiis solaribus, ut argilla fi-guli ab igne. Quam ob rem diffidentes, ne forte perfectam adhuc non acceperint soliditatem, in arbores percutiunt capita sua, et probant cornua per con-tactum et compulsum, an firmiter ere-cta sint.

34 Quod autem prudentia provida insit non uni alterique animalium, sed cun-ctis, omnium eorum generis tempus sta-tutum, maxime pro propagatione, ma-nifesto argumento est. Unum enim isto-rum excelsos montes petit, alterum profundum Scythiae Libyaeque deser-tum lustrat, tertium in nemoribus con-densis immersum latebras sibi assumit: multaque in terra invenies ripas[1] obside-re, quum ob procreationem huc con-ducuntur. Haec sane indicia sunt sa-pientiae redundantis, ac una cum pru-dentia providam diligentiam in generan-do ostendentis.

35 Verum mirificum quoddam operantur Palumbes, ac perfecta ratione dignum.

---

[1] Ad verbum, *rivus*, vel *in rivis germinans*; quo et *stagnum* potest designari, vel *arundinetum*.

Նախ նա առաջին ձ ուր ծնանինն
և ձարակին, այլ յայլ տեղի փոխին
յայսմանէ յղղագս ընելունելոյ . քանզի
բազումք են` որ սպասեն նոցա գտորմն
գտանել . յայսմանէ յղղագս զգուշուԹէ
ի բաց փախչին` զի մի գտցեն ինտրրակէին
Ձ. ապա իբր զղան` Թէ ոմանք յորս և
եռալ են յղղագս ձագուցն ,յոյժ արու
եստական կռուոյ փոխառակ մամոֆին . ի
վերայ Թռչելոֆ, և ի բաց փախչելոֆ .
և գարձեալ Հանգարոֆիկ Հանգարոֆիկ
գուղ գուղ երԹալոֆ . և փորք մի սակա
ի բաց Թռուցեալք , մինչ գորորդագն
փոյԹ տեսեալք Հողորբեացեն, պարապ
առեալ ձագուցն փախստեան ժամանա
կէ Ձ յակյորժամ գգայեն ի բաց Թռու
ցեալ, սպասեալով առ գարշաւարորն ,
կարի յոյժ քամնՀաբար արիեճցագ լլ
նին Ձ Ձակ նորա են պակուցեալք ի ձա
դուէ խաբեալք ի պատրանաց, որպէս
գեղ է ՝ գմֆղակուԹեան երևալք Ձ

36 Ձեէն որք գսեմելայն պատմեցին գ
բնուԹեանն և գծոֆայինն , չարադու
Թիւամբ իմն գիլի որումն ֆարել . Ձան
զի եԹէ ոչ կարասցէ զաա որ ի սկգբան
է բազմ որբինակէ մարգկանէ ի չարաՀար
արուեստուֆն գղուզուԹին առնել . իսկ
աֆսոցիկ որ ի ֆամֆձանն է անսգուստ
գործեղոֆ գյարձակմանն Ձ Ձանզի
կարԹ ոֆինչ սպանՀելի խաոճիք սպա
տեալ իբր գկերակուր կլանէ . և ֆարեն
սպաՀրանս երիս փախստեան Ձ Ձան
գի ի ֆերայ յարձակեալ և ընԹացեալ
գլարն ուտէ , երկուս իձս օգսակարա
առնելոֆ , իւր փրկուԹին անձին , և
տասնֆանս այսմ որ սկիգե . ծեռաց անե
բացեաց Ձ

37 Ձանգի Հրաչափառագ ձֆս Թեէրես

Nam in primis non ibi, ubi nascuntur,
pascuntur, sed in alterum locum se trans-
ferunt, ne capiantur; plures enim sunt,
qui machinantur, ut inveniant earum
latebras: itaque illinc ob cautelam effu-
giunt, ne persecutores inveniant se .
Deinde vero quum sentiunt quosdam
venatum exiisse pullorum, nimis artificiosa pugna accedunt in congressum,
supervolantes, et fugientes, et iterum
lente suspenso gradu obrepentes: atque
paulo longius volant; ut dum studium
venatorum vident, ipsos liberent, spatio
temporis pullis dato pro opportunitate
fugae capessendae: quos quum sentiunt
longe evolasse, expectantes adhuc modicum vestigia venatorum, demum de-
lusis eis, ipsi aërem petunt excellenter;
ita ut isti stupore agantur, derisos se
deceptosque cernentes, et exacerbentur,
ut par est .

Dicunt, qui Semelae [1] res narrave-
runt, quod genus est marinum cooperatione daemonis cujusdam usum; quod si
nequeat sub initio evitare multiplicem
hominum industriam malignam, postre-
mam eorum irruptionem irritam reddit.
Quippe quod hamum nulla exquisita esca obductum velut cibum inglutit: malum tamen effugiendi remedium invenit;
invadens enim et currens sursum, funiculum comedit; geminum emolumentum
ferens, et sibi salutem comparans, et
punitionem in eum, qui manu injurio-
sus fuit .

Incredibilem utique censeat quis hi-

---

1 Quid in Gr. lectum fuerit ab Interprete sic
istud nomen exponente , nos latet: σμίλη , σμιλίον , potuit esse nomen piscis, sicut scalpello
funiculum amputantis, vel ζῶον μέλαν . animal
nigrum, aut aliquid simile . Si vero legas μύξον , Lat. mugil, haec de eo habet Plin. L. xxxii.

c. v. « Scit et mugil esse in esca hamum , insi-
diasque non ignorat: aviditas tamen tanta est,
ut cauda verberando excutiat cibum ». Sed haec
et similia non ita facile credas ad praesentem locum pertinere . Varii sunt pisces, qui funiculum
hami abscindunt.

կարծեցէ որ զպատմութիւն զայնցիկ
որ 'ի Թրակէն են բազէք . քանզի և ես
յորժամ զառաջինն լսէի՝ Թերահաւա-
տէի , մինչև բազումք 'ի գաւառակա
նաց , և այնցիկ որ առ մեզ ոք եկն յօ
տարաց՝ որ զանմռունթեան 'ի մնի և
զեւալք կրթէին՝ առ վկայէին , զոր ինչ
զինէին յաղագս նոցա ։ \Բանզի աստ
ցին , Թէ միաբանութեամբ գործեն , Հա
ւուն՝ որսորդացն օժնականութիւն մա
տուցանէլ . և վաստակէն յիրս , և 'ի
վայելութիւն այնցիկ՝ որ ընդունեն
զեռալ լինեն ։ \Բանզի առնեն զնակոն
թեան իւրեանց յանտառս՝ որպէս վայել
է , 'ի Թանձր և 'ի խիտ մայրիս , ուր Հա
ւուց աղգքն շրջին յաճաս . իսկ այնցիկ
որ յորս գան՝ գործակից լինեն , այնցիկ
որ նախախնամութեամբ մատչին , առ
այսուիկ որ ոչ խորեն , և առ յոյս միզա
նակին զուղաանմատութեան , զոր վայելէ
'ի մարտակցացն մարտակցացն աներ ու
նել ։ \Սլ օրինակ զորավիզօ օժնական ու
լինեն այսպիսի ինն ։ \Բանզի յառաջ
քան զորսորդն 'ի սկզբան անդր ծառքն
շարժեալ լինեն ։ \Սլ զշարժումն փա
նաքի և զուզաակբաց Հաւքն ոչ կա
բեն տանել , 'ի բաց Թոթափեալբ եր
կուցեալ լինեն , և Թոշել 'ի բաց կամին ։
\Սկ բազէիցն զեր 'ի վերայ նոցա երևե
ցելոց՝ փոխանակ բռնադատեալբ 'ի վե
րայ երկրի իշեալ ստածին կոցելով , պատ
բատելով զիւրորս զորսն այնցիկ որ
յայսմ եկին ։ \Բանզի աստ մատոցս ուզգքին
զղզունին նոցա , և որ ստասիկ եղեալ լք
են 'ի շարժմանիկ կայլքիակք , և յայսուա
նէ որ 'ի վերույս նոցա բերեալ լինեն մի
զումին , միանգամային և աՀ, և աՀզգքին
պախուումին գործէ իբր զերան առ 'ի
յրմենունիմ լինել, ոչ միայն յորսողացն
'ի ձեռս մատնել ։ \Սկ նոքա ինդուԹբ
տան մասն յայսցանէ՝ զորս ունին , է
ինչ որ վասն գործակցուԹեանն փոխա
նակ Հատուցանելով , և է ինչ որ առ
Հանդերձեալ աւն որս Հրաւիրելով ։

storiam accipitrum Thraciae; siquidem
et ego ut primum audivi, minus fidem
praestabam, donec complures indigena-
rum, (quos inter unus, qui ad nos venit
peregrinorum, homo simplex,) testaban-
tur quidquid de illis noverant. Dixe-
runt enim, quod concorditer laborant,
ut opem ferant avium venatoribus, et
operam dant negotio et fruitioni capto-
rum. Quoniam habitationem sibi pa-
rant in sylvis, ut par est, inter densas
frondosasque quercus, ubi versantur a-
vium genera frequenter; et eis, qui venatum veniunt, cooperatores existunt,
iis inquam, qui provide accedunt, quique nihil fraudant circa spem aequalis
praemii, quod licet expectare socios a
sociis belli. Modus autem auxilii prae-
stiti ita se habet; primum (vel, ante) venatores sub initio arbores commoventur.
Hunc motum exiguae ac parvulae aves
non sufferentes, disjectae expavent, et
avolare volunt. Accipitribus tamen super se visis, retro coactae humi descen-
dunt, et rostris terram pulsant, (aut
ab accipitribus rostris laeduntur,) atque
ita venatio paratur facilis aucupantibus.
Quoniam ramorum vehemens commo-
tio, et violentus ex motu saltus, atque
ex superstantibus impulsus, sicut etiam
timor, et ingens horror facile reddit
aucupium, quo magis incidant in manus
venatorum. Isti vero laetabundi exhibent portionem ex iis, quae ceperunt,
partim cooperationis gratia retributio-
nem facientes, partim ad futuram venationem invitando.

38 Իսկ այսչափ աճելով յղղացան առ յա‐
լեւանումութիւն ինասատութեան, մինչ զի
'ի Հարկաւորապղտան` եւ քան զմարդիկ
 անցոյց ինասատութեամբն ։ ` Քանզի մեր
իւրաքանչիւր ոք յորժամ Հիւանդասցի,
այնօրիկ որ երթան յատոռչութիւն ոչ
գիտացեալ, բժիշկս կոչէ` որք գուցցենոն
զիւրկականն որ 'ի բժշկութիւն երթայ ։
Իսկ (ասկայն 'ի կենդանին) ոչ մեայ ուրուք
այլոյ ամենեին կարոտեալ` բժշկական
իմն իմննունիմն, եւ իմնասարդապեռ
սասրար ընկալեյոց ։ `Քանզի եղջերուք
առ իսանձս փաղիժիոն իմնգան փոսա‐
սակ դեղ իւրեանց իցզիտեթ, որոյ կեր
ասկուրն անղեն բժշկէ ։ Իսկ 'ի կրետէ
յորժամ այծ նետիւ Հարցի, այս որ կոցի
ողկուզուին իմնղրէ, եւ կերեալ` դիւրաւ
մերին բժշկեալ լինի ։

39 Ասեն, թէ` եւ կրայք յանյաղութենէ
որկորութեանն առցեալբբ,յիմիդ` ու‐
տեյ զուիրակն 'ի վերայ երկիւղէն մահու,
քանզի մեայն այսպէս 'ի թունաւերացն`
որ մնասա լինին` ճողոպրել կարեն ։ `Քան
զի աշա ունանց ոչ իսիկ յատաձայոֆն Հա
ւատալ սովորեցին առանց իսապարտրի,
կամեցեալ սասուդապօֆն յանղիսանութէ
զասացեյոց ընդունել ։ Օ կրայ աս‐
սեալ առցեալ յիժէ` զայս որ մօա առ‐
նոն բուսեայն էր զուիրակ, նողիմբբ
իմնն արմասնդֆն 'ի բաց կորձեայ` զեա
աննունէր . իսկ նա 'ի վերայ յարձա
կեա, ցուն 'ի զգասանէ սորբ` մեասնեք ։

40 Ոչ միայն բժշկել Հիւանդութիւս,
կամ որ իցէ իսկ աշա յուսերֆ, եւ կամ
որումֆ անինկալութին իցէ, այլ և ինչ`
որ եւ ցաս սասարածեայ պատճառ, զի
մի բնանդատեղ` յայս լինի ։ `Քանզի
ասէն, թէ արիսաողիսանֆ` այսողիկ որ
վարեցան առաբինութեամֆ'ի վարս բար
զաբին յայսանապէս երելեայ 'ի յաթէ

1 փաղիժ, in marg. փազանֆ, videtur innue‐
re Gr. vocem φάλαγξ, vel φαλάγγιον, ut aranei
speciem, sive tarantulam. Vide Phalangium apud

Adeo autem abundanter effluit copia
sapientiae (animalium), ut in apprime
necessariis superet vel hominum sapien‐
tiam . Quandoquidem unusquisque no‐
strum quum valetudine laborat, ignorans
quae conducunt ad sanitatem, medicos
advocat, qui indicent salubria ad salu‐
tem proficua. Ubi animantium nullum
umquam alterius opus habet, medicam
artem per se ipsum callens, et suo ar‐
bitrio professor compertum . Siquidem
cervi a phalangio [1] morsi, norunt pro
remedio cancrum (sive ostreum fluvii);
cujus esu statim curantur. In Creta ve‐
ro quum caper sagitta feritur, herbam
racemulum (vel dictamum) vocatam
quaerit, quam vescens facile sanatur ab
vulnere .

Ajunt quod etiam testudines ex avida
gula saturatae viperis, ob metum mor‐
tis origanum comedunt; hoc enim tan‐
tum modo ex veneniferorum discrimine
se expedire possunt. Verum enim vero
nonnulli qui nihil credere consueverunt
nisi exploratum, voluere certius argu‐
mentum dictorum capere. Videntes ita‐
que testudinem expletam vipera, origa‐
num prope eam natum radicitus evel‐
lentes, expectabant exitum rerum: illa
vero appetens, et non reperiens ut ede‐
ret, mortua est .

Nec solum curare morbos in aliquo
repertos, aut suspectos, sed dolores etiam
in se fingere fallaciter ad evitandum la‐
borem, observatum est (in animalibus).
Dicunt enim, quod Aristogitoni uni in‐
ter eos, qui virtute claruerunt in Re‐
publica Athenarum, equus erat piger,
qui in hippodromo ultro fingebat clau‐

Plin. variis in locis. In sequentibus ողկուզուկ
refunditur in Gr. βοτρίδιον. q. d. botriculus,
vel racemulus .

*Նաս, Հեղդ չի մի էր, որ 'ի կրկնեին ըստ
ելլեցն կերպարանեալ պատահաւէր կա
դալ. միսլէ ընթունեցաւ ժամանակաւ
'ի բուծողացն, զի խաբէր, և ստատկե
յաւ. որոց մազգվեն գարձեալ քան բղ
գանաղական 'ի վեր եղև. Բանգի քէն
նութիւն յայսմէ, քանզի յայասարլեա
Հարկ էր ենեգաւորին կեղծաւորութէ
զգենուլ, բայց խմատու--թիւն միաց:*

*Ո՜չ վայրապար և այս աստ, թէ ան
զեորզաց ունին միտ խորհրդայ արջա
ռոյն դաստկէ, և Հոբանք այծարա
ծաց, և Հօբքն Հովուաց. և որ ինչ
միանգամ 'ի զաղանաց վլայբեակողին
են, այսոցիկ որ Հոգն տանին ։ Ալ սա
կայն և կերակրող և ընակելեաց և միա
բող, և միանզանան որ միանզանմ յող
տութեն կամ 'ի չեան են նողա որխն
նուաղ է Հաստմեն գիտութեան, մինէ զի
և ժողով բազմաց և զանազանեցող միաց
ունին: Ալ է այս արութետական և բա
նական: Ալ սակայն յարզա միատական
բարութեան ծառ լլցե միեա ընղ նոցայն
զատմէլ:*

*Ո միջինդ ոչ տեսանես, քանզի դող
նարէս խիմ մարմին կենդանոյ է, և
մեծամեծ Հողական խարգաշխանօ ֆն և
մնտքն զորձող: Ս ա ամարայն Հայթալ
թէ, զե մի Հեղեղեսցէ, կզուեալ Հանդե
պեցուցանէ խր որխ. և աշտացց ան
դէն վախճանեցելոյ 'ի ներքս սպրդեալ
լ, 'ի ցրտոյ և յանձրեաց գիշան գգուշանն
լով: Ս ի 'ի ձմերոցն զորս նայն զոր
ծեաց, 'ի ներքս արգելեալ է, և զիբր
վաստական Հանեարբեաց անդ, ուրախ
լենի և բակկոկ Հանեն զկաղածոցն պլոս
առատապէս ։ Իսկ յորժամ պակասէ
զեալ լլցե ապոցիցն, դարձեալ ելանէ
'ի դուրս առ 'ի ժողով զունմերեցլոյ Հար
կաւորագն կերակուրս պամճարելով յայն*

---

1 Aequivocum est in Arm. *ստատկեցաւ*, potest
enim intelligi etiam, *deperditus fuit:* quod tamen
non patitur sequens periodus; ubi inaudita est

dicare; donec deprehensa fuit ab agasc-
nibus processu temporis fraus, et casti-
gatus fuit, [1] quo supplicio deinde supe-
ravit ignaviam; atque ita experimentc
probata fuit, quod evidenter fraudu-
lentus dolo usus fuerat, non aliundc,
quam sapientia mentis.

Haud inaniter et illud dicitur, quod 41
bubulcorum mentem et ingenium habent
bovum armenta, sicut et caprae capra-
riorum, et grex pastorum, et agrestio-
res belluae curatorum suorum. Verum
tamen ciborum quidem et potuum, prae-
sepiumque, et generatim eorum omnium
quae ipsis utilitati aut damno sunt, non
minor est peritia cognitionis ; ita ut
etiam copiam quandam teneant multa-
rum variarumque notionum, tam artifi-
ciosam quam rationalem ; ut proinde
absurdum sit maxime, inter se conferre
nostram et illorum providentiam.

Formicam non vides, exigui corporis 42
animal, et magnorum operum providum
per industriam ac mentem ? Haec aes-
tiva tempestate comparat necessaria ; et
ne forte inundationem patiatur, soller-
ter sibi parat sedem; autumno vero ver-
gente, obreptum ex frigore imbrisque
damnum cautius reparat; ut quae in hybernis a se constructis inclusit, et labo-
re proprio condidit, iis fruatur et gaudeat, horrea pro tempore insumens, et
copiose hauriens victum. Quando vero
accidit ut deficiant alimenta, rursum exit
foras cum suo exercitu, ut colligat copiam annonae, necessarium cibum procu-
rans pro tempore futuro. Ut autem malo-

vox Մաղովք, Graeco more, ut putatur, accepta
pro μάστιξ, *flagellum.*

ես ժամանակ ։ Զ գզարեացն ես դմին
դակութիւն 'ի բաց մերժելով գումի, զո
րբեանս և զարիս և որ ինչ միանգամ այլ
ինչ ժողովեսցէ, ընդ երկու հատանէ,
զի մի քսուեալք՝ անքատացի 'ի կերակ
րոյ ։

rum saevissimum expellat famem, triticum hordeumque, et quaecunque collegerit, in duo intercidit, ne pullulantibus illis penuria laboret.

43 Ո՛չ բաւեսցէ օր պատմելոյ զՀոմա
նման զայլոցն զՀոգս կենդանեաց, որոց
պատրաստութիւն կազմածոյ առատա
քար կերակրոյ վարի։ Քանզի ամենայն
որ ինչ միանգամ ցամաքային և կամ
ջրային, և կամ օդազնաց ինչ, զի մի՛ եր
բէք նոցա Հարկաւորքն պակասեսցեն,
նախախնամութեամբ ինն Հոգան ան
ճանg ։ Զ առանց այսոցիկ ո՞ ոչ գիտէ,
զի յմտութնդգաս միսա խոհՀրդոյ են՛
առնուլ զօգտակարսն և զանօգուսն,
զարժանաւորն և որ արտաքս քան զար
ժանն 'ի զանազանից ։ Իսկ սոցա Հա
սուին քննութեան ստուգագոյնք են, և
առաւել յօդացեալք՝ բանականին են
ումունքն ոգւոյ ։

Non sufficiet dies ad enarrandas consimiles caeterorum curas animalium, quorum apparatus oeconomicus abundanti gaudet cibo. Quandoquidem omnia terrestria, aquatilia, et volatilia per aërem, ne forte aliquid necessarii desit ipsis, provide sane procurant. Praeterea [1] quis nescit, quod intellectu consilii praeditorum est, eligere utilia et inutilia, convenientia, et inconvenientia cum distinctione? Istorum vero peritia nimis certa est (apud animalia), et solidata disciplina, animae rationali propria.

44 Բայց սակայն և այս աս արժանա
պէս, զի առ ընդդէմսն ոչ նոյնակէս վա
րի ։ յայս է՛ զի 'ի մարտնչողացն նոյն
պէս շարժեալք։Քանզի ջերմ կամ Հով
կամ քաղցը և դառն, կամ սպիտակ կամ
սեաւ, կամ մեծ և փոքր, և կամ որ ինչ
միանգամ յայսոսիկ որ ընդդէմն խառ
նուծք միաբանութեան են, միրը խոր
Հրդոյ այլակերս դնէ 'ի նոսա՛ առ 'ի կա
մել և ախորժել զՀեշտն զարաբրականն
Հեշտ ցանկութեան յորս փութայ, և
փախստեայ է 'ի ցաւագնացն՝ յորմէ 'ի
բաց դառնայ ։ Զ այսպիսի առաւելու
թեան իմաստութեան վարի, մինչ զի
և բազում ինչ 'ի խօսականէս յինքն
ձգեալ տանի՝ զխորՀրդականութնն
տեսակս մեկնել ոչ կարացեալ. վասն զի

Attamen illud etiam rite dicitur, quod oppositorum non idem est usus; quoniam, ut patet, a contrariis tales proveniunt motus. Quoniam calidum ac frigidum, vel dulce et amarum, vel album nigrumque, vel magnum et parvum, aut quidquid eorum, in quibus contraria sunt passim temperamenta, diversas pro sententia consilii in iis agit voluntates; ut animans quisque velit ac appetat quod placet et affert voluptatem ad quam properat, et effugiat dolorosum, quod aversatur. Et tanta copia sapientiae utitur, ut etiam quamplurima ex loquendi facultate hauriens praeseferat, consilii ideas quidem explicare non valens; quia linguam non habet articu-

---

1 Առանց այսոցիկ, apud Interpretem nostrum merus est Hellenismus, χωρὶς δὲ τούτων, praeter haec, exceptis his, sine illis, pro թող զայսոսիկ, բաց յայսցանէ, և հա, praeterea, adhaec.

Sicut etiam in sequentibus illud առանց հոտ-ութեյ, aeque potest reddi, praeter odoratum, vel sine odoratu.

լերգուն ոչ է յոդուաոր . բայց յայսմանէ դոյն քան զձայն՝ զգործոյն Հշմարտուն֊
թիւն այսոցիկ որ սէպ կարէն տեսանել, յանդիման կացուցանէ :

Շուն՝ զգազանն Հետ էր . Հասեալ 'ի փոս խոր՝ առ որում շաւիղք երկու, մին յաջակողմն՝ և միւսն յաՀեակ, առ սա կաւ մի գտեղի կայեալ, ընդ որ երթալ արժան էր՝ քննէր . և ընթացեալ յաջ֊ կոյս, և ոչինչ գտեալ Հետ՝ դարձեալ անդրէն ընդ միւս եմն զնայր : Իսկ եթր ոչ յայսմ յայտնապէս երևեալ ինչ նշա֊ նակ, վաղեալ զզխոսիւն՝ Հետաքննէս լլ֊ նէր, առ անց Հոտոտելոյ՝ փուխթոլ զնացս՝ ազգելով, զի վարեցեալ ոչ Հանդիպողի՝ այլ առաւել Հշմարտութեան քննութէ խորՀրդոյ :

Իսկ զխորՀուրդս մնածութեանս այ֊ սորիկ տրամաբանականքն կոչեն ապա֊ ցուցական բացդէք Հնգերորդ : Քանզի կամ յաջ զգազանն փախխաւ, կամ յաՀ֊ եակ, և կամ վազեաց : Իսկ արդ ար֊ դեօք այսպ և այսպիսի ձևք բանիցս մեկ֊ նին 'ի մարդկանէ . բայց իմացեալ լլնի ոչ զունելաբար և յայլոցն անստութ՝ եստվպաս 'ի լաւագոյն 'ի մասին գործոյ բանիցն իւրոց. և քանզի թեպէտս և ձայ֊ նի անբառ է յոդուորի՝ անստունիչ :
Քանզի առեղծուած ոչ է առաապելք այսն՝ որ զոյ 'ի նոսա իմաստունութիւն և զխոսութիւն 'ի բազունս : Քանզի աՀա յառակս, թէ պարտ է, և մարդկան զ(լսոյմն 'ի խորՀել՝ կոչել սովորութիւն է Լ'ոււեսս : Լ'յսպէս կենդանիչ այդ 'ի պայծառութեան և 'ի բազմորինակէ զ՛՛մակութեան մնացն զարմանալի է բազմաց : Քանզի և կապկաց և յիմն բաղինն խորամանկագոյն է ձեռնընդեֆ լութեամբ և խաղու կախարդելով . և յորժամ որոց կամէն յաղթեսցէ՝ զպատ֊ բեկեալսն ծաղր առնէ :

Բայց են և այլ ինչ բազունեք պատ֊ մել այսոցիկ որ իմաստունթիւն, որոնք կենդանեգ վարին . բայց բաւականն են

| latam; manifestius tamen, quam voce, operum veritatem eis, qui acute pos-sunt videre, exponens. |

Canis quum persequebatur feram, **45** perveniens ad fossam profundam, juxta quam duae erant semitae, una ad dex-tram, altera in sinistram; paululum se sistens, quo ire oporteat, meditabatur. Currens autem ad dexteram, et nullum inveniens vestigium, reversus per alte-ram ibat. Quando vero neque in ista a-perte appareret aliquod signum, transi-liens fossam, curiose indagat, praeter o-doratum cursum accelerans; satis decla-rans non obiter haec facere, sed po-tius vera inquisitione consilii.

Consilium autem talis cogitationis dia- **46** lectici appellant demonstrativum evidens quinti modi. Quoniam vel ad dextram fera fugit, vel ad sinistram, aut demum transiliit: et quidem haec et similes for-mae verborum explicantur ab homini-bus; verum intelligitur non obscure et apud caeteros sine mendacitate Aesopus quoque in parte maxima operis fabel-larum suarum. Nam etsi vocis expers sit articulatae animal, non tamen fictio est fabulae illa, quae inest eis sapientia scientiaque in multis rebus. Ecce enim in proverbiis etiam acutiores inter homi-nes in cogitando solent vocari Vulpes: adeo bestia ista ex claritate multiplici-que versutia mentis admirationi nota apud multos est. Etenim simiorum etiam stultissimus praestat agilitate manuum, et ludi praestigiis; ita ut postquam vi-cerit quos voluerit, deceptos ipsos irri-deat.

Caeterum multa sunt et alia relatione **47** digna de sapientia, qua utuntur anima-lia; sufficiunt tamen jam dicta. Adden-

և աստացեալքս․ զի ոչ միայն խմստու֊
թեան , այլ և պարկեշտութեան կցորդ
դուիթիւն ունին՝ ասեքի է․ Բայեթէ և
բանի պէտք են իրային , նոցա ձայն ար֊
ձակեաս ։ Քանզի զի՞նչ են առ նոսա
խոՀակերութիւնք իւրն , և կամ ո՞ր գինեն֊
զինութիւնք , և կամ ո՞ր ինչ արբեցու֊
թիւնք․ և կամ ո՞րք խոհարանապաց , և կամ
Հացագործաց ալելապործութիւնք թշ֊
շուառական որովայնի արուեստք այսու֊
ցիկ՝ որ առ 'ի Հեշտ ցանկութիւնս ընդ
մեզ են ։ Արտոքի մրըկեալս Հպարտու֊
թեամբ ընդ ո՞րչափս այսուսիկ զլերանս
բացեալ ոչ ամաչեմք․ այլ յորժամ ման
նասամք գյունդ քուրասեալ , և զվերս 'ի
վեր ՀանելաըՁեալ , անասունս խոսրորդովք
առյցեալք , անապատը և թափուրք մի֊
տոք , թոյլ վերմանեք Հանբարձեալք ,
կարծեցուՄք՝ որպէս ասեն քերթողք ,
եղեւս աստուածոց մերձասետ եսՄք և
ձօասաւորք , 'ի կերակրոյ 'ի դուվմասերյ
և յրմպելեյ 'ի գետնն անկեալ դնիմք
պարտեալք 'ի վերուստ լերկինե անկա֊
նիմք ։ Արդ աՀա ամղեն իսկ , իբր զի
պարտապան եմք ամօթոյ , և դարձեալ
և մսրՄինոյ տկարութե ։ Քանզի գանա֊
դուիթեան և գամբռութեան զՀետ եր֊
թան Հիսանդութիւնն․ զուզմապեացն՝
նուացք , և մեծին՝ անբոյժք ։

48 Իսկ գձձութիւնն կերակրոյ (ա֊ եկե-
դանիս ,) որք յաառաջին իսկ 'ի ծնունդենե
'ի մտի եդին ընդունել , դալարակերք
լինին , և դուվմապեացն շատին ։ Որոց ընդ
պելի ջուրք է , ոչ յաՃախս , այլ յասուրս ,
էյորժամ և բազում ամաց սուռեալ ըզ֊
բոլորս ։ Արդ ումանք զձՄեռնասպեանն ոչ
ըմպեն յերջանակս , փոխանակ ջրոյ շա֊
տեալ օդային զրսսութեամբ , և դալարա֊
կերութեան Հիսթն գիջուսթ ։ Այլ և
առանց այսոցիկ որ 'ի խոնաւն , և այսք
որ ընդ որովայնիսն Հեշտ ցանկութեսն

---

dum esset, quod non tantum sapientiae,
sed etiam sobrietatis participes bestiae
sunt. Quod si opus erit verbis pro rebus eis necessariis, vocem emittunt.
Quis vero apud illas apparatus dapium,
aut quae vinolentia, aut quae ebrietas,
aut quae coquorum pistorumque opera
superflua, artes miseri ventris eorum,
qui nobiscum voluptati dediti sunt? In-
super superbiae procellis nos jactati,
quanta sane aperto ore mirari non eru-
bescimus; praesertim quum supercilia
contrahentes, cervicesque extollentes,
belluinis sententiis tumidi, deserti vacui-
que mente, remisso fastu elati reputamus
nosmetipsos, ut Poetae dicunt, Diis
proximos factos fuisse semine, et pro-
pinquos generatione [1]; verum a cibo
exiguo potuque superati, humi jacemus,
ab alto caelo cadentes. Ecce illico ob-
noxii sumus non solum pudori, verum
etiam corporis infirmitati: quia aviditatem, et gulam sequuntur morbi, exi-
guam quidem minores, maximam vero
irremediabiles.

At bestiae, quae parcitatem alimen-
ti ab ipsa nativitate statuerunt sibi ad-
sciscere, herbis pascuntur, et exiguis con-
tentae sunt. Quarum potus aqua est,
non tamen frequens, sed semel in die,
aliquando etiam post multos menses per-
actos. Nam quaedam hybernis tempe-
statibus vix bibunt, aquae loco satis habentes frigus aëris, et viridis pastus hu-
morem humidum. Ad haec prurigo illa
ex humiditate orta, cupiditasque sub
ventre existens nimium minor et exigua

---

1 Locus usurpatus iisdem fere verbis Armeniis apud Chorenensem nostrum. Lib. III. cap. 65. vide in Praefat. nostra.

Էս, կարիքաշ յոյժ մեղմացեալ նուագին։ \Բանդի ումանք դարնային, և ումանք աշնային բառ զուգաւորութեանն, և էս ումանք որ միանգամ վկրանան յամի ։ Լ< եզք մինչ 'ի յղութիւն միայն յանձն առնուն վկրանալոյն, և ապա յարուցին 'ի բաց փախչին, բնութեանն կատարէ լով զօրէն' որ յապագա լինելոյ և ապա կանութեան զարուին առ եզն միառա նունութիւն և խառնս եզին ։

Լ<ծ արդ ո՞չ արդեւոք արժան է մարդ կան պատկառել՝ զիւրեանց զանարգել խառնակութիւնն Հակառակ զնելով սոցա ժուժկալութեան կրօնից ։ \Բան զի զո՞ր տարւոյն ժամանակ զաղարկէք յաստղկան ցանկութեանն, 'ի սովորու թիւն և զանխառնակացն զՀետ երթալ մեղուցեմ ։ Ո՛չ շատեալք միայն ամուս նովք' արտաքոյ շուրջ Հայելով, և 'ի պոռնիկս մխեալք, զեամ ծինս պայծառ ո՛չ պաշէն, անարգութիւն և աղքատու թիւն, առ այտրիւք և գվքանու լա փո խանակելքին ։ Լ< յաւադդ պատանեկացն ումանք զմահ քանսՀեալք' զՀետս զանկ կութիւնքան զկեանս ընտրեցին ։ Ո այ լոյն ամունութեան օրէնս լուծին, և ո՛չ յախնածու այզելուն մարդկան ամ շեղեամք պատառաել գին'ի դատաստանէ ։ Լ< ո՛չ որ 'ի վերայ շանց օրէնք կան' եր կեան ։ Լ< ո՛չ յանգգամ և յանշրամա բելլ սրամունութեն արանց մահ սպառ նողաց, որոց յապաՀուլոյն տայ աշնան ամփոփ ընդ վայր սպանանելին ։ \Բանդի 'ի վերայ եկեամք ումանք շարունթեան և առ ամենաագոյն խառնակունիւն կատարեցան, մանկրնն սերմանել, ա բ րոց ո՛չ եւ զբաղաբայն, այլ աՀա և բղ բնունեան կործզելով փոխանակ զՀռա մանս ։ \Բանդի քակտելով զօրէնս ան շարժ, 'ի նոյն ինքն 'ի Հշմարտունենէ յանղիմանին՝ անտարժանա գործելով, վ ամն սերմանելոյ անկատարս, և սեր մանս թթելոյ և վկծեղուցանելոյ ։

Լ< այլոյն կենդանեաց պարկեշտու

Թեանն՝ յայտնի արդեանցն ցոյցք են, և
նախանձ խանտապոյ առ այնոսիկ՝ որ
անյագքն լինին 'ի խառնակութիւն լռր
բաքար , անդէն վաղվաղակի վանել
բաղձացեալք . ապէթէ ոչ առժամայն
կարացեն , ժամանակաւ սպասեն ։ Ո՛չ
միայն ձեռնընդելացն և որ ընդ մեզ մնաւ
նին , այլ և յայնոսիկ որ Թուլին առ 'ի
պարկեշտութին և այնճն ընկալեալ ։
Ա՛րդ բազում ցոյցս յայտնիս պարտ է
այնոցիկ՝ որք Թերա—ատտեն՝ մատուցա
նել . բայց պարտ է միայ յիշատակաւ յան
դխմանութին զոլ և լռեքելոյն ։ Ո՛ր
յեզդպատոսկոկորդեղեա—մարդակէր կէն
դանին և երկակենցաց , յորժամ Հան
դերձեալ է առնել զխառնս , Հանեալ
շրջէ զէգն 'ի վերայ ափանցն , քանզի որն
'ի վայր ընաւորեալ է մերձենալ . և յետ
խառնակութեանն զարձեալ ձեռօք վէ
րացուցանելոյ վնա անդրէն շրջեալ ։
Իսկ յորժամ զզ—ացէ խառնակեալ և
յղացուցեալ , քանզի և կեղձաւոթէ , և
կերպարանս բիւն ձևոյ առնէ , իբր Թէ
զարձեալ 'ի խառնակութին վարարել
զօրէն սենեկին տա—թեգելլ զՏեսս երԹայ
և սովորակ օրինակաւն խառնակութն
շրջէ , և նա անդէն վաղվաղակի կար
ծեօքն ստուգեսցէ , Թէ Հատուցեալ և
Թէ այլով իւիք՝ չշմարին ինչ , Թէ ընդ
դէմն արդեօք . քանզի ունի 'ի ընուԹէն
զանցայտիցն զխտութին ։ Իսկ զտա 'ի
միաժանգ անշարկուԹիին Տանիխպողու
Թեանն անստուԹեամձ իրացն զորձե
լով Հատուատեալք , մածդոգն պատա
ւեալ զզեստերսն , լակէ , զի կակող է ։
և միանանցն Թեփ անարգել աղեւեալք
և կաղեալք , և զերկաԹն սրեալ բԹէն
'ի վերայ Հասուցանելով ։

51   ·Քանզի յաղագս պարկեշՏութան
այսպէս . զի՞ պիտի յաղագս արուԹան
պատմել . քանզի վարագաց կամ զուաււ
բակաց կամ փղաց և կամ այլոց և—ս յայ
ոցիկ վայրենեացն զօրուԹին և զ—ո—
զնութին ՞է ոչ , որ ոչ զիտէ , յան

ribus patet, et zelum contra libidinosos,
quos impudentes statim repellere nituntur : ita ut si pro tempore nequeant
compotes esse, opportunum praestolen-
tur tempus. Id autem cernitur non solum in mansuefactis, et apud nos de-
gentibus, sed quoque in belluis, quae
videntur continentiam ultro sibi adsci-
visse. Plura quidem oporteret in medium exempla ferre diffidentibus fidem
factura, verum satis erit unius mentione
facta similiter argumentari et de caeteris sub silentio . Crocodilus hominum
vorax, amphybiumque in Aegypto animal, coitum facturus, educens faeminam
revolvit ad ripam fluvii, quia dorso su-
pino solevit concubere: post autem coitum rursus manu elevans versat eam .
Quum autem suspicatur eam jam com-
mixtam aliis concepisse, quia subdolus
est, simulat se aliquo modo, quasi iterum
appetat coitum; tum tamquam amasius
cupidine actus sequitur eam, et solito
modo revolvit ipsam . Ut autem statim
certus redditur de opinione, odorando,
aut aliquo alio modo, verum ne sit, aut
contrarium, habet enim ex natura scientiam incognitorum; certior factus evidenter ab ipsis rebus de concubitu, un-
guibus lacerans intestina depascitur, tenera enim sunt; cum ex adverso mu-
sculi sint squamis solide connexis colli-
gatisque, ita ut vel ferrum acutum hebe-
tent contactum .

Quum hactenus de sobrietate actum
sit, quid opus sit de fortitudine referre?
Quandoquidem aprorum, aut taurorum,
vel elephantum, caeterorumque his fe-
rociorum robur vimque nemo est qui
nesciat ingentem esse, et loco exigente

դիմանս կանն խորեալս, և ոչ տեղեաւ
մեծ և ուժգին և առ 'ի մարտ արուեսա
տաւոր սպանչելադոքն ։ Քանզի աՀա և
կեալ 'ի կռիւ դսա տեսի ոռիւքն զորէն
րմբշի Հող Համելով, խեթիի Հայելով՝ շա
ռագոքն աչօք · և զեղջերսն խոցող կաղ
մեալ սրատումութեամ եռայ. և ամենայն
Հայեցուածով քն տեսանելոյ ։ Իսկ խոզ
յորժամ տեսանիցէ զորսորդս , ասեն
յանտառս որք երիման որսայ, յառա կա
ցեալ ոչ յառաջագոյն դիմել 'ի զորու
թիւն, մինչ չև յառաջ առ մերձաւոր
ծառան սրել զատամունսն ։ Իչ արջբ
բածանեալբ րնդուներն, և դսիչոյ թուռն
Հարեալ ուժգնապէս զարկուցանեն , և
պատեն ռռիւքն արուեստապէս կարթե
լով ։ և որ ինչ մնանգամ մանկավարժից
սովորութիւն է՝ մարթանօք դիստութն
գամենայն ինչ գործեն. որոց գիւրաբան
չիւրոքյատուով աւելորդ է թուել ։

Քանզի եռանող՝ վիշապակին առ իմա
եգիպտացւոց գիռին տեսեալ՝ ոչ սակաւ
Հիացեալ եղէ. մինչ գի 'իմնյ իսկ իրացն
Հարկ եղե Հաւատալ իձ Ճշմարտու
թեանն յադագս նորա,քանգի թամ իւր
թէ բանականք րնութեանն մասին՝ և
այլոց կենդանեաց շնորՀեալ մեծապես ։
Քանզի թունաներ իձն գայ 'ի Հաֆա
ռակութին իւր երկոյս և երկկերպիվ
եղեալ ; Քանզի գտտունն յոյ շաչմանն
դնդամ թարշեծ. իսկ առ 'ի լանշաչն
մինչ 'ի գլուխն 'ի վեր Համբարձեալ վա
գէ,քաշասյան, և լայնապարանցի,վէ
ցեալ լն սրատումութեամ և թունիւք.
որպես այլք՝ ծեռամե օրինակին, այր
անդր աձելով գլեգուն, սուլմանե յոյ
անգգասււ ։ Իսկ նա 'ի վեր Համբար
ծեալ կարի յոյժ գագին վիշապակն 'ի
վերայ շարժե գայն, պատրեւի, գի իւր
առ Հակաաակակայ կենդանին պատրեալ
ձգեսյծ, և յետեան 'ի դուծ իջեալ յար
հակի 'ի վերայ. և Հոստացեալ իՇն
բոլորովին լռվ գերեան 'ի վեր ունի՝ անն
ունելով և սպասելով ։ և երկուսեալ թէ

magnum fortemque animum demonstrare mira in pugna arte. Ecce enim istum
(taurum) vidi in luctam venientem, pedibus, gladiatoris instar, pulverem susci-
tare, torve aspicere oculis rubescentibus,
et cornibus ad feriendum paratis furore
fervere, totoque aspectu horrendum ap-
parere. Aprum autem dicunt, qui venatum exeunt in sylvis, quum viderit
venatores, non prius viribus armari,
quam in arbores vicinas acuerit dentes.
Ursi vero secedentes impetumque facientes cervicem immittunt sub medio
adversarii, et retorquentes violenter
sternunt in terram, et pedibus constringunt artificiose velut laqueo captivantes.
Et quidquid paedagogis (vel ludorum
magistris) in usu est, per astutiam pe-
ritiae, totum ipsi exequuntur. Singula
singillatim dinumerare superfluum est.

Ecce enim nudius tertius dracunculi
cum viperis Aegyptiacis luctamen videns, non modicum stupefactus fui: ita
ut ex ipsis rebus necessum habuerim
credere veritati de hoc, quod rationis
particeps esse, haud secus ac rationali-
bus, caeteris quoque animalibus conces-
sum fuerit. Vipera enim venenosa venit
ad congressum, velut geminae naturae
ac biformis facta; quippe quae caudam
nimio strepitu in forma globi retrahe-
ret; de pectore vero usque ad caput sursum erecta accurreret pectore forti, et
larga cervice, inflata, plena furoris vene-
nique; sicut alii quoque quasi digito indi-
cabant, huc illuc versando linguam cum
sibilo immani. Draco vero elevans altius caudam suam, removit eam, ut vipera decepta, in ipsam tamquam in ad-
versarium sibi animal irrueret. Ipse
dein super faciem pronus serpebat totus
humiliatus, oculis solis sursum directis,
et expectans praestolabatur; veritus tamen, ne forte illa praeveniret, et vulnus

Ս` զուգէ յառաջագոյն ժամանեալ Հա
բումածու ընկալեալ կարեօրս, և մխան
դամանն և պատրաստելով դեմ յարձակ
մանէ։ Իսկ յորժամ զզուացէ, թէ Հաւբը
մարտին իւրդ լոտ կաճսից եղէն, և իմ
իբր առ դագանն առ ազգին լրջաբար Հա
կառակիցէ, յանկարծ օրէն յարձակեալ
կորովի խածմանէ զպարանոցէն բուռն
Հարկանէ և Հեղձուցանէ ։

53 Բայց վարքն կենդանիքս այս և լաւ
  Հրամանաւք քան զզարդասապետոս և զգաւ
  ասպետաս՝ ոչ կարօտելաբ առ 'ի մխմանցն
  մխիթարութենէ. քանզի առաանց որլոյ
  'ի գործել պատրաստուտոյն է, 'ի ձեռն
  որոց յաճախ գործէն զարթուցեալ լե
  նին, և սրէ՝ ոչ սոսեալ թոյլ դանգա
  դէլ, այլ առ 'ի զորութիւն յառնել ։

54 Քանզի և ես յորժամ Հրէշտակ լոյաս
  'ի Հռոմ՝ յաճախ տեսի 'ի դագանա
  մարման գորէն քաջ սուրՀանդակաց յա
  ռաջագոյն սուրՀալոյ, և այլուաս կող
  մանէ իբր խարացեալ` ապեօբն զբա
  մանքն և զկողոն ձեծելով'ի յորդորուին.
  զոր և զայս կառավարք 'ի ձիընթացն
  ասեն թէ առնեն ։ Քանզի ոչ միայն 'ի
  պատիժս պատուՀասի զոանՋանն 'ի վե
  րայ Հասուցանեն, այլ և առ յիշատակ
  ոտից ։ Ոյս և քերթողն դիտացեալ
  ստոյգ ասէ .

  « Զգեաւ զկողս և զբամակ յերկր
  յոսից կողմանց ձեծէ, արդ եթէ զնա
  զրգռէ կռուէլ » ։

55 Արդ այսպէս արութեան ցանկուլթ
  'ի Հզորագոյնսն է, մինչ զի և տկարագոյն
  զորութեամէն քան յօժարութեամէն
  մարտնցէ յամօթ առնելով զագոււմն,
  ոչ անյայտ է ։ Քանզի ասէն թէ առիւծը
  յորժամ 'ի Հակառակս'ի բազում Հա

sibi adferret periculosum, parabat se in-
simul ad impetum faciendum. Quando
itaque sensit insidias luctae suae ad li-
bitum cessisse, quum nimirum vipera
cum cauda sua serio dimicare videbatur,
subito invadens eam, acri morsu pre-
hendit cervicem ejus, et suffocavit.

Porro haec animantia utuntur etiam
ordine meliori, quam cohortis praefecti
ac centuriones, minime indigentes ad-
hortatione mutua; quoniam prompta sunt
ad operandum sine excitamento, quo
necesse est homines identidem ad opus
impelli, quod acuit audientes, nec eos
cunctari sinit, sed excitat ad virtutem.

Etenim ego etiam quando per lega-
tionem adii Romam [1], crebro vidi in be-
stiarum congressibus, quod more opti-
morum cursorum prius huc illuc discur-
rebant, et ex altera parte tamquam fla-
gello caudis propriis dorsum et latera
percutiebant, ut sese hortarentur, quemadmodum in hippodromo aurigae faciunt. Quoniam non solum supplicii loco
flagella adhibent, verum etiam monen-
dis pedibus (ut se se moveant). Pro-
pterea et Poëta certior factus dicit:
«Caudà latera dorsumque ex utraque
parte
Percutit, ut irritet ad certamen se-
ipsum ».

Tanta itaque fortitudinis vis in robu-
stioribus existit: ut quae vires debiliores
desiderio habent, erubescant, quod igno-
tum non est. Dicitur enim, quod leones
quum in congressu ob multam oppugnationem debellantur, obtuentes secedunt

---

1 Alexander, vel Lysimachus satis hic se mon-
strat legatione Romam petiisse, haud secus ac
Philo. Si quis autem nolit id concedere, necesse
erit dicere, quod aliqua interlocutus sit etiam

Philo ipse. Verum integrum responsum Philonis
ad omnia adversarii dicta paulo inferius clare
videbis .

DE ANIMALIBUS ADVERS. ALEX. 243

կառավելութեւնն Թուեցին 'ի պարտու
թիւն մամանել, Հայելով 'ի թաց զնասցեն
Հանդարտիկ դոշելով. և այսր անդր շրջ
շելով գողցս անoթոյ մատուցանելով մի
լռութեան, և շրշեալ զերևան և ձայս
պախեալ և թաբուցեւալբ՚Հաւանեցու
ցին ինդութեամb գվանդիմանութիս
որ մաանցոյց առնեն, մի շոգապետել
կամեցեալբ, և յետոյ դարձեալբ երբ
յարձակարանե ինթանան :

Իսկ ձիոց ազդ՚ այսպէս է պատուու
սեր, որպէս զի աշա անդլեն ինք յելլա
դա, և 'ի խումձռոմձ ազզի, որպէս
մարդկան՚ և ձիոց մարտակցան ընթայն
մրցանակ երլեալ լինեն : Լ ոչ միայն
նուազ յայլին որոց զգզն է թան, այլ
և 'ի մեՀենական մարտա նաՀատակու
թեան, յորս 'ի տիեզերաց երկրէ զան
ժողովին բստ զանազան տեսակլաց սպ
տոյից, ումանք յաղագա մարտոցին յաղ
թութեան յումմվ. և ոմանք տեսլեանն
ցանկութեամե, և ոմանք բստ վաճա
ռականութեան : Քանզի պատրաստա
գոյնն կամբ 'ի ժողոսմ ;ոնիգն Հաւեցող
աnoթ Համարելով՚ երե ոչ անդր և
լեալբ 'ի տուն ինչ թերգեն :

Բացեալ զուգաճնայից տեսլարանաց
և ձիոց, յոլոմպիադեին, 'ի նեմեայմն, 'ի
պիթիեանդ, բատ նեոցին, և զծարաn
ծրից և տեսանել փոյթս, և դմդա
կութէ Հոգս անճառա, անրնայ ձախա,
ստպատսան կազմելով,կերաիլ պատ
բատելով, մարդեա 'ի վարձու ունելով,
որբ զանանցն զենական արադութիս
դրդսեալ օրեսգե ինատնե զարտսոյ
Ժորժա մարան Հասնե, յատաճոս
անդ զզայ (գ. անդզայ), տեսանելով
զարծակարանն, ուստի յատա;ագոս
նդատուցեալ լինի յեին ընդերեւա հ.
արագդիճթացն. և աղաս 'ի դոռս թո
դեայ, և երկայն երկայն և յա
ձախատես վացեբ տատնապել փու
թան :

Քանզի երանդ՚ և ՀՀամայլ տեսել
20

lente vociferando, et huc illuc ambulan-
tes pudere se de errato manifestant;
atque aversis vultibus, et voce remissa
se se abdentes, persuadent, quod ultro
reprehensionem indicantium se digito
volunt evitare, et retrorsum reversi velut ex stadio currunt.

Equorum autem genus adeo cupidum 56
est honoris, ut in ipsa jam Graecia, atque apud Barbaros, sicut hominibus, ita
etiam equis certatim currentibus prae-
mium apponatur. Et non solum in mi-
noribus congressibus, quorum exigua
est ratio, sed et in sacris certaminibus,
ad quae ex orbe terrarum veniunt con-
gregatim ob varias necessitatis species,
aliqui propter collectationem spe victo-
riae, quidam verò desiderio spectaculi
visendi, ac nonnulli ob negotiationem;
quoniam paratiores sunt voluntates ad-
venientium in coetum solemnem, pudori ducunt sibi, si eò egressi nihil in do-
mum adduxerint.

Apertis itaque heroicis theatris, cur- 57
sibusque equorum, ludis Olympicis, Ne-
meaeis, Pythiis, Isthmiis, agasonum
quoque studium videre est, et diras curas ineffabiles, impensas sine parsimonia,
stabula construentium pro equis, cibos
parantium, locantiumque exercitatores
(vel instructores), qui animalium velocitatem naturalem excitent acuantque
per curam et alimentum. Quando cer-
taminis tempus venit, jam jam rem sen-
tit, videns circi carceres. Unde primum
irritatur ad resiliendum velociter disru-
pto claustro; deinde foras emissus evo-
lat, atque longius per longitudinem fe-
stinato cursu satagit properare.

Ecce nudius tertius mirum spectacu- 58

եղեալ՝ բազում՛ ըիւր մարդկան դար
ձոյ . քանզի էր մարտ քառաձիոց եօ
թանց ընթացելոց ։ Քանզի էին որք 'ի
վերջ՛ մնացին , բայց երկուք՝ որք և արա
գութեամբ առաւելեալ էին , վիճէին
ընդ միմեանս Հանդիսաբար ։ և կառա
վարացն՝ զի թէպէտ և ոչ մարմնովք ,
այլ գոնեա յաղագս ոգւոյն յօժարու
թեան , որպէս զեալ էր , ընդ նոսա ընդ
թացելոց ։ և ոմն 'ի պատառաւ[ ]ԹԵԵ
յափշտակեալ անկաւ ոչ գիտացեալ ։ Իւ
նոքա անսպատ մնացեալք 'ի կառավար
բէն , ինքնաՀրամանք Հանդերձ յօժա
րութեամբ ամենայն սրտմտութեամբ
ѕկրտէին , ձգտէգուցանէին գընթեանս
առ 'ի յընթացս փութոյ . օրինակ իմն
'ի կառավարէ զարագութիւն գուցա
ѕեւլոյ Հանդերձ տեղեկութեամբ . գոմա
ոչ վայրապար փոխէին , իբր թէ այսր
անդր անօգուտ յածելով ընթացիցն ,
այլ իբր 'ի վերայ գծի ընթանալով ուղ
ղակի , անխափան կացեալք 'ի մէջ , և
այնոցիկ որ գՀետն ընթանային , զի մի
ժամանէալք յառաջ վարէալք անցցեն
'ի շրջանակին , յայն տեղւոյ ուր բազում
սպառանեն լինէն , 'ի ձահ և 'ի դէպ շրջ
ѕեկեալք Հանդիպէգուցանէլով ։ Ընդ
միտ ածէալք՝ ընդ շրջանակէն մօտ Հպէ
զի սեռն , և գյաջեկեն ձիև , զի մի ըր
լորակ մեծաց պարագութեան ընդար
ձակութիւն ունէլ այնմ որ գՀետն գայ ,
և որ այլ ևս ։

59   Իւ գինչ եւ՝ և գյաղագս գազանա
գունէն կէնդանւոյ փղէն պատմէն , որպ
ասէ՛ 'ի սկզբանէ , և գամէնայն Հնումն
յիշատակ մատուցանէլ , այնց որ առ անն
տիոքաւն եղէն իբր ։ Ասիացւոց իշխանն՝
երամակ փղոց բուձանէր առ 'ի ձիամարգ
տութին Հակառակակից օգնականու
թին . քանզի 'ի ճայնէ փղաց սարսէալ
երկչոտ ձի՝ իբր յերկիւղէ ընդակէլով
փախչէ ։ Իւկ այնոցիկ որց վէհ էն ըստ
մարմնոյ առոգութեան , և զայն որ առ
'ի գործէն գօրութին է , անուն եդ ըդ

դիւցազանցն անուանս ։ Օնաֆրապա
սոււււթիւնն՝ էանոսվյ սււււեալ, որ քան
զամենայս էր առաբինի երեւեԼ և դող
ծէԼ։ Ւնդ զետ Համզերբ̇ծալ անզանէԼ
զորու, եաս՝ 'ի վերոյ ափիսււն անզեալ
կայր՝ յանզլէս երկուգեալ կարծեմ յոր
ճանայն ։ Ւսկ որբ դՏեւաս դայֆս, որպէս
վլայեԼ էր, անեբկիւլ եդեւււբ՝ զզաման
զզուշեաւ պաՀէֆս, մֆսչե Թազաւորս
առաբեալ կոչեւազ զզուծառոզին ․ Բ̇աս
զի ժամանակն եբազութեւասբ էր, մեբ
զանակ կբկֆս զնեբ առաջ̇ի ափսոզֆկ̇ որ
յաւււա̇ քան զզյան անզգէս, որ 'ի վե
րայս է արծաթ, և փֆֆն̇ սաֆսապատու
ււււթֆււ ։ Բ̇րդ վասն զի պատրոֆդու
զանց աբաբ վաբեաԼ 'ի զեւււ , քանսՀա
բաբ աբաբ զանզն, զորոյ Հետ ամեսեբֆս
անզֆն այլբս, և պատււֆ աբժանֆ եզև ։
Ւե զժււաբեաԼ զֆ մֆ զրկեազֆ 'ի սա
խաապատււււթեսեն̇ եսան , և զժււււ
բեալ զզֆայ̇բ խաս̇ս̇ազսղորբեԼ յուզեաԼ
զֆզարդ զււււսեն , մֆսչ զֆ ււակա ֆֆս պե
սոյ էր քան զզեԼւամսիֆասյ ււրդֆս վասո
Թաբ սֆսեԼ՝ յազոզա ափԼեւււյ զֆււււֆն
սսֆս ււբֆսաՆ ս̇ֆղ̇եաԼ ։ Բ̇անզֆ խաս̇աւււյ
և բսդզֆզեԼււ̇Լ և յււււֆ̇բՆ և 'ի զււրււՆ
անււմֆֆբեբֆԼ էր , յամանգզանազ փասււււ
զեալ, և պատււււ̇ բազՆ̇ազեալ ։ Ւ́բ
մււբզււ այսապէս, ււչ աբււււթեան մֆււֆս
զդՏեււ է զււււսեԼ․ այԼ և որ զդՏեւաս զայ
քաջււււթԷ , բււբեֆֆաբււււԹԷ փափագէ ։
Բ̇անզֆ աֆս ււււմֆււ̇ բ̇ կեսզււււն̇ ասււււ
ււււֆս̇ Հասեբ̇ս̇զծ ասււ̇ազեզֆֆբ̇ ււււ̇աբ̇ֆււււ
Թեււֆբ̇էբ̇ և զաբզււււբ̇ււււ̇թֆււ̇ կբեզֆս ,
զււ̇զււ̇աֆֆբ̇, և զասս̇ս̇աֆֆ̇եբ̇, և ււղ̇ասֆււ̇ազբ̇ ։
Բ̇անզֆ աֆս ււբ բ̇սա ծււֆււււն̇ 'ի ափ̇ս̇ս̇ս̇
ււֆս̇ և 'ի ափ̇ս̇ս̇աապատ̇եֆֆ̇ս̇ Հասււ̇արաֆււ̇
թֆս̇ս̇ մֆբ̇ս̇աֆււււււթ̇եան̇ յայււ̇ է ։ Բ̇աս̇
զֆ կզււ̇բ̇զ̇ււ̇ււ̇թֆս̇ււ̇ ււււ̇ֆֆ̇ կեբ̇ս̇ֆֆ̇բ̇զ̇այ, և զււզ̇
զււ̇ա բ̇ս̇բ̇ւււ̇եղ̇եալ սււ̇ս̇ֆ̇ ։ Ւ̇բ ծֆս̇ււ̇ֆս̇ զայււ̇
զււ̇բ̇ււ̇ֆ̇ֆ̇ֆ̇ււ̇ււ̇ււ̇ս̇ ասսֆ̇ս̇եֆ̇զ̇ս̇ս̇ սււ̇ս̇ււ̇եֆ̇Լ , և բ̇ս̇զ̇
ափ̇ս̇ս̇ֆ̇ֆ̇ֆ̇ֆ̇ււււ̇ս̇ʼ խասււ̇ս̇ււ̇ֆ̇մ̇ս̇ս̇ֆ̇ս̇ս̇թֆս̇ս̇ս̇ ասււ̇ս̇բ̇
ծ̇ս̇ֆ̇ֆ̇ս̇ʼ ս̇ս̇ֆ̇ս̇ ււբ̇ 'ֆ̇ Հս̇ս̇ս̇ս̇ս̇ս̇ս̇ս̇բ̇ս̇ս̇ֆ̇ֆ̇ֆ̇ս̇ս̇ս̇ մ̇ֆ̇ս̇ս̇
բ̇ս̇ս̇ււ̇ս̇ֆ̇ֆ̇ֆ̇ս̇ս̇ս̇ ս̇ ։

didit ex Heroum nominibus, praeferentiam concedens Ajaci; illi nempe, qui
prae omnibus fortis videbatur, et operi deditus. Quum ergo flumen rex transiret cum exercitu, Ajax (nuncupatus
elephas) ad ripam morabatur, transitum
timens, ut puto, ob gurgitem. Qui vero
sequebantur, ut decebat, impavidi reperti, ordinem (tamen) diligenter servabant. Donec rex misit vocavitque altores, quia tempus erat accelerandi,
bravium duplex pollicitus transeuntibus
ante alios, insidenti nimirum argentum,
et elephanti dignitatem primi honoris.
Quoniam vero Patroclus transfretavit
desiliens in amnem, ita ut audacter
transitum fecerit, quem secuti et alii
transierunt, ideo ipse honorem assecutus est. Aegre id tulit Ajax, ac ne privaretur dignitate, tam molestus, et agitatus, deturbare totum ordinem castrorum voluit, ita ut videretur conqueri se
deteriorem evasisse Telamonis filio ob
arma Achillis similiter furentis: siquidem
livore plenus, et rebellis, in itinere ac
in castris intolerabilis erat, ignominiam
effugiens, et gloriam petens. Utinam
et homo talis fuisset, ut non solum fortitudini acquirendae studeret, verum
etiam sequelam virtutis feliciter optaret!

Ecce enim quaedam animantia prae- 60
ter jam dictas virtutes, justitiam etiam
praesetulerunt, natantia, terrestria, et
aërea. Quando quidem jam in marinis
Pinnae et satellitis ejus aequitas in societate manifesta est; contubernium enim commune cibi habent, et aequaliter
eum distribuunt. Idem facere et Trochilum ac Pompilum in confesso est apud
omnes, qui minores iis comperiuntur in
vita communi.

61 Իսկ 'ի Թռչունա՛ բուն զվերրին զար
դարութիւն՝ արագիլ ցուցանէ, զծնողս
փոխանակ կերակրելով, և միանգամայն
ընդ Թևաբուսէն ոչինչ այլ նախ գործ
առաջի դնելով յառաջ քան զշնորհՏ
Հատուցանել, և այնուցիկ որ բարին ա.
բարին ոմա՛ փոխարէն տալ: Բայց Հա
ասար ումքն և զաանապափնչն ումանք որ
գործեն զեզ է: Օ ի Թէպէտ և անեչ
ոչ ունիմք զամենեցուն ընուԹիւնս, և
ոչ տեսանել և ոչ ստուղել կարենէ, բայց
Հաստատւլ պարտ է վկայից այյմնեւաց
յաղագս անեբւոԹիյս: Խանզի ար
ժան՛ ոչ ումանց, այլ անեներցնց մասանցն
զաշխարՏն զղմանւլ. և սակայն յոր ան
կանի արղարութՏն և անիրաւուԹիւն,
ամէնայն իրոք և բանիւ: Սամն զի և
բան երկաբանդլերող. և որպէս մարդ
կան, այսպէս և ասացելող կենդանեւագս
բաժանեւալ տացի: Յիրաւի այնոցիկ որբ
կերակրեն զծնողն արագ իլբ՝ պատւիս
իւն որոշեալ տան յառաշկեկերցացն, և
կամ պատիժս: Խանզի գովեն և սլինա
սպատակացն բաշեկւլով պինաայցն զկերա
կւրն զզգ, և զֆուրդերն՝ պարասանաց
և ամբաստանւԹեան արժանի առնեն,
Թէ որ զասացեանն միայնբ վառբեցսն:
Ւ-c սակայն որբ մինագւած'ի պապա ամ.
զւացg աշխատին, զեշահեղուն ֆասաւ
կավա գորcdեց1g լիղրֆրստ Հանբեւալ՝
ասսակէն: Ս ամն զի որպէս ասէ եսիո
գոս, « զայլոց վասատակս յիւբեանg որո
վայս մուծին »:

62 Տեսևալբ առաՏֆէին զառաբնււթիս
մինչև յայլ կենդանֆն Հասեալբ՝ 'ի
մարմնակերուԹենէ Հրաժարեցֆն. և
մինչ'ի բազութ ժամանակս սովորուԹիս
յեչլայզ և 'ի խունֆաղուcd ազգֆն ասեն
կայեւալ: Ի ծննդողն ապա լւծանֆլ
որբ 'ի փափուկ կեանս բարեկեզդանու.
Թեան Թաւաւլեgան: Որբ նոբոզ զււեւլ

1 Haec usurpat cum aliis Sergius doctor Ar-
menus in comment. II. Petr. Serm. I. pag. 529.

Inter volucres autem vel ipsam supre-
mam justitiam ciconia demonstrat, parentes nutriens, [1] et statim ut alata fuerit, nullum aliud opus sibi proponens,
antequam gratias retribuat, et benefa-
cientibus sibi vices reddat, Verum hoc
pariter et terrestrium aliqua agere per-
suasum est. Nam etsi nequimus omnium
ipsi naturas proferre, nec lustrare, neque certius indagare, oportet tamen fidem praestare testibus manifestis in iis
quae ipsi non videmus. Aequum est quidem non nonnullis, sed universis partibus mundum compositum esse; verum
partem eam, in quam cadit justitia et
injustitia, illam omnino praeditam esse
etiam ratione: utraque enim ad rationem pertinet. Itaque sicut hominibus,
sic quoque animalibus memoratis distri-
buatur (ratio) necesse est. Jure sane
(homines) ciconiis adultis parentes a-
lentibus honorem certum exhibent, aut
contumeliam; laudant et satellites Pin-
nis aequaliter distribuentes cibum; ex
adverso vero, opprobrio et accusationi
dignum censent, si soli consument dictum (cibum). Qui vero laborant in e-
xamine apum, fucos nocivos favo elaborato existimantes deperdunt; quoniam
ut ait Hesiodus, « aliorum laborem in
ventres suos induxere ».

Videntes antiqui virtutes usque ad
bruta animantia devenisse, a carnis esu
se se abstinuerunt; ac ad multum temporis consuetudinem talem inter Grae-
cos et Barbaros obtinuisse ajunt, donec
prisca temperantia labente, in delica-
tam vivendi rationem cum voluptate de-
clinarunt: novis enim cupiditatibus ad-

* Արագիլ արդար, զի զծերացեալ հայրն կե
րակրէ || : Similia occurrunt etiam apud alios.

դզանկութիւն` արտորինագոյն ՀԷշա
գանկութեամբք դ,քաղաքս լցին, որոց
սկզբունքն են փձանք. և սպանմանքն
ոչ մաքրելով եղեն։ Որւոք աղգ ասֆոնցիկ
որ 'ի խրատոււ են, որք նախանձաւորք
պիթագորեան իմաստութեանն են, և
ոչ եթէ կարքի ինչ լլցի, 'ի մարմնա Հպեն
ցին, վասն կրօնից ժուժկալութեանն,
և 'ի Հեռանալութեանց խուաւմնէ։

Քանզի ունի իմն գործ բնութեամբք
մարմին ող,ոյ, ոչ միայն վասն ընտանու
թեան որոց են ընդելք մնայն, այլ և
վասն փոխանակ Հատուցանելոյ մտերբ
մունթեան. քանզի սիրեցեանն և փոխա
նակ սիրէ, և որոց միանգամ ինամ ինչ
դարմանոց արասցուք բաղձայ գՀանդիան
օգտութիւն փոխանակ մատուցանել։

Իսկ Հաստատութեան և արդարու
թեան մեծ ցույցք, և նորա վարքին. այնամ
ոչ առաւել նուազ քան գմարդիկ։ Շ
խանն կացուցանեն, և դ,ասա` յորում
կարդցացի՞ ոչ երբեք լբեալ. և դ,որս
ընտրէնն ի՞նման` ամենայն Հասանու
թեամբ դ՞Ստ երթանկոնցա։ Իս Հաստ
ասացեայն Հաւատի ութսանելիես։
Քանզի է դ,ուարակ յանդեայս, և Հա
րանաց` քաղ, և խոյ Հօտից Հեղինակ։
Իսկ 'ի վայրենին շուրբս մի ինն 'ի Թագա
լորականացն առաջիկայ եղեալ, որոյ
մեծամեծք և բազում կենդանքը դ՞Ստ
երթան, աստուարապէս դ,Էայ եղէ լլինէլ։

'Ի պատա մեղուաց անկբեդ,կայս կա
գեալ լբէնն Թագաւոր. որոյ խնամունց
առաջակացութեան ընկալեալ, երկինչին
'ի՞նման, և սապահեալ են։ Իս ապատ Հասա
կբէնն որք ընդ նոսայ են սպատաւորքն
ամենեբքին։ Իս եթէ իցէ և աճեցեալ
պարաբն` առաւելեալ լցբէնն գորէն բազ
մմայր քաղաքէ, դ,աղթեւբք յայյլ վայս
փոխսին. իբր մանկից իմն առաքեալ լլ
էնն։ Իս եթէ իցէ որք գիշխանութիւնն
առինն` վերակացու լբեել ոչ կարացեալ,
յանիշխանութենէն իբր 'ի բազմիշխա

նութեանն յոզդողղեալ.ք քանանալ.ք
գրուին . Վանդի աՁս և քաղմանութիւն
Հաւոլց տեսի տողապար՝ զի թռչին ա
ռաշնորդուԹեանՁ ումանց , և այլցն
գՋետերթ[Թ].Ꝁեանն , և '[գնաս և '[
Ꝁրաւեղ
՟
ꝏ , և '[ծալս երաՁ ꝅանꝝ
[ꝏ ալճս իՂն ուղիզ '[ՄինուՁ զծի ꝝիꝣ
երկար՝ զորՂն զորու յերկար ꝝսունեալ
'[զպրաՂսալ.եսաց և '[ꝣսասսպ.ꝏաց :
Ꝫ,յսսյՂս '[քաղաքական սսուԹՂ
ꝅꝛսնꝝանսꝝան սնսուꝁք ꝅցոր.դութ.ꝣꝏ ոս
սꝣꝏ , և ꝝելՂեալ ꝝ'զ[և Ꝡսուսսꝅուսսուꝛ	
ꝣꝏ : Ꝫ,յսսրꝅ ꝣꝏ յססꝣ.ꝝ ꝅꝛꝣ.ꝣꝏ
նուԹꝣꝏꝣꝏ քսꝛꝛ.ց ꝣꝛսսսꝏ և ꝣ.ꝏꝝ
Հꝏꝝ.Թꝣꝏ :

66 Ꝫ,.յց սսկ.ꝛꝝ ꝣꝏ ꝛ.ꝅꝝ.ꝛꝅꝣ.ꝛ ꝣꝛ.ս
քꝛꝏꝏꝅꝏꝏꝛꝛ՝ և ꝛꝛꝣ.ꝝꝝꝛ քꝏꝏꝅꝛꝛ
քꝛꝣꝝꝅꝏꝣ ꝣꝏ ꝝꝛꝅꝏꝝꝅ.ꝝꝣꝣꝣ : Վꝏꝣꝣ
ꝣꝝꝝ.ꝛꝣꝝꝏꝅ սꝏꝅꝏꝝ ꝣꝏꝝ սꝣꝣ.ꝣ ꝝꝏ ꝝꝏ.
ꝏꝏꝣꝣ՟ , ꝝꝛꝏꝏꝏꝏꝏ՝ և ꝝꝏꝛꝝꝅꝏꝣ և սսꝣꝏꝝ
ꝅꝣꝣ.ꝏꝣꝣꝏꝏꝏ ꝣ ꝛꝏꝣꝛꝝꝏꝣꝏ.ꝝ[ꝣ.ꝣ՟ : Վꝏꝣ
ꝝꝣ սꝣꝝꝏꝝ ꝝꝣꝝꝏ.ꝝꝣꝣ ꝣ ꝣꝛꝣꝅꝣꝏꝝꝝ՝ և ꝣꝛꝅꝣꝣꝏꝝꝏ ꝣ
ꝏꝣꝏꝣꝛꝏ.ꝝꝣꝣꝏ՝ և ꝏꝛ ꝣꝝ.ꝣ սꝏꝝꝏꝅꝏꝝ '[
սꝏꝛꝏꝏꝣ՝ ꝛꝛꝣꝝꝣ'ꝣ ꝝꝣꝏ.ꝝ סꝏꝝꝏꝣꝝꝣ Հꝏꝏꝝ
ꝛꝏꝅ ꝣꝛꝣꝣ քꝏꝛꝏꝏꝝꝣ և ꝛꝏꝝꝝ.ꝝꝛꝣ , և χꝛꝛ.
սꝏꝛꝏꝝꝏ.ꝣꝏꝝꝣ Թꝏꝝꝝꝏꝣꝣ ꝣꝛꝣꝣꝏꝝ ꝣꝣꝣꝣ
ꝝꝏꝝꝣꝣꝝꝣꝝ : Оꝛꝛ ꝣꝛꝛ ꝣꝏꝛꝅꝀꝣ ꝏꝝꝏꝝꝝ ꝝꝏ
ꝛꝏꝝꝏꝝꝛ , ꝣꝏꝣꝛ և ꝏꝣꝏꝣꝛꝛ : Ꝭꝛꝛ ꝣꝏꝛ
ꝝꝣꝏꝛ ꝣꝣꝣ ꝣꝏꝏꝏꝝꝏꝛꝅꝛ , ꝛꝏꝛꝛ և ꝏꝣꝏ.
Ꝡꝛ . քꝏꝝꝛꝛ ꝣꝝꝏꝝ ꝣꝏꝣꝏꝛꝛꝛꝏꝝ սꝏꝛꝛ.ꝝꝣꝣ
ꝣ , և ꝛꝏꝛꝛꝣ ꝛꝏꝏꝣꝛ սꝏꝛꝛꝛꝏꝛꝛꝏꝛꝛ ꝝꝛꝛꝣꝝ՝ .
ꝝꝏꝝꝣ ꝛꝣ ꝝꝣ ꝣꝛꝛꝣ'ꝣ ꝝꝝꝛꝛꝛꝝ Ꝁꝏ.ꝝꝝꝏꝛꝛꝝ ꝛꝏꝣꝏꝛꝛꝏꝣ
ꝣꝏꝛꝏꝝꝛ ꝝꝛꝛꝣ ꝝꝝꝛ սꝝꝛꝛꝛꝝꝏꝏ : Վꝏꝝꝣꝝ
ꝏꝏ.ꝝꝏ.ꝝꝣꝣ ꝣꝏꝛꝣꝏ.ꝝꝣꝏꝝ և սꝏꝝꝏ և
Հꝏꝏ ꝏ.ꝝꝛꝛꝝ '[ꝝꝝꝛꝏꝝ ꝏꝝꝝꝛꝛꝏ , ꝛꝝ ꝝꝣꝏꝝꝝ
ꝅꝏꝛꝛꝏꝝ ꝝꝝꝏꝛꝛꝏꝝ '[Հꝏꝝꝝꝛꝛꝣꝝ , ꝏꝝꝣ
Թꝝꝣꝏꝝꝝꝛꝏꝛ ꝛꝛꝣꝛꝛꝏꝝ ꝝꝏꝝꝛ ꝏꝝꝝꝏꝣꝝꝝꝝ :
Վꝏꝝꝣꝝ ꝛꝏꝛꝛꝏꝝꝝꝣꝏꝅ ꝅꝏꝛꝝꝝꝛ ꝝꝏꝛꝛꝣꝅ
ꝛꝏχꝏꝝꝛ և ꝛꝏꝝꝛꝛ ꝛꝏꝝꝛꝛ ꝛꝏꝛꝣꝏꝝꝛꝏꝝ ꝅꝛ.
ꝛꝛꝅꝛꝏꝣꝛꝛ և ꝏꝝꝏꝝꝛ ꝝꝏ Հꝛꝛꝛꝝꝣ՝ ꝣꝏꝛꝛ '[
սꝏꝝꝏꝛꝏꝛꝅꝛꝛꝣ ꝅꝛꝛꝝꝛꝛꝏꝝ , ꝛꝛꝏꝝꝝꝛꝏꝛꝛꝣꝏꝝ
ꝝꝏꝛꝏꝛꝛ ꝛꝛꝛꝝꝝꝛꝛꝛ , ꝣꝛꝛꝛꝏꝝꝛ ꝛꝛꝝꝏ ꝏꝣꝛꝏ
ꝛꝏꝛꝛꝏꝛ ꝛꝛꝛꝝꝛꝛꝏꝝ՝ և ꝛꝏꝝꝛꝏꝛꝛꝏꝝ ꝛꝛꝛ
Ꝡꝛꝛ· ꝛꝛꝏꝛꝝꝛ քꝏꝝ ꝛꝝꝝꝛ ꝛꝛꝛꝏꝛꝛꝛꝛꝛꝛ ꝛꝛꝛ

tur . Ecce enim et multitudinem avium
vidi , gregatim volantem , quibusdam
praecedentibus , aliis postsequentibus ;
necnon in fluviis, stagnis, et maribus
coetus piscium tamquam rivus rectus in
una linea protrahitur, exercitus instar
per longitudinem extensi a praefectis le-
gionum magistrisque militiae . Hoc modo politicae virtutis animantium genera
habent participationem ; jamque osten-
sum est, quod dispensatione gaudent;
quae sunt argumento moralis disciplinae
et providentiae .

Porro non minus vitia quam virtutes
sunt indicia naturae rationalis . Pronum
enim mihi est demonstrare, quod per-
versitas peccati tam hominibus, quam
animalibus inest. Nam avida luxuries,
timiditas, injuria, et istiusmodi, in una fere specie communiter multiplicatae sunt
usque ad insaniam ; probitate autem
vacua sunt saepe animantia. Quaedam,
exempli gratia, sunt callida, ut lupi et
vulpes. Item alia luxuriosa, ut hirci et
columbi; adeo ut ob vitium veneris mas
soleat ova frangere, ne foemina super
ova sedens impediat sibi coitum. Cupi-
ditas enim libidinosa praevalet etiam in
animalibus modo violento, ita ut non sit
eis satis ferri ad cognatos, verum etiam
ad inimicos propter aviditatem. Nam
quemadmodum ut putatur, homines per
frequentes multiplicesque cibos, et avi-
das dapes, quibus de more dediti sunt,
novas cupidines sibi fecerunt, et meretri-
ces alienas adinvenientes, incitati sunt
insolita perpetrare; sicut sermo habetur
de Pasiphaë Minois Cretensis filia, quae
fertur adamasse taurum, peperisseque
monstrum Minotaurum; simili modo in
homines nonnulla (animalium) exarse-
re, sicut cantatus ille aries Glauci. Alio-

Նմկայ դուստրին կրեստացեալ Հարեալ 'ի
զլու ծնանել խառնաղզական ինն՛ մնա
վլայ գուլ։ Ո այս օրինակ և 'ի մարդկա
են ումանք Հնարին ցանկալ, որպէս այս՛
որ երգի խոյս զեղլալկայ։ Լ՛այս եթէ
ոչ, ուստի՛ այլազգէ այլահերպ և Հրա
շածին ինչ դէպ եղև ծնանել՛ եթէ ոչ
յանննմանագրեացն 'ի խառող խառնա
կութենէ։

Լ՛է յաքայիա սակայն աստեն 'ի Հինան,
թէ զերվիքս՛ մանկան, որ առաւելեալ
եր զեղեցկութեան երեսոք, ցանկութէ
ըմբռնեալ, բազում անգամ'ի ծովեզե
րէն 'ի ներքս 'ի խոր 'ի վերայ բառնա
լով մկանանցն վլերամեարծ բերէր 'ի
վերայ ալեացն։ Լ՛ բազում անգամ'ի
ներքրոստ աննի մինչև 'ի ծովեզրն գալ
բերել առանց վնասի։ և լուծանէլ
զերկայն ասպարէսն՛ յորժամ և ըզ
կենանն. քանզի իբր մեռաւ մանուկն 'ի
մանկութեան տոս, զղուարեալ դ վլիննն'
ընդ նմա և զոգին երեք Հանել, հաս
կան կապանացն 'ի մարմնոյն սնդեալ
Հաստատապէս բռնութեամ լուծեալ։

Լ՛է են ումանք՛ որ այսչափ լի են եր
կիւղիւ, մինչ զի տկարագունիցն տեղի
տալ, և առ ստուերս զորաւորացն պա
խուցեալ է։ Լ՛է է զի յայըս՛ որք ընդ
լերամբ են, և 'ի ձորս և 'ի թանձր ան
տառս սարասեալ թաքչի։ և է զի 'ի բարձր
և 'ի թևէ մայրիս, և զերկոսն այսր անդր
շուրջ ածէ, անն ունի, ոչ մինչս'ի Հաւուց
թոչմանն երկուցեալ։ Ն կայեն եղջ
րւք բանիս. զորոց անարունեան յա
ռաջագոյն բազմիստ եղեալ ընութեան,
որպէս երեկ, վլիժակացն զործեաց 'ի
նմա բազմութիւն. ոչ որպէս այլոց կեն
դանեացն կապելով'ի զլուին երկին եղ
ջերս, այլ առատագոյն ուղէշ այս
անդր բուսեալք զորէն ոստոց 'ի սաե
զուսն, զի կարոցագոյն յոգւոյն առ զո
րութիւն առաւելատացուցէ զինոս
բութեան Հանգխասցի։ Լ՛այց զի՞ն՛ս
օգուտ է անարեաց ապառգինութիւն

quin unde alieni variique generis monstra
portentosa evenit nasci, nisi ex dissimi-
lium prava commixtione?

In Achaïa ipsa, ut dicitur apud an- 67
tiquos, delphin pueri excellentis in pul-
chritudine vultûs desiderio captus, multoties ex ora maris intus in pelagus
deferebat super dorsum elevatum supra
undas, et multoties ex pelago usque ad
ripam seducebat sine discrimine. Et
tunc demum cessavit a stadio hoc longo,
quando et a vita: mortuo enim puero
in puerili aetate, moleste id ferens delphin, cum illo et animam visus est ex-
spirare, ligamen spiraminis cum corpore
constantius connexum violenter dissol-
vens.

Sunt etiam alia tanto timore plena, 68
ut infirmioribus cedant, et ad umbras
fortiorum trepident. Nunc in speluncis,
quae sub montibus sunt, et in vallibus,
condensisque sylvis, cum tremore lati-
tant; nunc in altis frondosisque saltibus:
ubi oculos huc illuc versant haesitabun-
da, ipsum volatum avium metuentia.
Testes sunt his dictis cervi: quibus fortitudinem cum negarit natura, ut videtur, instrumenta quibus se tuerentur
concessit, et haec ipsa ramosa; haud sicut in caeteris bestiis in capite sitis ge-
minis cornibus, sed abundantiora germina hinc illinc pullulare faciens velut
ramos in truncis, utpote animo forti
carentes, saltem per excessum armatu-
rae aequalitatem sortirentur. Sed quid
refert ignavis ex armatura fortitudinem
acquirere? nec enim mulieres deformes

յարութենէ սառնացեալ ։ Քանզի ոչ զար
դու՛ կանանց տղերաց 'ի զեղեցկակեր
պութիւն . յման որոյ պայծառացեալ
զարդու ամօթ ներանաւոր է ։

69 'Ի մարտի երկիւղածութեան ոչ երբ
եք տարցի նապաստակ զերկրորդութ
եղջերուէ ։ Քանզի դողայ միշտ կեն
դանիս այս, և սարսեալ է . ոչ ուսկը հա
ստատ ամենեսին, ոչ ասեմ գազանաց,
այլ և բնութեան մասանց ։ Քանզի
երկնչի և 'ի բխմանէ վճմից աղբերց, և
'ի ծովու շղփմանէ (կամ շղփմանէ)
և 'ի հողմոյ և յօդոյ . և ամենայն որ
ինչ վանգամ 'ի վերայ երկրէ է . զի յառ
բիւրս, և 'ի պրակս, և 'ի չանեաց տերևա
կոյաս 'ի նեքրս գալարել (կամ ամլարել :)
Ա ասն որոյ ինձ թուի՛ քերթողական
ազգն ոչ վայրապար զողակ զնապաս
տակն կարծել վման զողէ զին և պախե
լոյ ։ Ա ասն այսորիկ և ամենեսուն օգնա
կան բնութիւն զօրակիցն ձեռնտրական
լութիւն երկիւղէ շնորհեաց կենդանւոյս
արագոտութիւն՝ հանգխտարաս թռոց
գործեցեալ զոտս, ոչ եբր այլոյցն հա
ստատեաց յանփուտ և յանսեխ և 'ի
թանձեալ մորթէ շ, բայց զոբէն սպունգի
անգայտացգց, զի օդ բազում ընդունել
լով՝ ընդ նմա թեթեագոյն համբարձ
գէն յընթացս ։

70 Ա գորչափ սոքա լի են պակեղով,
այնքան ոմանք առ անամօթ և լիրբ յան
դգնութիւն աձեգուեեալ զարգացու
ցանեն զինքեանս ։ Զի զի՞ պիտի զվարա
զաց, կամ զընձոց, կամ զառխծոց
պատմել զՎարստութիւն, որք առ ա
նիրաւել և զզեք պատրաստք են . քան
զի թէպէտն և ամընգակից մարդկան կեն
դանին շուն՝ յորժամ յումեքէ յանրն
զել երբաաս բաժեցոյս, մոլի և կատաղի .
և եթէ երևեցի հեռագոյն՝ յաճախ
կռնչէ, և յարձակեցեալ անխատարած հա
ոէ . և մօտ երեալ՝ 'ի վերայ ընթանայ,

ornatus juvat ad formositatem, immo claritas ornatus ruborem inducit.

In certamine timiditatis nunquam ferat lepus secundas cervo concedere. Namque tremit semper haec bestia, et timet cuiquam fidere, non dico belluarum, sed etiam partium naturae. Quoniam timet et scaturigines fontium salientium, et strepitum maris, venti aërisque, omniaque, quae sunt super terram; ita ut circa fontes, dumeta, et siccorum foliorum congeriem contrahatur. Quare ut mihi videtur, poëtarum genus non sine ratione *latitantem* vocat leporem, eo quod lateat trepidetque [1]. Ideo omnium auxiliatrix natura adjumenti loco contra timorem concessit huic bestiae velocitatem pedum parem celeritati alarum; pedes ejus non ut caeterorum firmans in imputrida ac glomerata pelle, sed adinstar spongii dilatans, ut nempe multum aëris recipiendo, ope ejus levius elevarentur ad cursum.

Quantum vero isti pavore sunt pleni, tantum nonnulli in impudentem audaciam feruntur. Quid enim oportet aprorum, vel pardorum, aut leonum referre praepotentiam, qui ad injurias violentiasque faciendas parati sunt, quum vel connutritus inter homines canis quando obtingat vultum sibi infamiliarem, furit praeter modum? qui si ex longinquo appareat, crebro murmurat, et cum impetu impavide latrat; appropinquantem vero invadit, cunctis imminentibus malis contemptis. Nam supersiliens vehe-

---

1 Λαγὼς enim lepus est, et λαθεῖν, latere.

պատճառին զ՚ի վերայ արկածս քաման՝հէ
լով․ Քանզի անզեալ ուժգնագոյն սաս
տիկ արիւնրուշտ աչօք հայէլով՝ վէրր
վիրով քերանան լէ, ոչ միայն առ քարանց
հանդուրժէ ձիգս ընդդէմ կալով, այլ եւ
առ տիգաց եւ նետոց ընդդէմ յարձակէ
լով հարուածողս, եւ առ ՚նոսին միէլով․

Արդ այլ եւ մասին Հասին, ոչ միայն
մարդ, այլ եւ․ այլ կենդանիք նմաց խորհ
հրդականութեան ժառանգութիւն ըն
կալան, ոչ անյայտ է․ այլ՚ի կցորդու
թիւն առաքինութեան եւ չարութեան ՚ի
նոցանէ մատուցեալ լինէ Հաւատ․ Այլ
է ինչ որ անկլեր այսպիսի քանեձն է՛ յինչ
քենէ ոչ ուսեալ, անդիտութիւն խրատ
ունելով՝ սպասանութիւն ալերս կրէ
լով մատեալ․ Իսկ այսոցիկ որ Հանդերձ
բնական պատճառօք՝ Հանճարոյ սկրզբ
մասնած չնորհեաց, եւ ընկալան եւս զա
՚ի քանէն վարդապետութիւն, արժան
է սատել, կամ իբր խօթասրաց, եւ
կամ իբր թշնամեաց ճշմարտութեանն․

ԼԻՒՍԻՄԱՔՈՍ·

Այսոքիկ են, ով պատուականն փիլին,
զորս ալեքսանդրոս եղբօրորդին մեր շա
րագրեաց, եւ ՚ի մէջ անցեալ եցոյց․

ՓԻԼՈՆ·

Աքանչելապէս, ով լիւսիմաքոս, եւ
մէ՛ծ է քան զժամանակս Հասակին․ եւ
Հիացուցին ոչ միայն զերշկաց, այլ եւ ի
մասաասիրութեան սոլորական առնել
կարէ․ Արդ ես ոչ առանց կրթութեան
եմ այսպիսի քանից, այլ ՚ի մանկութենէ
ընդ այս խրատս մեծալ առ ՚ի գիտա
ստուգութէ, եւ անուանց զեղեցկութէ,
եւ մեկնութեանն արագութեան ոչ չա
փաքեր ընկալայ, այլ անշուշտ եւ ճշմա
րիտ զիտեմ․ Քանզի ոչ անդիոս ընեէր
զու զոլմանեք ալդող ՚ի կիր առեալ, եւ

---

menter oculis sanguineis, oreque spuma pleno, non tantum lapidum projectionem suffert, se se opponendo, sed etiam hastarum sagittarumque immissarum ictus obviat, et interserit se.

Itaque reliquam etiam partem, non 71 solum homines, verum etiam bestias sortitas esse intellectûs, quasi haereditate accepta, haud obscurum est: cujus rei participatio virtutis ac malitiae satis fidem facit. Sunt quidem aliqui, qui nequeunt haec audire, per se minime edocti; sed ignorantiam pro disciplina habentes, contrarium omnino tuentur. Quibus tamen una cum naturali ratiocinio principia intelligentiae Deus concesserit, atque doctrinam ex verbo acquisiverunt, istis licet adversarios increpare veluti insanos, aut veluti inimicos veritatis.

Haec sunt, o honoratissime Philo, 72 quae Alexander noster ex fratre nepos conscripsit, et in medium veniens publicavit.

Mirifica sane res est, o Lysimache, 73 majorque aetatis tempore, ita ut stuporem facere possit non solum rusticis, verum etiam instructis philosophià. Ego itaque non careo studio hujusmodi verborum, sed ex juventute in hac nutritus sum disciplina ad inveniendam veritatem; atque nominum venustatem, interpretationisque celeritatem non mediocriter percepi, sed certe et vere novi. Quoniam haud ignoranter te rem percipere monstrabas, dum declamares

 անարկելով անընդ Հատ գլխովկ . և
գինաչ թուի քեզ կրել՛ զի եթէ լուեալ
եր քո նմա, զի ընթեռնույ. իսկ թուէիր
րմՆենել գործէն ածարելոց և կոկոզ
զանծ, քանզի նոցա նոքա՛ ոչ զնոյն յայտ
նութիւն աղզեն առ գառողացն և մեկ
նչայն ասացեալս : Ս աստ որոյ է խնչ որ
մէր 'ի վայր է Նախասացութիւն լեզուէն,
և սաՀմանօք բերանոյ եզերեալ . և կ
խնչ որ յոգւոցն յառաջնորդականէն
ծզտեալ՛ զմայինն գործի զեղեցիկ յար
մարեալ պատկանաւորաբար բախէ :

74 'Բանզի Հոր կամ մօր առ ծնունդս
անսա' ոչ է քատակեալ : Ֆայց ճշմարիտ
ուրենն ծնողք՛ որք ըստ ՀանՃարոյ են
իմաստուն, ք, յորժամ արդեօք պատմեկ
գեն զիւրեանց զիւման, խառնեն իննն ընդ
քաննան աննպատմելի ապա ապուցիկ՛ որ
պատուՀէնն լսելքը : Ս՛չ գդցն ընդ ան
ուաննն և ընդ բայան գանծեալ լիննն,
ասես : Ս աներրաբար և դկզ. բայց ընդ
մեկնչքն քո յոմարութիւն Հիացեալ եմ.
քանզի զբանն այսպէս Բուիս իննծ պատ
մել, մինչ զեներքն իսկ գրողքն՛ յորժամ
կարդայլոին յուցանէը, իննծ Բուի Բէ
ոչինչ 'ի բաց մնացեալ լիններ:

75 Իսկ արմայեալ գպատանեայն ճնորող
իննդիրան, ճնորս և զբազմեքննակզիւման,
և զանունանցն դառ իրաքանչելւրզ գոր
յայտնեքննն դպարմարականնն, ոչ ընդ նոսն
Հաւանեալ եմ, ընդ որս սակաւամիտքն.
որը զիւրաւ գաֆն որ առ ձեռանն է իբր
Բուելռւ Բումանքը իննն կաասարզելէ՛ սոփլ
րեալ են մեզաՆեալ : Ֆայց ճշմարիտ ըստ
որուՆ բնասւորեալ ոււնել յուզելով քննե
ցեզ : Տ ինա , առաւել Հարցափործեալ՛
աննենեցունն քան վարդապետուցգ . քանզի
վիրզ է իննծ ոչ Հաւանական բանագոււ
Բեաննքը մլշտ պարտիլ : Օ ի ոչ զեղզոր
որդին մեր՛ որով ը զրեաց յառաջքան
զզնողզեման ստոյզ ուանել, յառաշա
գոֆն Հաասատալ : Ս՛չ պայստպիկ եթէ պա
րապես , արդէն իսկ յիշեցուցեզ պատ

verba citata, nutu continuo capitis tui;
quid quaeso putares, si ipsum audires
legentem? Mihi videbaris rem altius con-
cipere adinstar divinantium sacra quadam insania; quoniam illi non eamdem
revelationem se accepisse dictis decla-
rant, quam faciunt auctores et com-
mentatores . Quapropter aliud est,
quando huc illuc versatilis comperitur
praedictio linguae, et per terminum oris
finita; et aliud, quando ex anima mo-
deratrice producta, instrumentum vocis
pulchre coaptans, decenter pronuntiat.

Verum enim vero patris aut matris
ad natos affectio non habet parem: veri
tamen parentes, qui secundum scientiam
sapientes sunt, quum referant suas inventiones, admiscent cum verbis inenar-
rabilem quamdam affectionem ad auditores. At non pauca una cum nominibus verbisque conduntur velut iu thesau-
ro, inquis. Decenter ac convenienter:
ego tamen erga commentatorem incli-
nationem tuam miror; ita enim verbum
mihi narrare videris, ut etiam ab ipso
scriptore, quando legens publicabat, nihil differre videreris.

Miras autem adolescentis quaestiones
recentes, novas multiplicesque adinven-
tiones, et nominum ad singulas res rite
exponendas accommodationem, non eo
modo probabiles mihi fateor, quo homines modici intellectûs solent adire eum,
qui in promptu habet quasi per incan-
tationem allicere auditores; sed veraciter dicta, ut reipsa se habent, indagan-
do perpendam, immo magis ventilatum
universos verbum edocebo; cura enim
mihi est, non persuasoriis verborum semper adinventionibus vinci; neque etiam
nepoti nostro ex fratre fidem praebere
licet in iis, quae scripsit, ante quam op-
posita certius sciret. Haec itaque, si va-
cat tibi tempus, jam jam rememorans

Մելով․ ապե՞ս զընկենունս յապաղելով,
Հրամարեալք՝ դարձեալ 'ի ժամու աստ
գունք ։

ԼԻՒՍԻՄԱՔՈՍ:

Ո՞թէ ո՞չ կարծես զիս, ով փիլոն, ամեն
նայն իրիք որ զինչ և իցէ արհամարհան
առնել՝ ցանկութեամբ խրատու և Հրշ
մարտութեամբ բաղձանօք : Ա՞ դայս
սիկ պատրաստադոյն ուսուցեալ քո` յա
ժարագոյն կամեցեալ:

ՓԻԼՈՆ:

Ա՞ սակայն, ով լիւսիմաքոս, որ ինչ
յինէն է պատրաստ : Բայց ընդ մեղ
ած այսպէս Հարցանելով և քննելով՝ և
թէ ո՞չ վրիպեմք 'ի սուրբ խորհրդոյ :
Մեղուք և սարդք կարծես զի ոմանք ու
տայնանք են, և ոմանք խորհիս մեղու
աստեղծանեն, բատ արուեստի՞ ինն մար
թամոց Հանճարոյ, թէ բատ առանց բա
նի ընականի դործոյ : Քանզի 'ի վերայ
ասցա առանց Հանդամանաց՝ թէ պատ
է զձշմարիտն ասել, խնամեալ դարձա
նէ ։ Որ 'ի Հանդամանք՝ յորժամ ո՞չ յա
ռաջադոյն իցէ Հասաումն գիտութեան,
որում սկիզ ելք դոլ արուեստից սկիզբն:
Քանզի ժողով միաբանութեանն Հաստ
նեցալ է արուեստ ։

Արդ գործոյնակ թռչունք օդայանդ
են, իսկ ջրայինքն դուլեն, և գնայուն
են ցամաքայինք : Միամինե արդեօք,
և ո՞չ բնութեամբը երբափանելերը յա
սացեծոյս դործէ զայս : Այսն օրինակ և
մեղուք անուսին, բնութեամ զլու
ըխանն դործեն։ Ա՞ սարդք ինքնուսումն
նակի զայսն ոբք յանկանեն՝ զբարակա
դործութեան կաղմեն : Ա՞ եթէ ոք
յեղլոյս խաբեցոյ Հահանութենէ, եբ
թէ առ ձառս, և Հայեցոյ տեսցէ զլու

254 PHILO AD LYSIMACHUM

թեանն : Օ ի բազում ին՚ կ ե ՛ի սաա
առանց արուեստի արուեստաւոր :

79 Օ որթ՚դ ոչ տեսանես , յորժամ յա_
լուրս գարնայնոյ գպաուցն ձնանիցի ·
քանզի գառաջինն տերեւովք՚ն ձածկէ·
ե ապա զորէ՚ն մօր , առ սակաւ սակաւ
մնուցանե ա՚ծգուցանելով :]չ փոխ_
խու՚մն զարձեալ յազզելեն խազոէ զոր_
ձէ, մինչե ամեներն Հասսւցանել զլիրզ՚ն
]չ սակայն կրթեզա՚ե Հոզացեալ. ե ա
մենասարանց ել՚ոլ ե աւազասսրծ՚ու բնու_
թեամ՚ն, ոչ միայն որ փիսասագււ՚ն էր
զպաուցն ձնանեզ այլ ե զարզարեալ
զատեզն 'ի վայելեու՚թէ, երկայն ձձմանըի
ազոլեազ , պասասանումէ զ՚ուակազ ,
տարաձեալ տերեւովք՚ն , զաձմր սապալ
թիէ,տեսակզօրձեոց: ե եզ՚ն երեւան պայ
ձառազոյնք , ե զարմանալ՚ի խսասարն:
ե զեզզեզկուելթիէ՚ն սռա ոչ զազս միայն,
այլ ե զունզունու՚ե յ՚ինզն կոզզելլ զորու_
թեամ՚ե ձզեր , շնչմանէ սբ 'ի ՚ եոզանե
զմեբձաւոր զոզմ՚ն ին՚եաՀա զորձել , որ
Հանզարտ՚իկ՚ն շնջելոյ զտեզխ՚ն լ՚ առ_
՚ել՚եր ամու_զաՀոուՀթէ. ե զՀասու՚ն ասխր
ձեզ՚լ ե կամ՚ակ Հե՚շութէ լզուցաներ:

80]չազու_՚ն ե յազպազս այլոզ ա՚ն՚եցե՚ն
ասել կարոազեալ, ոչ վա՚ն զանզազելոյ
լու_եմ, այլ ջայ՚ի միայն, յ՚ա՚եյազ երկայ_
նաբանու_թէուք խոզս սուեեալ: այլ զայս
ոչ զխու_մ, զի բազում 'ին՚ա զլաեզելոն՚ն
սմիոյ կարձեմ Համեմասու_թեան օրինա
կաւ ձա՚նուզեազ լ՚ին՚ի : Հասգի ա՚ն՚եր
ե զանեբելոյ զմու_թեա՚ն, զի կազզեա՚ն
ե արուեստաւորն ա՚մ՚եդ՚ զ ապրեզուզա
՚ել՚ն զոգ բանա՚ակա՚եբա, ե որբ ոչ չասա_աո
ե՚ն ամե՚ նելե՚ն :]լ սակայն ոչ ուռ՚ա՚ ե ոչ
սաազեզոզ կե՚ազանեազ՚ն · բանզի կասու_
բե՚ն զլերապա՚նյլլ՚ր ին՚ ոչ ՚ նախասխնա
՚ու_թեա՚մէ Հոզոլ ե խորՀզզուլ, այլ լսա
ակա՚ամ բին զոզձոյ 'ի վեզ բերե՚ն 'ի
կազմա՚ծոյ առա՚նե՚աս՚ոռու_թիսսս : Օ ի
՚ եոֆսորֆ՚ն՚ակ Հաւ_փայզ 'ի ՚ այեզ փախ՚ի՚ն,
ե ձազբ՚ն 'ի սոզի՚ոզ երկ՚ի՚ին: ե ձոזայ՚ի՚ն

istis multa sunt, absque arte, artificiosa.

Vitem non vides, dum in diebus vernalibus germen producit? Primum enim foliis illud operit, deinde more matris paulatim nutrit, et id augmentum ducit; postmodum remutans ex acido uvam efficit, donec omnino maturet fructum. At instructa ne in his fuit, cura adhibita? Certè omnino per mirabilem, operique praesidentem naturam par fuit non solum utilissimo fructui ferendo, verum etiam adornando trunco decore, extensis per longum palmitibus, volutatis capreolis, dilatatis frondibus, viridibusque foliis eleganter; conspectus erat ejus clarissimus, et species mirabilis: pulchritudo vero ipsius non oculos solos, sed etiam nares vi attractiva ad se trahebat; exhalatione vero spiritûs circumvicinum aërem fragrantem reddebat: lente enim flante aura adimplebat spatia odore suavi, odoratumque grata placidaque voluptate explebat.

Plura et de caeteris plantis cum dicere possim, non morae taedio taceo, sed tantum solum, nimiam verbositatem evitans. Etenim puto, quod sub silentio multis relictis, unius analogiae modo totum innotescit. Dicimus enim et de invisibilibus naturis, quod quum ordinatae et artificiosae sint, integrum servant esse rationales, etiam illae, quae omnino carent anima. Revera tamen non ista dantur et in praedictis animalibus. Cuncta enim peragunt non provida procuratione ac consilio, sed secundum irrevocabilem operationem eliciunt ex propria constructione proprietates naturales. Namque consimiliter palumbes fugiunt vultures, ac pulli timent reptilia; et marini murices dum tanguntur

զաղտակուրբդ յորժամ հպեցին ոք, պնդ-
դեն, բուռն հարեալք յորում են զվէ-
մաց իբր երկոքումբք ձեռօք կծկեալ. և
դժուարաւ 'ի բաց կորզեալ 'ի միաս-
րութիւն բռնադատեն. խորհրդակա-
նութեամբ, ո՞չ. բնութեան յատկորէն
ինն կազմածոյբ՝ այն որ առ յարմարա-
կան գործս իւրաքանչիւրոք հրաւիրէ։
Վասն որոյ թեւաբուսեալք հաւք անդէն
թռչին. և ցուլք եղ ջիւրս բուսուցեալ
անդէն եղ ջիւր աձեն, և դարձեալ որ 'ի
կրիւ գան՝ ընդդիմ կայ։ Դարձեալ ո՛չ
խորհրդով, մի՛ գոյ կարծեր, զի կարիճ
'ի վերայ կանգնեալ զխայթոցն շարժէ՝
դատեալ յնկէան ընդրութեամբ որ յա-
ոձադոյն զրկեցա։ Եւ ոձիգն ապդբ զայնոսիկ որ յարձակին, զնայ 'ի վերայ խաա
ձանել թունաբէր զորութեամբն. և
այսմ որում զեկ լիցէ հարելոցս սատա-
կուցին լինել գիտացեաւ։ Դ՛ց դպաձա-
րանս կերակրոյ, և հիւանդութեանց
բժշկութիւնք, և որ ինչ սմանգ ման սոցա
նմանորէն է, բնութեանն դարձեալ են
զիւրք՝ բունք կենդանւոյն աներև գո-
րութեբ. և ստանան զիւրաբանչիւրոք
յայնցանէ առ իւր փրկութ, իսկ այնցիկ
որ յարձակին 'ի վերայ վնաս բերելով։

Ո՛չ եթէ ուսեալ մուկն 'ի ձակ և
յորձս իւր սուզանի, որ ինչ մնանգաման
ո՛չ զ մնունն յանայ քան անհունն ընդունել 'ի ներբս սպրդել ջանակայ։ Սց սա
կայն և ծիծառանց ձագք զմարսկացեալ
կերակուրն յետս դարձեալ 'ի բաց ընկենուն. և ստոյգ երբէք օգտակարին զգա
յութեան, այլ կենդանոյն յախափ ինչ
փուձոյ առաբինուբեանն եղև. և զայն
որ բաշ տևէ՝ զիմեցին գործեւ նախախ
նամութեամբ, զի աւելի ինչ յեղելոցն
աճձան իցէ։

Ծանրգ ծանրագոյն է և կարծել ա-
մենախորգ խարդախութեամբ, և աւելորդ ինն խմանութեամբ վարել անաս
նոցդ։ Ծանրգ խմանութիւն որբ սոցա
վկայէն՝ ստոյգ զիմեցեանս ո՛չ ինանան,

---

ab aliquo, tenaciter constringunt lapidem, cui haerent, velut ambabus manibus tenentes, et aegre divulsi vim faciunt pro unione. Deliberatione forsan? Minime: sed naturae peculiari quadam dispositione, quae singula ad opus coaptatum invitat. Quamobrem aves statim ut alae eis natae sint, illico volant; et tauri, cornibus natis, illico cornua movent, et inpugnantes oppugnant. Praeterea, haud consulto, (neque enim id putes,) scorpio attollens aculeum dimovet, ac si per electionem statuerit debere ulcisci eum, qui jam injuriam sibi fecerit. Sicut etiam genus serpentium invadentes se nititur mordere vi venenosa; anne scire etiam poterit quem ex morsis mori contingat? Item comparatio victûs, et morborum curatio, et his similia, iterum naturae sunt inventiones innatae animalibus invisibili virtute. Et haec omnia possident, ut salva se servent, et ut vindicent se ab invasoribus.

Non enim edoctus mus in rimas suas 81 antraque immergitur, et quid quid sit frugi, et natos suos antequam incrementum sortiti sint, intus inferre studet. Pulli tamen hirundinum cibum digestum retro versi deponunt. Num certae cujusdam utilitatis sensus alius fuit umquam animali, aut continuum studium virtutis? An ad aliquid diu durabile operandum provide se verterunt, ut aliquid amplius augmenti sortiantur facta?

Gravius quippe est putare callidissima 82 fraudulentia, et superabundante quadam sapientia uti bestias. Qui enim his sapientiae concedendae testimonium perhibent, ipsi certe non noverunt se pla-

թէ ոչինչ ոչ գիտենն ։ Ո ի թէպլէտ և է
ինչ 'ի սոսա ընդունելութիւն, աղօտ ա‐
ղօտ և անճամ և դիւրոստոյց ։ Ս ասն
զի այսոցիկ որ դեռևս տղայ մանկունիցն
են, հասանէք գիտութեան իբր տղայոց
անյայոք են և անճառասանք, քան թէ
'ի բանս կատարելոյս առ միտս ։

83 Իսկ եթէ արիստոգոնայ ձին ոչ ինչ
խոտեալ ոտիւք կաղայր, ոչ խաբէր իբր
թէ կերպարանէր ինչ գխաբկանան. քան
զի խաբելդ բանական է ։ այլ վասն զի
յառաջագոյն թերևս ցաւեցեալ ոտքն,
հանգստեան հասանէ յիշեցեալ ։ ապա
եթէ ոչ քանիցն երբեմն հանգարութեան
դադարեցան ինչ աղօտ աղօտ անյայ
է. քանզի մինչև ցորչափ ժամանակ յև
շատակէս և 'ի նոսա արձարձեալ պա‐
հին.

84 Բայց սակայն տապարզելլ է և գվարկս
զայսոցիկ որք կարծեցին զորսորդս շունս
Հետամար լինելով 'ի Հնդերորդում
յեղանակին առնուլ 'ի կար 'ի նմանէ ։
Ո այն ասել զգաղատակրաքաղ̀այն, և
կամ այլ ինչ որ յուզէն. ցուցականն գՀետ
երթան երևմամբք 'ի տրամաբանական
տեսէնէն, այլ և ոչ երազ (կամ յերազ)
երեքք իմաստասիրութեան գգածեալք.
քանզի պարտ է ասել 'ի վերայ ամենայ‐
նի որդ ինչ որ և խնդրէք, զի ՀՆդե‐
րորդն յեղանակին վարեն ։ Կամ այղ
եկ, յայլոց վայրի արդէոք այսոքիկ և
այսպիսիք գլացմանք ապիրատութն
են, Հագաձաբաներ և իմաստակել գրան
սոփերեցին առաւել̀ քան թէ գձշման
բեան, որպէս բնաւորեալ ուն̀ի քննել ։

85 Բանզի ասեմք, զի վայելչականացն
և բարեացն որ անկ է նոցա, և բազում
այլոց որ ինչ միանգամ գործակից առ 'ի
փրկութիւն և աւելուն յարութեան՝
ունէլ առաջք ունելեանց. և ընդՀանուր
ամենայնի Հասմանց 'ի բաց բարձեալոց՝
զայն որ 'ի տեսականն է ստուգութիւն ։
Բայց բանական ունակութեան 'ի Հալ
կէ անբաժ է. և բանական ունակութի‐

ne nihil scire. Nam etsi detur in istis perceptio aliqua, obscura tamen et insulsa est, leves animo faciliter irritans; quoniam eis, qui adhuc juvenilem aetatem agunt, comprehensio scientiae velut pueris incerta est ac inconstans, non ita perfectis in verbo et intellectu.

Quod si Aristogoni equus illaesis pedibus claudicabat, non decipiebat, quasi fngens fraudulenter; deceptio enim rationalis est: sed quia fortassis antea dolores in pedibus senserat, quietem acquirere recordabatur; at quantum temporis esset ex quo quieti se dederat ac otio, obscurum est et incertum: nam et in illis per aliquod tempus tantum memoria rerum servatur.

Proscribenda et opinio eorum, qui canem venaticum bestias persequentem autumarunt quinto argumenti modo uti. Idem dicendum de collectoribus conchyliorum, deque quaerentibus quidquam; indicia enim rerum sequentur apparenter sub specie dialectica, verumtamen nec per somnium quidem philosophentur: alioquin dicendum esset de omnibus aliquid quaerentibus, quod quintum illum modum usurpent. Caeterum ex aliis observa, quod isti et tales feruntur in negationem pravitatis suae, soliti praetexere sermonem fuco, et sophistice de rebus agere potius, quàm verum, ut res se habet, examinare.

Nos enim dicimus, quod ex decentibus bonisque sibi convenientibus, multisque rebus juvantibus ad sanitatem perseverationemque valetudinis habent (animalia) appetitionem (aut imaginationem); et universali comprehensione universorum carentes, eam possident certitudinem, quae in propria specie cernitur. Verum tamen rationalis habi-

ժողով յոչէից համանանգ է, ըստ որում
մինչ յաղագս այ, յաղագս աշխարհի,
յաղագս օրինաց, յաղագս դաւառային
սովորութեան, յաղագս քաղաքի, յաղագս քաղաքավարութեան, և այլոց ևս
անթուից. զորոց և ոչ մի ինչ ոչ ընդու-
նին անասունքդ։

Իսկ զգլուխ համբառնայ ի վեր ձի, և
փռնգայ. և երկիտա երշջերու և նապաստ
ակ են, և շաքակարծ աղուէս. և մանն
կասէքը և նապխանձաւորք և բազումք։
Ո՛չ ըստ ոգւոյ ինչ սրատեսութէ ստու-
գութեան ուրեիին համատ մնաց, այլ ըստ
այնմ որ ի բնութենէն կազմած համբից
իւրաքանչիւր ումեք ի խահ և ի դեպ
տուաւ, իբր մասունքը մարմնոյ և ոգւոյ
իւրաքանչիւր յատկութեան առ ինքնւոյն
և առ քոյորին կատարելութիւն գործե-
ցելոց։

Ահա զի և առիւծք տեսանելով ի
պարտութեան յայտնապէս ի բաց անցեալ գնան, և ոչ տեսանելով փախչին
առանց յետս դառնալոյ. փառասիրու-
թեան և ոչ դոյզնաքեալ ինչ զգացեալք։
բայց յետնեձ լլմել ուսեալք, իբր սրեալք
առ ի զօրութին առանց վարդապե-
տութեան. իսկ զօրութին ոչ առնու
յանձն դայլոյն արհամարհան։ Ա՛յս ինքն
թուի ինձ կրել և ձիք և երշջերուք, և
որ ինչ մնանդամ այլք դասին կենդանիք
փառամոլեալ։ Քանզի ոչ ինչ մխա մ
ձեալ նախագահութին, այլ ընտանա
ցեալք դարմանումբ` ընդ որս կրթէ ոք,
զՀետ երթան։

Բայց ոչ յուսահատելի է և սերմանս
մեծարողդ լինելոյ ի սոսա յառաջադրոս
Հարկանել. զի առ իւրն յատուկ առա
քինութիւն իւրաքանչիւր ոք խաղա
ցեալ` մի ումբատ լցցի յարժանաւոր կա
տարմանէն։ Որդ ի դէպ և սիրամարգդ,
որը զպայծառ և զնախշան թևանորպ
թեանն գորէն ոստայնից տեսանելով, և
ամենագեղեցիկ տեսիլ ցուցանէ տեսա-

---

tûs necesse est illa nullam habere participationem. Rationalis autem habitus est syllogismus ex apprehensione entium, quae minime adsunt; ut intellectus de Deo, de mundo, de lege, de patrio more, de civitate, de politica, deque aliis innumeris, quorum nihil percipiunt bestiae.

At caput efferens equus hinnit, et pa- 86 vidi sunt cervus lepusque, et perfida vulpes, alii puerorum amasii, invidi, et id genus multa. Non utique secundum acumen oculorum animae, certitudinem quamdam comprehendunt intellectus, sed secundum illud, quod ex natura apparatus necessarius singulis conveniens datus est, sicut partes corporis et animae, singulis proprio modo ad sui totiusque perfectionem operantibus.

Ecce et leones videntes se evidenter 87 debellatos, abscedunt, et nolentes quemquam videre fugam capessunt, quin retrorsum se vertant: ambitionis nihil quidquam vel modicum sentientes, sed quia petulantes esse consueverunt, tamquam acuti ad robur exercendum sine doctrina: robur autem non acquiescit contemni ab aliis. Idem mihi videntur pati et equi, cervi, caeteraque animalium, quae ambitione affici putantur. Etenim haud de primate cogitant, sed assuefacta alimento, ea in quibus exercitantur ab educatore, imitantur.

Nec tamen desperandum supponere 88 jam in istis semina curae habendae, ut unumquodque ad virtutem sibi propriam progrediens, non destituatur a debita perfectione. Convenienter et pavo, qui cernens in se claram variegatamque alarum formam texturae similem, pulcherrimam praebet speciem videntibus, non ostentatione utitur, sed naturae suae

դայն, ոչ գոյց առնելով, այլ քնութեանն
իբրյ առապքնութեան տեսանատ տե_
զի տուեալ:

89 Բայց կարծել իբր թէ արդեոք որ
յամոնոքական փղիյն անուանեալ եղե
էան, զրկեալ յաա_ջաապատունֆէն ան
դէս մելողնեաց, ամենաձդռ առնելլ
է, ապարատունֆ ե ընֆպըֆ4քունֆ վայ_
րֆնե դագանֆն. Թեֆրֆս արդ_որ_ վան
յապելոլ ե դայն որ յամ_ օ_ինմամֆն
բաղումֆփափկունֆ կերակրոֆ 'ֆ ֆմ_
եղ_ֆ_ փառատուունֆ նախխանմ_ ձ_
անռ_գ_ան_ե_:

90 Բայց ֆնֆ_ֆ_րդ, ե գֆր_ը, ե ու_լ_ը,
ե _ած_ֆ_գ_ ադ_ը, ե որ ֆնֆ_ մֆ_աֆ_դ_ամ_
առ_ ապ_անֆ_ե_լ_ագ_որ_ձ_ֆ_ է 'ֆ_ տ_ե_ս_ար_ան_ու_
կան_ գֆ_յ_ց_ֆ_ն_ առ_աց_ե_ա_լ_ մ_աանոնֆ_, մֆ_
կարձ_եր_ երֆ_ե_ը_ բ_ անֆ_ֆ_ ո_սնֆ_ե_լ_. բան_
զֆ_ կ_ե_ր_ակ_ր_գ_ կ_արոտ_ունֆֆ_ե_ն_ ե_ յ_աֆ_նֆ_ն_
_ անֆ_ֆ_ անֆ_ առ_ ապ_ատ_ֆ_ն_ դար_թ_ոֆ_գ_ա_
նե_ն_: Բանֆ_ անֆ_ն_աֆ_ն_ ֆ_ն_ ֆ_ որ_ ֆ_ն_ մֆ_անֆ_
դ_ամ_ մ_աֆ_ն_ ո_ ֆ_ ո_ղ__ո_յ_, ֆ_ն_ դ_ ա_ան_ձ_ան_ը_
անֆ_կ_ե_ա_լ_ է_, ե_ ֆ_ն_դ_ ս_ո_լ_ո_վ_ ա_ան_ձ_ան_ա_գ_ ֆ_ն_դ_
_ ար_կ_ա_ո_ր_ օ_ր_ֆ_ն_ո_ք_ լ_ծ_ե_ա_լ_:

91 Ֆ_սֆ_ ե_ թ_ե_ կ_ար_ձ_ե_ա_ ա_յ_լ_ մ_աաֆ_կ_ար_ար_ա_
կ_անֆ_ն_ մ_ֆ__ֆ_ե_ն_ ե_ մ_ե_դ_ո_ւ_, ո_չ_ մֆ_ ք_ար_ար_ա_կ_ա_
բ_ո_ֆ_թ_է_ ք_ա_ն_ք_ ե_ն_, թ_ո_յ_լ_ տ_ո_ւ_ֆ_ դ_ֆ_ մֆ_ա_ն_դ_ամ_
ա_ա_ե_լ_ դ_ֆ_ ս_ա_ց_ե_ա_լ_ ֆ_ն_դ_ ո_ր_ո_ֆ_ ա_ն_կ_ան_ֆ_ ֆ_ն_ֆ_:
Ֆ_ո_ւ_թ_ֆ_ թ_է_ ո_չ_ ա_ն_դ_ֆ_ ա_ կ_ե_դ_ե_ ա_ք_, դ_ֆ_ ն_ո_
գ_ո_ֆ_ն_գ_ է_ մ_ա_ա_կ_ար_ար_ո_ֆ_թ_ֆ_ն_ն_, ե_ ք_ա_դ_ա_
ք_ա_վ_ար_ո_ֆ_թ_ֆ_ն_ն_ մ_ֆ_ո_ յ_ առ_ ապ_ք_ն_ո_ֆ_թ_ե_ա_ն_ն_
ձ_ն_ո_ֆ_ն_դ_ը_. ք_ա_ն_զ_ֆ_ ո_ր_ք_ ֆ_ն_դ_ ո_ւ_ս_ա_ո_ք_ դ_դ_կ_
ե_ն_, ե_ ա_ն_դ_ո_դ_դ_ մ_ե_ծ_ո_ֆ_թ_ե_ա_ն_ֆ_ը_, վ_ա_ն_ն_ ո_ր_ո_ֆ_
ե_ ս_ո_ւ_ն_ ե_ ք_ար_ա_ք_ա_ ա_յ_լ_ դ_ֆ_ թ_է_ ո_չ_ ք_ար_ա_
ք_ա_վ_ար_ո_ֆ_թ_ֆ_ն_, ե_ ո_չ_ մ_աա_կ_ար_ար_ո_ֆ_թ_է_
'ֆ_ վ_ե_ր_ա_յ_ ս_ո_ց_ա_ ա_ս_ե_լ_ֆ_ է_:

92 Ֆ_սֆ_ դ_ֆ_ն_ֆ_ ո_ր_ ա_ա_ա_ց_է_, դ_ո_ր_ յ_առ_ա_ֆ_ա_
դ_ո_ֆ_ն_ դ_ա_մ_ա_ձ_ե_ա_լ_ Հ_ա_մ_ֆ_ա_ր_ե_ մ_ֆ__ֆ_ե_ն_, ե_ յ_ա_
ո_ա_ֆ_ա_դ_ո_ֆ_ն_ դ_ն_ե_ ֆ_ր_ դ_ֆ_ ր_ա_ Հ_ա_մ_թ_ա_ր_ո_յ_:
Բ_ա_յ_ց_ առ_ ա_յ_ս_ո_ֆ_ֆ_ք_ ե_ս_ ա_ֆ_ ա_ա_ս_ո_գ_ո_ֆ_, ե_
ո_ր_ 'ֆ_ դ_ո_ր_ձ_ ե_ր_թ_ա_յ_ ձ_ա_ղ_ֆ_կ_ս_ ժ_ո_ղ_ո_ֆ_ե_լ_ո_վ_
մ_ե_դ_ո_ֆ_ն_, ե_ դ_լ_ո_ֆ_ր_ֆ_ս_ա_ դ_ե_դ_ե_գ_ֆ_կ_ ս_ա_ե_դ_ձ_ա_
ն_ե_լ_ յ_ո_յ_ժ_ ս_ ք_ան_ֆ_ե_լ_ա_ո_լ_ե_ս_ մ_ե_դ_ր_ա_դ_ո_ր_ձ_ե_ա_լ_:
Բ_ա_յ_ց_ ա_ս_ե_մ_ ս_ա_կ_ա_յ_ն_ ե_ ե_ս_, ք_ա_ն_զ_ֆ_ դ_ո_ա_ա_

virtutis desiderio cedit.

Putare autem, quod unus Antioche-
norum elephantum Ajax nominatus,
quum privatus fuisset dignitate princi-
pali, furore captus fuerit, irridendum
omnino est una cum procacitate inso-
lentiaque ferae agresti. Fortassis ob
saturationem, fruitionemque cibi multum delicati diligenter alteri comparati
invidia ambitiosa accessit.

Catulos vero, et onagros, haedos, si-
mias, et quaecumque mirificas in spectaculo ostentationes exhibent, noli putare ullo modo rationis esse documenta;
nam cibi indigentia, et crebre flagellum ad haec excitant: siquidem omne,
quodquod animae partem habet, subje-
ctum est tormento; et famis tormentum
sub lege necessaria redigit.

Quod si censeas alia praedita esse oe-
conomica quadam prudentia, ut formica
et apis, cum nemo istarum tamen politica gaudeat, concedas oportet posse
falsum esse in specie, quod verum sit
in genere sub quo cadit. Ut puto, non
ignoras, quod eorum est oeconomia,
quorum et politice: unius enim utraque
virtutis proles, etsi sub specie pares,
impares tamen magnitudine, ut domus
et civitas. Atqui deest eis politice, neque ergo dispensatio dicenda est de illis.

At quid quis dicere potest (inquies),
quum antea collectione facta condit in
thesauros formica, et horrea sibi prius
parat? His utilius in opus exit apis, et
collectis floribus, alvearique pulchre con-
structo, mirifice mel facit. Ita quidem
ego etiam dico. Dico tamen haec providentiae, non animalium ratione ca-

աասեմ եթէ՝ նախախնամութեամբ՝ ոչ
կենդանին անասնոյն, այլ այնր որ զդ
լորն մատակարարէ բնութիւն՝ խոստո
վանել. Քանզի այսն ոչինչ գործէ մտօք,
իսկ անա Հոգ է զանասական իրաց զպարբ
չականասն դպարձակվմունն, իւրաբան
չիւր ուրոյք սրկել Հրապնդ. արարեալ՝
առ որ եղեն՝ յայս որ անեն է կատարէ
լութիւն.

Իսկ ՚ի վերայ պիննային և պիննոսպա
սեկին ասացեալք՝ առ միաբանութեան
Հասարակաց ՚ի ցոյց. և ասէ ոչ վայրա
պար այնոցիկ՝ որ Հաւանեալ ՚ի գիւտ
ճշմարտութեան՝ յայտնութեամբ Հնա
րին արուեստիւ. բայց առ ոչինչ յայն
ցանեն, որք դիտեն խնատութիւն՝ դիմաս
տութիւն յանդիմանել.

Իսկ որ միանգամ յերկբայս է, ուսցի
՚ի ծառոց և միանգամայն ՚ի անկոց մի
պատրիլ. Քանզի այսոքիկ զի թէպէտո
և ոգեյ սման ոչ ունին, բնատութեան
և օտարութեան այլ ոչինչ Նուազ յայտ
նութիւն. շարժի և աճ է, իբր սիրական
Համբուրիւք ողջունին զմիմեանքք պատու
տի, որպէս զձիթենեօրդ բաղեղն, և
կամ զորդարբոք որթ. Զ և ինչ յորմէ
խուսէ և ՚ի բաց զառնայ, ոչ միայն յան
դիմանս կացուցանելով ՚ի ձեռն յայտնի
երեսացն. զի ՚ի բաց փախչէր, թէ ոտք
լիեալ էին. և պաաստ ոչինչ Նուազ մօ
տաւորագոյն մատուցեալ. քանզի անա
ցեալ լինին իւր առաշին բոյսքն. և եղեալ
բոդորձք դեռեւս կան մնան Նորածինք.
և զարձեալ անեերելյ՝ աս արարեալ գո
սացուցէ, և դայլն առ սակաւ սակաւ
ճնջեցուցանէ.

Ո ի կաղամբէ յորթոյ խուսափին, և
սարոյ. այլ ոչ զոք կարծեմ այսպաԷ լի
յիմարութեամբ, մինչ զի երժնել ասել,
իբր զի եթէ ոչ ՚ի սոցանէ մետբ մտ
կամ թշնամութիւն թնակրեցաց գոր
ծել. այլ վերնագոյն բնութեան բանի,

rentium, sed ejus qui universam mode- 93
ratur naturam, esse tribuenda. Illud e-
nim (animal) nihil agit intellectu. Isti
vero (Deo) cura est variis de rebus;
ut nimirum sicut Creator impetum singularum creaturarum constringit, ac
corroborat ad id, propter quod factae
sunt, sive ad eam quae singulis convenit perfectionem.

Quae vero dicuntur de Pinna, et sa-
tellite ejus, communem societatem de-
monstrant; quaeque non gratis dicta sunt
eis, qui persuasi de inventione veritatis, manifestare rem nituntur artificiose;
nihil tamen juvant illos, qui sapientiae
ducunt esse reprehendere sapientiam.

Quisquis autem haesitat, discat ex 94
arboribus plantisque, ut non decipiatur.
Hae enim etsi nullam habeant animae
partem, tamen familiaritatis abalienatio-
nisque non minorem praeferunt manife-
stationem. Moventur, et crescunt, atque
tamquam osculo dilectionis salutando
amplectuntur se invicem, ut olivam hae-
dera, et ulmum vitis; aliquas tamen non
solum aversatur (vitis), verum etiam evi-
tat, manifeste ostendens in aspectu suo,
quod in fugam se verteret, si pedes ha-
beret: atque languet non modicum, si
propius accedat; quoniam steriles red-
duntur primae ejus germinationes, et
germina producta adhuc restant ut no-
vella; ita ut invisibiliter exsiccetur, alia
quoque paulatim mulgendo.

Vitem evitant populus (vel cram- 95
be)[1], et lauri. Sed neminem arbitror
adeo insanire, ut audeat dicere, haec
ex fidelis amicitiae aut hostilitatis animo oriri, sed supremae naturae ratione
quaedam in unum adducuntur, alia dis-

1 կաղամբ, vel կաղամբուք erit, vel կաղամբք.

Է ինչ որ 'ի մի վայր աձեալ լինի, և է
ինչ դարձեալ՝ որ անշատեալ մեկնի, ոչ
յարելով 'ի միմեանս : Բրդ վայս օրէ
նակ կարձեմ և կենդանեացն որ ինչ
մ[ա]նգամ մահ[…] է, առանց մար-
դյ, Հակառակութիս և միաբանութէ
և որ ինչ այլ յայս գ[…] 'ի բանական[…]յար
մ[…]ում, առ[…] է : Քանզի ամենեքին
այսոքիկ բանիւ և մ[…]ոք ընաւորեալ են
[…] դնել : Իսկ գնանութիւն, և
երր գկերպարանս 'ի կենդանեին առ
բաղձրի գ[…] երև . և արհամարՀանաց
փոփոխմանե, և քանզի պատ[…]ց և յո
ժարութեան, և շնորՀակա[…]ութեան և
նմ[…]ցն այսոցիկ երևեալ լինին ոմանք
աղոտ աղոտ կերպարան տպաց, որ ոչ
են առ Ճշմարտութիս : Քանզի բուն
տ[…]ոքն և Հաստատուն ձև[…]ն յոգֆան
են մարդկան :

96   Վասն որդ և արագել ոչ փոփոխակ
կերպկրելով գծնողն անիրաւութեան
պարտող ոչ ինչ լ[…]ցի, թայց թ[…]նան
ին առնել անիրաւելյ, վ[…]ն զի ոչ 'ի
կ[…]ից : Իսկ ոչ իշ[…]կզ[…]ք գերի մ[…]
գ[…]ց խանգարելով լինին պարտա
պ[…]ք, զի ոչ կ[…]ք առ[…]ել՝ ք[…]ն կե
բակրդ ց[…]ութէ գործձն : Ո՞չ տես
ս[…], զի յոյժ տղայոցն ոչ ոք ոչ 'ի վերայ
մի[…] երբեք գործեցելոց ոչ զոք բազդ[…]ն
ամ[…]ին : վ[…]ն զի խ[…]ն Հասատկ
[…]է և մ[…]ն ընկ[…]եալ : և սակայն մա
[…]ի զի թ[…]կ[…]ն և անկատար է, այլ
սակայն վ[…]ն զի մ[…]դ է ընուխ ք[…]
[…], […]րդ այս ինչ ընկ[…]եալ գ[…]ան
ութեանն մ[…]ն : թայց այժ[…]ի[…] կա
կերպարանեալ, թայց փ[…]ք մի սակաւ
ապա ձ[…]ե : վ[…]ն զի սի և Հ[…]ն
[…]ելով սեր[…]կան քանք գ[…]ն կա
ձ[…]ան խ[…]ի ըստ ժամ[…]կաց ընդ
ն[…] ախ[…]ն, […]ն Շարկ[…]ե և յարել
սե[…] : Իսկ այլոցն ոգիք աղ[…]ե մ[…]ց
ոչ ունելոյ, 'ի խոր[…]գոյ են անս[…]քը :

97   Քանզի բազ[…]մ և այլ ինչ ասի այս[…]
գ[…] որք յա[…]ք[…]ութիւս և 'ի չ[…]ու

junguntur, haud sibi invicem convenientia . Hoc itaque tenore existimo falsum
esse animalia mortalia quaeque, praeter
hominem, contrarietatem societatem-
que, et quae ad istas referuntur rationali cum harmonia, praeseferre . Haec
enim omnia ratione et mente solent con-
ciliari; similitudines vero, et tamquam
imagines in animalibus delineari conti-
git; et speciem contemptûs, vel honoris,
studii, gratiarum actionis, et consimi-
lium rerum exhibent aliqua subobscura
delineamenta impressa, quae non perti-
nent ad veritatem; quandoquidem propriae, certae, et solidae formae in animis sunt hominum .

Quapropter et ciconia non enutriens
genitorem, iniquitatis rea nullatenus e-
rit, etsi aliquid simile injuriae facere videatur ; quia non procedit ex volunta-
te . Neque fuci laborem apum devastan-
do rei evadunt; non enim voluntarie id
agunt, sed potius desiderio victûs . Non
vides, quod parvulum infantem nemo
de ullo opere facto accusat umquam,
eo quod prudentis aetatis nondum particeps fuerit? At puer quamvis imperfectus comperitur, quia homo tamen est
rationalis naturâ, paullo ante acceptis
sapientiae seminibus, quamvis nondum
formare rem potest, paullo tamen post
oriri faciet: etenim aura recepta, seminales vires ad modum scintillae in syl-
va juxta tempora cum illo crescentes vi-
gere ac adhaerere debent. Aliorum vero (animalium) animae non habentes
fontem mentis, progressu deliberationis
carent .

Multa et alia de illis dicunt ad virtutes et malitiam pertinentia explorato-

Թիւն երթան, զոր իմդրեաց չ[...]ել
յաղագս կենդանեացն պատմութեանն:
որոց սոֆորութիւն է 'ի դատարկութեւ
անիորձ եւ անընտիր սոֆորութեամբ
յերկարել բանիւք: Ի՞այց մի զել֊յառաւ
ջաղգ[ն] ասացեալ է զեչուրդդ ո՞չուն
թիւն այսոցիկ ամենեցուն որ ձեռնարկ֊
կեն ֆակառակել, ֆարկաւոր է եւ ժ[...]
յանսանող զխոութեամբ ընտրութեանն
ֆանդերձ նախախնամունեամբ որ զ[...]
եւ գործիցէ: այլ թէկ 'ի նոցանէ լինի
րատ ունանունեան մարդկային գործոյն,
առանց մնայ կատարիլ, րատ առաչին
ընունեան կազմ[...]ս իւրաբանչիւր
ուրուք ձնունդ բաշեբեալ եւ ընտանե֊
բար:

Արդ բաւական է որ յաղագս 'ի խոր֊
ֆուրդս բանին կայ խոսեցեալբ զգոյոֆն
յետս այսորիկ ընենկացուք: Ի՞անգի կեռ֊
նեւրբ՝ եւ ագռաւը եւ պապկայք, եւ որ
մնանգամ ֆոֆանանանբ, զի թէպէտ եւ
զանազանագոյն բարբառեսցեն, յոդա֊
ւոր ոչ երբէբ եւ ոչ իւեբ ձայն կարաւ
գեն ֆանել: Իյլ որպէս կարձեմ առ
բախողարարող ձակբ ֆաւասոի Հիգբեր
տութեան մաս ունելով ոչ բանաւորս
ֆաւասոի Հիչմնւբ են անեբարբ, եւ
ոժ[...]ս յայտնի կարբեն յանչկխան կացու֊
ցանել: Ոյնոււնեալ եւ ասացեալ կեն
զանեացն են աննչանբ եւ տարապարբբ
ձայնբ, զբանամձեն ոչ բառիւ ունել զձ֊
մարտունիւ յայտնունեան առաւել
բան ֆամնորական:

Իււ յարագայ բեղ փող եւ բնար, եւ
որ ինչ մնանգամ երբաշշականն գործի
է: Ի՞անգի բախելով 'ի ձեռն սոֆա օդ֊ք՝
Հնչմնման կատարել նմանեցով մարդկային
ֆայնից: եւ են նորա անկյավբ, Հասատա֊
տունն ընշանակեան ոչ կարեն ձեցագու
ցանել: վասն զի զիւբ այսմ որ կամի՝
պատկանել ընիբ զանագանունիւն եւ
վեճող բաււ, որ նմին կայանի ֆանգի֊
տունայ. եւ Հասատատեն եւ յայանի նշա֊
նակ առցէ որ յայլընդ[...]յլ լսողացն ա[...]ն

res historiae animalium, ii nempe, quibus mos est ob otium improbâ et minus
laudabili consuetudine verba protrahere.
Nos autem jampridem asseruimus ex
diametro argumentum adversus praesumentes contradicere, nihil urgere cur
dicamus animalia per scientiam deliberativam provido consilio agere quidquid
faciunt; sed etsi aliqua ab illis fiant ad
similitudinem humanae operationis, sine
intellectu peragi primae naturae impul-
su, quae apparatum singulorum in ipsa
nativitate distribuens, illum donavit eis
velut innatum.

Hactenus satis locuti de ratione exi- 98
stente in intellectu, locutionem nunc
examinemus. Siquidem merulae, et cor-
vi, et psittaci, et consimiles, etsi varie
vocem proferant, articulatum tamen
numquam, et nullo modo vocabulum pro-
nuntiare queant. Sed puto quod quemadmodum in instrumentis musicis fo-
ramina quamvis habeant portionem veritatis constantis, non tamen rationales
sonitus sunt constantes, sed forma carentes, et consequenter nihil manifeste
exponere possunt; ita et praedictorum
animantium voces sunt significatione carentes et deformes, veritatem formae ser-
monis non vocabuli modo exprimentes,
sed per cantilenam.

Exempli loco sint tibi tibia et citha- 99
ra, cum aliis instrumentis musicis; quoniam pulsatus ab his aër sonos perficit,
imitando voces humanas: qui tamen in-
certi sunt, et nequeunt constanter formare significatum. Facile est enim ei,
qui velit, innumeras fingere differentias,
et vocabula contentiosa, quae eidem
sententiae aequivaleant: argumentum
autem rei firmum et manifestum reci-
piet quis ex eo, quod auditores vario

ունէր՝ եւ ոչ վնոյս ամենեքյուն ։ Քան
զի ոչ եթէ զորէն քաղթնատաց թո
թով բարբառ-բդ՝նցգա. բայգ աղչամիշ իւ
լի են ամենայն որ ինչ սիանդամ՝ի Հաիաս
րակեալ եւ 'ի յզկեալ ձայնէ սառւգու
թեամբ զրկեալ է ։

100 Արդ դադարեցցւ՞ք տղգող լնելով
զեսուլ թենես , եւ աեբարշտել ով՝ Քան
զի վնչ 'ի մարզկան ագգ Հուշակելով ։
եւ չանարդ-եւ՞գ՝ն բաշխել լով տալով , ա՞ն
Հանգեա ա՞նիրաւււ թեա՞ս է , պարաա
ւել զայնոսիկ որ՝ նախաստնունւ թեա՞ն 'ի
բնու թենէն Հասին,՝նազել-լի պարիկէշտու
թեն արկանել զա՞նՀոգալւ օջ՞ս եւ զա՞ն
եբւ-ւ-թիւք ։

modo concipiunt, et non eadem omnes;
nec enim adinstar blesorum balbutien-
tium est sonus istorum (animantium,
ac instrumentorum), sed tenebrosa comperiuntur omnia, quae politâ voce ac
diserte pronuntiante certo privantur.

Desinamus itaque obmurmurare contra naturam, et impietatem subire. Quando quidem usque ad gradum humani
generis extollere immerentia, et indi-
gnis distibuere aequalia, summa est injuria: tum vituperare illos (sc. homines),
qui dignitatem primi gradus ex natura
sortiti sunt [1], et egregiam sobrietatem
adscribere ignavis, ac exiguis minimeque
apparentibus bestiolis [2].

---

1 Plin. vII. 1. « Animantium in eodem (mundo) natura, nullius prope partis contemplatione
minor est, si quidem omnia exsequi humanus animus queat. Principium jure tribuetur homini,
cujus causa videtur cuncta alia genuisse natura ».

2 In codice subjicitur inscriptio, sive memoria, cujus ex parte mentionem fecimus in Prae-
fatione nostra. Verum varias Amanuensium in-

scriptiones variis in locis de scribendo Philonis codice, quoniam respiciunt opus integrum, dabi-
mus in fine ultimi tomi loco suo. Post itaque
tres hosce sermones erit locus publicandi Quaestiones et Solutiones Philonis in Genesim, et
consequenter caetera ejus opera, quae desideran-
tur in occidente .

ՎԱԽՃԱՆ ԵՐԻՅ ԲԱՆԻՅՆ ՓԻԼՈՆԻ.

FINIS TRIUM SERMONUM PHILONIS,

APPENDIX II

GREEK FRAGMENTS OF THE *DE ANIMALIBUS*[1]

Frag. 1, tit.
'Ο 'Αλέξανδρος ἢ περὶ
τοῦ λόγον ἔχειν τὰ
ἄλογα ζῷα.[2]

Ֆիլոնի յաղագս բան ունել
եւ անասուն կենդանեացն:

Frag. 2, § 6.
Τὸ ζητεῖν καὶ πυνθάνεσθαι
πρὸς διδασκαλίαν ἀνυσιμώ-
τατον.[3]

Բանցի խնդրել եւ հարցանել
ի վարդապետութիւն վ'ճարա-
գոյն իմն է եւ յաւգնւտ:

Frag. 3, § 7.
Διδάσκουσι μὲν οἱ τὰς
ἰδίας τέχνας μυοῦντες
ἑτέρους, ἑρμηνεύουσι
δὲ οἱ ἀλλοτρίαν ἀκοὴν
εὐστοχίᾳ μνήμης ἀπαγ-
γέλλοντες.[4]

Վասն զի ուսուցանեն այնորիկ
որք զիւրեանց զհմտութիւն
այլ ոց ուսուցանեն, իսկ մեկնեն այնորիկ որք զայլոցն
լուրս հանրիպողութիւն յիշատակի պատմեն:

Frag. 4, § 100.
Τὸ νέμειν ἴσα τοῖς ἀνίσοις
τῆς μεγίστης ἐστὶν ἀδικίας.[5]

Հանարդեացն բաշխելով տալով,
ան հանգէտ անիրաւութեանց է:

---

[1]Except for the title of the treatise (Frag. 1), these frags.
are reproduced from J. R. Harris, *Fragments of Philo Judaeus* (Cambridge, 1886), p. 11.

[2]Eus. *H E* ii. 18. 6. Arm. has Φίλωνος περὶ κτλ.

[3]Parisinus gr. 923, formerly Regius 923, fol. 230. For more on
this codex, see Harris, *Fragments*, pp. viii-xxii, 89-95.

[4]Vaticanus gr. 1553, formerly Cryptoferratensis; A. Mai,
ed., *Scriptorum veterum nova collectio*, VII (Rome, 1833), 99. The
excerpt is introduced as Φίλωνος ἐκ τοῦ περὶ τῶν ἀλόγων ζώων in Λεον-
τίου πρεσβυτέρου καὶ 'Ιωάννου τῶν ἱερῶν βιβλίον δεύτερον. The next
excerpt, introduced as 'Εκ τοῦ αὐτοῦ, *ibid.*, p. 100, is not from Anim.
The quotation may be worthy of note, even though it is of dubious origin:
'Επίστησον ὁ διδάσκων, ἐξέτασον ἀκριβῶς ἀκοὴν τοῦ μανθάνοντος· εὐηθὴς
γὰρ ὁ κωφῷ διαλεγόμενος, καὶ μάταιος ὁ λίθον νουθετῶν· καὶ σὺ ἔνοχος
ἁμαρτίας ἔσῃ ὁ μὴ ἐπισκεψάμενος ὅπως καὶ πηνίκα καὶ πότε δεῖ λόγον προ-
έσθαι.

[5]Parisinus gr. 923, fol. 208.

PHILO'S *DE ANIMALIBUS* AND PLATO'S *PHAEDRUS*

Parallels between the introductory and transitory dialogues preceding and following the discourse of Alexander in Philo's *De animalibus* (§§ 1-9, 72-76) and the introductory and transitory dialogues preceding and following the discourse of Lysias in Plato's *Phaedrus* (227A-230E, 234D-237A).[1]

Philo's De animalibus	*Plato's Phaedrus*
(1) PHILO: You remember the recent arguments, Lysimachus, which Alexander, our nephew, cited in this regard, that not only men but also dumb animals possess reason.	And meeting the man who is sick with the love of discourse, he was glad when he saw him, because he would have someone to share his revel (228B).
(2) LYSIMACHUS: Admittedly, honorable Philo, some differing opinions have been amicably brought to the speaker three times since then, for he is my uncle, and my father-in-law as well. As you are not unaware, his daughter is engaged to be my wife. Let us resume the discussion of this long, difficult, and wearisome subject and its absurd interpretation which does not appeal to me since it	He said the same thing two or three times, as if he did not find it easy to say many things about one subject, or perhaps he did not care about such a detail (235A). Well then, my dearest, what the subject is, about which we are to take counsel, has been said and defined, and now let us continue keeping our attention

---

[1]Since the Gr. of Anim no longer exists, the parallel passages are most conveniently given in Eng. translation. The translation of Plato's *Phaedrus* is that of H. N. Fowler (LCL I, 412-449, including the Gr. text *en face*).

affects the clear light by distorting the obvious evidence.

(3) PHILO: With regard to great assertions, it is agreed that one ought to listen to them carefully, for nothing else seems to be so helpful to good learning as to critically examine what the lecturer is emphasizing. Had he truly wished to continue learning, he would not have allowed himself to become occupied with other concerns. Tell me, why would he leave his other affairs and come merely to entertain a relative with useless words designed to tickle the ears?

Such an action would be considered neither kind nor appropriate by that person who has already rejected his former courtesies. Therefore do not anticipate receiving a particularly significant response to your request. You will not get very far with your request.

fixed upon that definition (238D).

No, such arguments, I think, must be allowed and excused; and in these the arrangement, not the invention, is to be praised; but in the case of arguments which are not inevitable and are hard to discover, the invention deserves praise as well as the arrangement (236A).

What was your conversation? But it is obvious that Lysias entertained you with his speeches (227B).

Believe this of me, that I am very fond of you, but when Lysias is here I have not the slightest intention of lending you my ears to practice on (228E).

He says that favours should be granted rather to the one who is not in love than to the lover (227C).

But when the lover of discourse asked him to speak, he feigned coyness, as if he did not yearn to speak; at last, however, even if no one would listen willingly, he was bound to speak whether or no (228C).

(4) LYSIMACHUS: Is not his want of leisure, Philo, the reason? You are not unaware of how many things are involved given relatives, the classes, and community affairs at home.

What Lysias, the cleverest writer of our day, composed at his leisure and took a long time for (228A).

(5) PHILO: Since I know that you are interested, indeed that you are always eager to hear new things, I shall begin to speak if you will keep quiet and not always interrupt my speech by making forceful remarks on the same matter.

Be careful and do not force me to say . . . "he yearned to speak, but feigned coyness" (236C).

(6) LYSIMACHUS: Such a restrictive order is unreasonable. But since it is expedient to seek and to ask for instruction, your order must be complied with. So here I sit quietly, modestly, and with restored humility as is proper for a student; and here you are seated in front of me on a platform looking dignified, respectable, and erudite, ready to begin to teach your teachings.

I concede your point, for I think what you say is reasonable. So I will make this concession: I will allow you to begin (236A).

So now that I have come here, I intend to lie down, and do you choose the position in which you think you can read most easily, and read (230E).

(7) PHILO: I shall begin to interpret, but I will not teach, since I am an interpreter and not a teacher. Those who teach impart their own knowledge to others, but those who interpret present from others information through accurate recall. And they do not do this just to a

I know very well that I have never invented these things myself, so the only alternative is that I have been filled through the ears, like a pitcher, from the well springs of another (235C-D).

268

few Alexandrians and Romans--
the eminent or the excellent,
the privileged, the elite of
the upper class, and those
distinguished in music and
other learning--gathered at a
given place.

(8) The young man entered in
a respectful manner, without
that overconfident bearing that
some have nowadays, but with a
modest self-reliance that becomes a freeman--even a descendant of freemen. He sat down
partly for his own instruction
and partly because of his father's continuous, insistent
urging.

(9) Eventually one of the
slaves, who was sent to a place
nearby, brought the manuscripts.
Philo took them and was about
to read.

(The MS of Alexander's
discourse is brought forth and
read [§§ 10-71]).

(72) LYSIMACHUS: These are
the matters, honorable Philo,
that Alexander, our nephew,
presented and discussed when he
came in.

(73) PHILO: Wonderful Lysimachus; time is longer than life!
These matters may interest not

I know very well that
when listening to Lysias he did
not hear once only, but often
urged him to repeat; and he
gladly obeyed (228A).

At last he borrowed the
book and read what he especially wished (228B).

(The MS of Lysias' dis-
course is brought forth and read
[230E-234C]).

What do you think of the
discourse . . . (234C)?

Is it not wonderful
. . . ? More than that . . .
(234D).

only the peasants, but also those
trained in philosophy. Now it is
not as though I was not taught
the things referred to; in fact
I was nurtured with such instructions throughout childhood, on
account of their certainty,
intriguing names, and easy comprehension. And it is not that
I studied them thoroughly, but
surely I do know them well. Nor
are you ignorant, as expressed
by the tone of your voice and
indicated by the constant nodding of your head. Since you
were listening to what was being read, what else would you
need? You seemed to be absorbed
like bacchanals and corybants,
whose self-proclaimed revelations
are not consistent with the reports of researchers and interpreters. On the one hand, there
is a diction which results from
the up and down movements of the
tongue and terminates at the
edges of the mouth; on the other
hand, there is that which stems
from the sovereign part of the
soul and, through the marvelous
employment of the vocal organ,
makes sensible utterances.

I have not at all learn-
ed the words by heart; but I
will repeat the general sense
of the whole . . . in summary
(228D).

Now I am conscious of
my own ignorance (235C).

As I looked at you, I saw
that you were delighted by the
speech as you read. So, thinking that you know more than I
about such matters, I followed
in your train and joined you in
the divine frenzy (234D).

(74) The affection of a father or of a mother for their children is unequaled. Even honest, wise, and knowledgeable parents blend with their words an indescribable affection when they relate their experiences to those who listen. They add quite a few nouns and verbs. That is fine and appropriate, you say. But from the interpreter's point of view, I admire your method. You appeared to present the subject much as the author himself would have presented it by reading. It seems to me that you have not omitted anything.

He has omitted none of the points that belong to the subject, so that nobody could ever speak about it more exhaustively or worthily than he has done (235B).

(75) As for the recent questions which the young man raised, the new and diverse discoveries, and the terms used to delineate everything that is being disclosed, I am not persuaded by them as the fickleminded, whose habit is to be easily attracted by any fascinating thing. But I will thoroughly examine the truth, as one accustomed to do so, and will make it known to everyone after analyzing it critically. I must not always be impressionable to persuasive

He appeared to me in youthful fashion to be exhibiting his ability (235A).

argumentation; otherwise what
our nephew has already written,

which is contrary to sound learning, would be readily believed.
If you want to concern yourself
with these matters, I will discuss them right now; but if you
want to wait, let us agree to
defer them to some other time.

(76) LYSIMACHUS: Do you not
realize Philo, that I hold in
low esteem all other duties for
the sake of my love for learning and hunger for truth? If
you wish to teach these matters
now, I would be most pleased.

(Clearing of conscience
at the beginning and at the end
of the refutation [§§ 77, 100]).

You shall hear, if you
have leisure to walk along and
listen (227B).

Don't you believe that I
consider hearing your conversation with Lysias "a greater
thing even than business," as
Pindar says [*I*. i. 1] (227B)?

(Clearing of conscience
at the end of the refutation
[242C-243E]).

BIBLIOGRAPHY

BIBLIOGRAPHY

Texts and Translations

Aelian, On the Characteristics of Animals, I-III. Translated by A. F. Scholfield. The Loeb Classical Library. Cambridge, Mass., 1958-1959.

Aëtius in *Doxographi graeci.* Edited by H. Diels. Berlin, 1879, pp. 273-444.

The Apocrypha and Pseudepigrapha of the Old Testament in English, I-II. Edited by R. H. Charles. Oxford, 1913.

The Works of Aristotle, I-XII. Edited by J. A. Smith and W. D. Ross. Oxford, 1908-1952.

Athenaeus, The Deipnosophists, I-VII. Translated by C. B. Gulick. The Loeb Classical Library. London, 1927-1941.

Biblia Hebraica. Edited by R. Kittel. 7th ed. by P. Kahle, A. Alt, and O. Eissfeldt. Stuttgart, 1958.

Cicero, De finibus bonorum et malorum. Translated by H. Rackham. The Loeb Classical Library. London, 1914.

Cicero, De officiis. Translated by W. Miller. The Loeb Classical Library. London, 1913.

Cicero, De republica, De legibus. Translated by C. W. Keyes. The Loeb Classical Library. London, 1928.

Cicero, De senectute, De amicitia, De divinatione. Translated by W. A. Falconer. The Loeb Classical Library. London, 1923.

Cicero, The Speeches, with an English Translation: In Catilinam, I-IV, Pro Murena, Pro Sulla, Pro Flacco. Translated by L. E. Lord. The Loeb Classical Library. London, 1937.

Cicero, Tusculan Disputations. Translated by J. E. King. The Loeb Classical Library. Rev. ed. London, 1945.

Dio's Roman History, I-IX. Translated by E. Cary. The Loeb Classical Library. London, 1914-1927.

Diogenes Laertius, Lives of Eminent Philosophers, I-II. Translated by R. D. Hicks. The Loeb Classical Library. London, 1925.

Epictetus, The Discourses as Reported by Arrian, The Manual, and Fragments, I-II. Translated by W. A. Oldfather. The Loeb Classical Library. London, 1925-1928.

Euripides, I-IV. Translated by A. S. Way. The Loeb Classical Library. London, 1912.

"S. P. N. Eustathii Archiepiscopi Antiocheni et Martyris *Commentarius in hexaemeron.*" Translated by L. Allatio. *Migne, Patrologia graeca,* XVIII. Paris, 1857, cols. 707-794.

Herodotus, I-IV. Translated by A. D. Godley. The Loeb Classical Library. London, 1920-1925.

Homer, The Iliad, I-II. *The Odyssey,* I-II. Translated by A. T. Murray. The Loeb Classical Library. London, 1919-1925.

Hesiod, The Homeric Hymns and Homerica. Translated by H. G. Evelyn-White. The Loeb Classical Library. London, 1914.

Hierokles, Ethische Elementarlehre. Edited by H. von Arnim and W. Schubart. Berliner Klassikertexte, IV. Berlin, 1906.

Iamblichi De vita pythagorica liber. Edited by L. Deubner. Bibliotheca graecorum et romanorum Teubneriana. Leipzig, 1937.

Sancti Irenaei Adversus Haereses, I-II. Edited by W. W. Harvey. Cambridge, 1857.

"S. Irenaeus, Εἰς ἐπίδειξιν τοῦ ἀποστολικοῦ κηρύγματος, The Proof of the Apostolic Preaching with Seven Fragments: Armenian Version." Edited and Translated by K. Ter Mekerttschian and S. G. Wilson. *Patrologia orientalis,* XII. Paris, 1919, 653-746.

Irenée de Lyon, Démonstration de la Prédication apostolique. Edited by L. M. Froidevaux. Sources chrétiennes, LXII. Paris, 1959.

Isocrates, I-III. Translated by G. Norlin and L. Van Hook. The Loeb Classical Library. London, 1928-1945.

Josephus, I-IX. Translated by H. St. J. Thackeray, *et al.* The Loeb Classical Library. Cambridge, Mass., 1926-1965.

Juvenal and Persius. Translated by G. G. Ramsay. The Loeb Classical Library. Rev. ed. London, 1940.

Macrobius, The Saturnalia. Translated, with an introduction and notes, by P. E. Davies. New York, 1969.

Martial, Epigrams, I-II. Translated by W. C. A. Ker. The Loeb Classical Library. London, 1919-1920.

Novum testamentum Graece. Edited by Eberhard Nestle, Erwin Nestle, and
K. Aland. 25th ed. Stuttgart, 1963.

Oppian, Colluthus, Tryphiodorus. Translated by A. W. Mair. The Loeb
Classical Library. London, 1928.

Origène, Contre Celse, I-V. Edited and translated by M. Borret. Sources
chrétiennes, CXXII, CXXXVI, CXLVII, CL, CCXXVII. Paris, 1967-1976.

The Oxyrhynchus Papyri, I-XLV. Edited by B. P. Grenfell, *et al.* London,
1898-1977.

Archiv für Papyrusforschung und verwandte Gebiete, I-XIV. Edited by
U. Wilcken. Leipzig and Berlin, 1901-1941.

Einige Wiener Papyri. Edited by E. Boswinckel. Leiden, 1942.

Philonis Alexandrini opera quae supersunt, I-VI. Edited by L. Cohn and
P. Wendland. *Indices,* VII. Edited by H. Leisegang. Berlin,
1896-1930.

*Philonis Judaei sermones tres hactenus inediti: I et II de providentia et
III de animalibus.* Edited and translated by J. B. Aucher. Venice,
1822.

*Philonis Judaei paralipomena Armena: libri videlicet quatuor in Genesin,
libri duo in Exodum, sermo unus de Sampsone, alter de Jona, tertius
de tribus angelis Abraamo apparentibus.* Edited and translated by J.
B. Aucher. Venice, 1826.

*P'iloni Hebrayec'woy čaŕk' t'argmanealk' i naxneac' meroc' oroc' Hellen
bnagirk' hasin aŕ mez* ("Works of Philo Judaeus Translated by Our
Ancients the Greek Original of Which is Extant"). [Edited by F. C.
Conybeare]. Venice, 1892.

Fragments of Philo Judaeus. Edited by J. R. Harris. Cambridge, 1886.

Philo, I-X. Translated by F. H. Colson and G. H. Whitaker. *Supplement,*
I-II. Translated by R. Marcus. The Loeb Classical Library. Cambridge, Mass., 1929-1962.

Die Werke Philos von Alexandria in deutscher Übersetzung, I-VII. Edited
by L. Cohn, I. Heinemann, M. Adler, and W. Theiler. Berlin,
1909-1964.

Les oeuvres de Philon d'Alexandrie, I-XXXVI. Edited by R. Arnaldez,
C. Mondésert, and J. Pouilloux. Paris, 1961- .

*Philo about the Contemplative Life, or the Fourth Book of the Treatise
concerning Virtues.* Edited by F. C. Conybeare. Oxford, 1895.

Philonis Alexandrini Legatio ad Gaium. Edited with an Introduction,
Translation and Commentary by E. M. Smallwood. Leiden, 1961.

278

Philonis Alexandrini In Flaccum. Edited with an Introduction, Translation and Commentary by H. Box. Oxford, 1939.

Philostrati minoris Imagines et Callistrati Descriptiones. Edited by C. Schenkl and A. Reisch. Bibliotheca scriptorum graecorum et romanorum Teubneriana. Leipzig, 1902.

The Odes of Pindar, including the Principal Fragments, with an Introduction and an English Translation. Translated by J. Sandys. The Loeb Classical Library. 2d rev. ed. London, 1937.

Plato, I-XII. Translated by H. N. Fowler, *et al.* The Loeb Classical Library. London, 1914-1935.

Pliny, Natural History, I-X. Translated by H. Rackham and W. H. S. Jones. The Loeb Classical Library. Cambridge, Mass., 1938-1963.

Plutarch, Moralia, I-XVI. Translated by F. C. Babbitt, *et al.* The Loeb Classical Library. Cambridge, Mass., 1927- .

Porphyrii Opuscula. Edited by A. Nauck. 2d ed. Bibliotheca scriptorum graecorum et romanorum Teubneriana. Leipzig, 1886.

Seneca, Ad Lucilium epistulae morales, I-III. Translated by R. M. Gummere. The Loeb Classical Library. London, 1917-1925.

Seneca, Moral Essays, I-III. Translated by J. W. Basore. The Loeb Classical Library. London, 1928-1935.

L. Annaei Senecae opera quae supersunt, I-III. Edited by E. Hermes, *et al.* Bibliotheca scriptorum graecorum et romanorum Teubneriana. Leipzig, 1898-1907.

Septuaginta, I-II. Edited by A. Rahlfs. 7th ed. Stuttgart, 1962.

Sextus Empiricus, I-IV. Translated by R. G. Bury. The Loeb Classical Library. London, 1933-1949.

Joannis Stobaei Anthologium, I-V. Edited by C. Wachsmuth and O. Hense. Berlin, 1884-1912.

Stoicorum veterum fragmenta, I-IV. Edited by J. von Arnim. Stuttgart, 1903-1924.

Theophrastus, Enquiry into Plants and Minor Works on Odours and Weather Signs, I-II. Translated by A. Hort. The Loeb Classical Library. London, 1916.

Thucydides, I-IV. Translated by C. F. Smith. The Loeb Classical Library. London, 1919-1923.

M. Terenti Varronis de Lingua Latina quae supersunt. Edited by G. Goetz
and F. Schoell. Bibliotheca scriptorum graecorum et romanorum
Teubneriana. Leipzig, 1910.

M. Terenti Varronis Rerum rusticarum libri tres. Edited by G. Goetz.
Bibliotheca scriptorum graecorum et romanorum Teubneriana. Leipzig,
1912.

Virgil, I-II. Translated by H. R. Fairclough. The Loeb Classical Library.
Rev. ed. London, 1934-1935.

Die Fragmente der Vorsokratiker. Edited by H. Diels. 5th ed. by W. Krantz.
Berlin, 1934.

Xenophon, Memorabilia and Oeconomicus. Translated by E. C. Marchant.
The Loeb Classical Library. London, 1923.

Literature

Aall, A. *Der Logos. Geschichte seiner Entwickelung in der griechischen
Philosophie und der christlichen Litteratur,* I-II. Leipzig, 1896-1899.

Adler, M. "Das philonische Fragment De Deo." *Monatschrift für Geschichte
und Wissenschaft des Judentums,* 80 (1936), 163-170.

_____. *Studien zu Philon von Alexandreia.* Breslau, 1929.

Akinean, N. "Yownaban dproc'ə " ("The Hellenizing School"). *Handes
amsorya,* 46 (1932), 271-292.

Amir, Y. "Philo Judaeus." *Encyclopaedia Judaica,* 13 (1971), cols.
409-415.

Arevšatyan, S. "Pĺatoni erkeri Hayeren t'argmanowt'yan žamanakə" ("A
propos de l'époque de la traduction en arménien des dialogues de
Platon"). *Banber Matenadarani,* 10 (1971), 7-20.

Arnim, J. Von. *Quellenstudien zu Philo von Alexandria.* Philologische
Untersuchungen, XI. Berlin, 1888.

Arnold, E. V. *Roman Stoicism.* London, 1911.

Awedik'ean, G., Siwrmēlean, X., and Awgerean, M. [J. B. Aucher].
Nor bargirk'haykazean lezowi ("New Dictionary of the Armenian
Language"), I-II. Venice, 1836-1837.

Baer, R. A., Jr. *Philo's Use of the Categories Male and Female.* Arbeiten
zur Literatur und Geschichte des hellenistischen Judentums, III.
Leiden, 1970.

Balogh, E. and Pflaum, H. G. "Le 'concilium' du Préfet d'Égypte. Sa composition." *Revue historique de droit français et étranger*, 30 (1952), 117-124.

Bauer, W. *Griechisch-Deutsches Wörterbuch zu den Schriften des Neuen Testaments und der übrigen urchristlichen Literatur*. 5th ed. Berlin, 1958.

Behm, J. "γλῶσσα." *Theologisches Wörterbuch zum Neuen Testament*, I (Stuttgart, 1949), 719-726.

_____. "ἑρμηνεύω." *Theologisches Worterbuch zum Neuen Testament*, II (Stuttgart, 1950), 659-662.

Belkin, S. *The Alexandrian Halakah in the Apologetic Literature of the First Century C. E.* Philadelphia, 1936.

_____. *Philo and the Oral Law: The Philonic Interpretation of Biblical Law in Relation to the Palestinian Halakah*. Harvard Semitic Series, XI. Cambridge, Mass., 1940.

_____. "The Alexandrian Source of *Contra Apionem* II." *Jewish Quarterly Review*, 27 (1936), 1-32.

Bischoff, [E.] "Isthmia, 1." *PRE*, 18 (1916), cols. 2248-2254.

Bogharian, N. *Grand Catalogue of St. James Manuscripts*, I-VII. Jerusalem, 1966-1974.

Bousset, W. *Jüdisch-christlicher Schulbetrieb in Alexandria und Rom. Literarische Untersuchungen zu Philo und Clemens von Alexandria, Justin und Irenäus*. Forschungen zur Religion und Literatur des Alten und Neuen Testaments, Neue Folge, VI. Göttingen, 1915.

Bréhier, É. *Les idées philosophiques et religieuses de Philon d'Alexandrie*. Études de philosophie mediévale, VIII. 3d ed. Paris, 1950.

Brink, C. O. "Οἰκείωσις and Οἰκειότης: Theophrastus and Zeno on Nature and Moral Theory." *Phronesis*, I (1955), 123-145.

Bruns, J. E. "Philo Christianus: The Debris of a Legend." *Harvard Theological Review*, 66 (1973), 141-145.

Bultmann, R. *Der Stil der paulinischen Predigt und die kynisch-stoische Diatribe*. Göttingen, 1910.

Burr, V. *Tiberius Iulius Alexander*. Antiquitas: Abhandlurgen zur alten Geschichte, I. Bonn, 1955.

Christ, W. von. *Geschichte der griechischen Litteratur*, I-II. Handbuch der klassischen Altertums-wissenschaft, VII. 6th ed. München, 1920.

Christensen, J. *An Essay on the Unity of Stoic Philosophy*. Copenhagen, 1962.

Christiansen, I. *Die Technik der allegorischen Auslegungswissenschaft bei Philon von Alexandrien.* Beiträge zur Geschichte der biblischen Hermeneutik, VII. Tübingen, 1969.

Cohen, B. "Betrothal in Jewish and Roman Law." *Proceedings of the American Academy for Jewish Research,* 18 (1949), 67-135.

_____. *Jewish and Roman Law: A Comparative Study,* I-II. New York, 1966.

Cohn, L. "Einteilung und Chronologie der Schriften Philos." *Philologus Supplementband,* 7 (1899), 387-435.

_____. "Zur Lehre vom Logos bei Philo." *Judaica, Festschrift zu Hermann Cohens siebzigstem Geburtstage.* Berlin, 1912, pp. 303-331.

Conybeare, F. C. "The Lost Works of Philo." *The Academy,* 38 (1890), 32.

_____. *Specimen lectionum armeniacarum, or a Review of the Fragments of Philo Judaeus, as Newly Edited by J. R. Harris.* Oxford, 1889.

Coste, J. "Notion greque et notion biblique de la souffrance éducatrice." *Recherches de science religieuse,* 43 (1955), 488-523.

Daniélou, J. *Philon D'Alexandrie.* Paris, 1958.

Dashian, J. *Katalog der armenischen Handschriften in der Mechitaristen-Bibliothek zu Wien.* Vienna, 1895.

Delling, G. and Maser, R. M. *Bibliographie zur jüdisch-hellenistischen und intertestamentarischen Literatur, 1900-1970.* Texte und Untersuchungen zur Geschichte der altchristlichen Literatur, CVI. Berlin, 1975.

Dey, L. K. K. *The Intermediary World and Patterns of Perfection in Philo and Hebrews.* SBL Dissertation Series, XXV. Missoula, Montana, 1975.

Dickerman, S. O. "Some Stock Illustrations of Animal Intelligence in Greek Psychology." *Transactions and Proceedings of the American Philological Association,* 42 (1911), 123-130.

Dillon, J. and Terian, A. "Philo and the Stoic Doctrine of Εὐπάθειαι: A Note on Quaes Gen 2.57." *Studia Philonica,* 4 (1976-1977), 17-22.

Dirlmeier, F. "Die Oikeiosis-Lehre Theophrasts." *Philologus Supplementband,* 30 (1937), 1-100.

Dörrie, H., Fritz, K. von, and Waerden, B. L. van der. "Pythagoras." *PRE,* 47 (1963), cols. 171-300.

Drummond, J. *Philo Judaeus: Or, the Jewish-Alexandrian Philosophy in Its Development and Completion,* I-II. London, 1888.

Dyroff, A. "Zur stoischen Tierpsychologie." *Blätter für das bayerische Gymnasialschulwesen,* 33 (1897), 399-404; 34 (1898), 416-430.

Eganyan, Ō., Zeyt'ownyan, A., and Ant'abyan, P'. *C'owc'ak jeragrac' Maštoc'i anvan matenadarani* ("Catalogue of the Manuscripts of Maštoc' Matenadaran"), I-II. Erevan, 1965-1970.

Ewald, H. *Geschichte des Volkes Israel*, I-VII. 3d ed. Göttingen, 1864-1868.

Farandos, G. D. *Kosmos und Logos nach Philo von Alexandria*. Elementa: Schriften zur Philosophie und ihrer Problem-geschichte, IV. Amsterdam, 1976.

Fauth, W. "Pythia, 2." *PRE*, 47 (1963), cols. 515-547.

Feldman, L. H. *Scholarship on Philo and Josephus (1937-1962)*. Studies in Judaica, [I]. New York, n. d.

Freeman, K. *The Pre-Socratic Philosophers: A Companion to Diels, Fragmente der Vorsokratiker*. Oxford, 1959.

Früchtel, U. *Die kosmologischen Vorstellungen bei Philo von Alexandrien. Ein Beitrag zur Geschichte der Genesisexegese*. Arbeiten zur Literatur und Geschichte des hellenistischen Judentums, II. Leiden, 1968.

Giblet, J. "L'homme image de Dieu dans les commentaires littéraux de Philon d'Alexandrie." *Studia Hellenistica*, 5 (1948), 93-118.

Ginzberg, L. *The Legends of the Jews*, I-VII. Translated by H. Szold and P. Radin. Philadelphia, 1946-1947.

Goodenough, E. R. *By Light, Light: The Mystic Gospel of Hellenistic Judaism*. New Haven, 1935.

————. *An Introduction to Philo Judaeus*. 2d ed. Oxford, 1962.

————. *The Jurisprudence of the Jewish Courts in Egypt: Legal Administration by the Jews under the Early Roman Empire as Described by Philo Judaeus*. New Haven, 1929.

————. "A Neo-Pythagorean Source in Philo Judaeus." *Yale Classical Studies*, 3 (1932), 115-164.

————. "Philo and Public Life." *The Journal of Egyptian Archaeology*, 12 (1926), 77-79.

————. *The Politics of Philo Judaeus: Practice and Theory*. New Haven, 1938.

Goodhart, H. L. and Goodenough, E. R. *The Politics of Philo Judaeus, Practice and Theory* with a *General Bibliography of Philo*. New Haven, 1938.

Goodspeed, E. J. *A History of Early Christian Literature*. Revised and enlarged by R. M. Grant. Chicago, 1966.

Gould, J. B. *The Philosophy of Chrysippus.* Albany, N. Y., 1970.

Grant, R. M. *The Letter and the Spirit.* London, 1957.

Grigoryan, G. "P'ilon Alek'sandrac'ow ašxatowt'iownneri Hay meknowt'iownnerə" ("The Armenian Scholia on the Works of Philo of Alexandria").
Banber Matenadarani, 5 (1960), 95-115.

Gross, J. *Philons von Alexandreia Anschauungen über die Natur des
Menschen.* Diss. Tübingen, 1930.

Gulak, A. "Deed of Betrothal and Oral Stipulations in Talmudic Law,"
Tarbiz, 3 (1931-1932), 361-376 (Heb.).

_____. *Das Urkundenwesen im Talmud im Lichte der griechische-
aegyptischen Papyri und des griechischen und roemischen Rechts.*
Jerusalem, 1935.

Hamerton-Kelly, R. G. "Sources and Traditions in Philo Judaeus:
Prolegomena to an Analysis of His Writings." *Studia Philonica,* 1
(1972), 3-26.

Hanell, K. "Nemea, 4." *PRE,* 32 (1935), cols. 2322-2327.

Harl, M. "Cosmologie grecque et représentations juives dans l'oeuvre de
Philon d'Alexandrie." *Philon d'Alexandrie.* Colloques Nationaux du
Centre National de la Recherche Scientifique, à Lyon du 11 au 15
septembre 1966. Paris, 1967, pp. 189-203.

Haussleiter, J. "Nacharistotelische Philosophen, 1931-1936." *Jahresbericht über die Fortschritte der klassischen Altertumswissenschaft,*
281-282 (1943), 107-116.

_____. *Der Vegetarismus in der Antike.* Berlin, 1935.

Heinemann, I. *Philons griechische und jüdische Bildung. Kulturvergleichende Untersuchungen zu Philons Darstellung der jüdischen
Gesetze.* Breslau, 1932.

Heinimann, F. *Nomos und Physis.* Basel, 1945.

Henry, M. -L. "Tier." *Biblisch-Historisches Handwörterbuch,* III
(Göttingen, 1966), cols. 1984-1987.

Hilgert, E. "A Bibliography of Philo Studies, 1963-1970." *Studia
Philonica,* 1 (1972), 57-71.

_____. "A Bibliography of Philo Studies in 1971 with Additions for
1965-1970." *Studia Philonica,* 2 (1973), 51-54.

_____. "A Bibliography of Philo Studies, 1972-1973." *Studia
Philonica,* 3 (1974-1975), 117-125.

_____. "A Bibliography of Philo Studies, 1974-1975." *Studia
Philonica,* 4 (1976-1977), 79-85.

284

Jennison, G. *Animals for Show and Pleasure in Ancient Rome*. Manchester, 1937.

Jervell, J. *Imago Dei. Gen 1:26f. in Spätjudentum, in der Gnosis und in den paulinischen Briefen*. Göttingen, 1960.

Kannengiesser, C. "Philon et les Pères sur la double creation de l'homme." *Philon d'Alexandrie*. Colloques Nationaux du Centre National de la Recherche Scientifique, à Lyon du 11 au 15 septembre 1966. Paris, 1967, pp. 277-296.

Keschischian, M. *Katalog der armenischen Handschriften in der Bibliothek des Klosters Bzommar*. Vienna, 1964.

Kiwleserean, B. *Ełišē, k'nnakan owsowmnasirowt'iwn*. ("Elisaeus: A Critical Study"). Vienna, 1908.

Klebs, [E.] "Baebius, 41-46." *PRE*, 4 (1896), cols. 2731-2734.

Koester, H. "Νόμος φύσεως: The Concept of Natural Law in Greek Thought." *Religions in Antiquity: Essays in Memory of E. R. Goodenough*. Edited by J. Neusner. Studies in the History of Religions, Supplement to *Numen*, XIV. Leiden, 1968, 521-541.

Kunsemueller, O. *Die Herkunft der platonischen Kardinaltugenden*. Erlangen, 1935.

Lagrange, M. -J. "Le Logos de Philon." *Revue Biblique*, 32 (1923), 321-371.

Leisegang, H. "Philon." *PRE*, 39 (1941), cols. 1-50.

_____. "Philons Schrift über die Ewigkeit der Welt." *Philologus*, 92 (1937), 156-176.

Lepape, A. "Tiberius Iulius Alexander, Préfet d'Alexandrie et d'Egypte." *Bulletin de la Société Royale d'Archeologie d'Alexandrie*, 8 (1934), 331-341.

Lesky, A. *Geschichte der griechischen Literatur*. Bern, 1957-1958.

Lewy, H. "The Date and Purpose of Moses of Chorene's History." *Byzantion*, 11 (1936), 81-96.

_____. *The Pseudo-Philonic De Jona*. Part I: *The Armenian Text with a Critical Introduction*. Studies and Documents, VII. London, 1936.

Long, A. A. "Carneades and the Stoic Telos." *Phronesis*, 12 (1967), 59-90.

_____. "Freedom and Determinism in the Stoic Theory of Human Action." *Problems in Stoicism*. Edited by A. A. Long. London, 1971, pp. 173-199.

_____. *Hellenistic Philosophy*. London, 1974.

_____. "Language and Thought in Stoicism." *Problems in Stoicism.*
Edited by A. A. Long. London, 1971, pp. 75-113.

_____. "The Stoic Concept of Evil." *Philosophical Quarterly,* 18
(1968), 329-343.

Macler, F. *Catalogue des manuscrits arméniens et géorgiens de la
Bibliothèque Nationale.* Paris, 1908.

Magie, D. *Roman Rule in Asia Minor to the End of the Third Century after
Christ,* I-II. Princeton, 1950.

Manandean, Y. *Yownaban dproc'ə ew nra zargac'man šrǰannerə* ("The
Hellenizing School and Its Development"). Vienna, 1928.

Marcus, R. "An Armenian-Greek Index to Philo's *Quaestiones* and *De Vita
Contemplativa.*" *Journal of the American Oriental Society,* 53 (1953),
251-282.

_____. "The Armenian Translation of Philo's *Quaestiones in Genesim*
[sic] *et Exodum.*" *Journal of Biblical Literature,* 49 (1930), 61-64.

_____. "Notes on the Armenian Text of Philo's *Quaestiones in Genesin,*
Books I-III." *Journal of Near Eastern Studies,* 7 (1948), 111-115.

Massebieau, L. "Le classement des oeuvres de Philon." *Bibliothèque de
l'école des hautes études, Sciences religieuses,* I. Paris, 1889, 1-91.

Mendelson, A. "A Reappraisal of Wolfson's Method." *Studia Philonica,* 3
(1974-1975), 11-26.

Meyer, G. *Index Philoneus.* Berlin, 1974.

Meyer, H. *Geschichte der Lehre von den Keimkräften von der Stoa bis zum
Ausgang der Patristik.* Bonn, 1914.

Miller, J. "Aristogeiton, 1." *PRE,* 3 (1895), cols. 930-931.

Mühl, M. *Der Logos endiathetos und prophorikos in der älteren Stoa bis
zur Synode von Sirmium 351.* Archiv für Begriffsgeschichte, VII.
Bonn, 1962.

Nikiprowetzky, V. "L'exégèse de Philon d'Alexandrie." *Revue d'histoire
et de philosophie religieuses,* 53 (1973), 309-329.

Nöldeke, T. *Kurzgefasste syrische Grammatik.* Leipzig, 1898.

North, H. F. "Canons and Hierarchies of the Cardinal Virtues in Greek and
Latin Literature." *The Classical Tradition: Literary and Historical
Studies in Honor of Harry Caplan.* Edited by L. Wallach. Ithaca,
N. Y., 1966, pp. 165-183.

Oltramare, A. *Les origines de la diatribe romaine.* Lausanne, 1926.

Otte, K. *Das Sprachverständnis bei Philo von Alexandrien: Sprache als
Mittel der Hermeneutik.* Beitrage zur Geschichte der biblischen
Exegese, VII. Tübingen, 1968.

Pascher, J. Ἡ βασιλικὴ Ὁδός. *Der Königsweg zu Wiedergeburt und Vergottung bei Philon von Alexandrien.* Paderborn, 1931.

Pease, A. S. "Caeli Enarrant." *Harvard Theological Review,* 34 (1941), 163-200.

Pembroke, S. G. "Oikeiōsis." *Problems in Stoicism.* Edited by A. A. Long. London, 1971, pp. 114-149.

Pohlenz, M. "Die Begründung der abendländischen Sprachlehre durch die Stoa." *Nachrichten von der Gesellschaft der Wissenschaften zu Göttingen,* Philologische-historische Klasse, Neue Folge I, 3 (1939), 151-198.

_____. "Philon von Alexandreia." *Nachrichten von der Gesellschaft der Wissenschaften zu Göttingen,* Philologische-historische Klasse, Neue Folge 1, 5 (1942), 409-487.

_____. *Die Stoa. Geschichte einer geistigen Bewegung,* I-II. Göttingen, 1964.

_____. "Tierische und menschliche Intellegenz bei Poseidonios." *Hermes,* 76 (1941), 1-13.

Pulver, M. "The Experience of the Πνεῦμα in Philo." *Eranos Jahrbücher,* 13 (1945), 107-121.

Reinhardt, K. *Poseidonios.* Munich, 1921.

Regner, J. "Olympioniken." *PRE,* 35 (1939), cols. 232-241.

Rengstorf, K. H. "διδάσκαλος." *Theologisches Wörterbuch zum Neuen Testament,* II (Stuttgart, 1950), 150-154.

Reynders, B. *Lexique comparé du texte grec et des versions latine, arménienne et syriaque de l' "Adversus Haereses" de Saint Irénée, I: Introduction, index des mots grecs, arméniens et syriaques, II: Index des mots latins.* Corpus scriptorum christianorum orientalium, CXLI-CXLII, Subsidia, t. 5 et 6. Louvain, 1954.

_____. *Vocabulaire de la "Démonstration" et des fragments de Saint Irénée.* Louvain, 1958.

Rieth, O. "Über das Telos der Stoiker." *Hermes,* 69 (1934), 13-45.

Rist, J. M. *Stoic Philosophy.* London, 1969.

Sandbach, F. H. "Phantasia Kataleptikē." *Problems in Stoicism.* Edited by A. A. Long. London, 1971, pp. 9-21.

Sarghissian, B. and Sarksian, G. *Grande catalogue des manuscrits arméniens de la bibliothèque des PP. Mekhitaristes de Saint-Lazare,* I-III. Venice, 1914-1966.

Schalit, A. *Namenwörterbuch zu Flavius Josephus*. A Complete Concordance to Flavius Josephus. Supplement I. Edited by K. H. Rengstorf. Leiden, 1968.

Scherling, K. "Pasiphae." *PRE*, 36:3 (1949), cols. 2069-2082.

Schmid, W. and Stählin, O. *Geschichte der griechischen Litteratur*, I-V. Handbuch der Altertumswissenschaft, VII. Munich, 1929-1948.

Schmidt, H. *Die Anthropologie Philons von Alexandreia*. Diss. Leipzig, 1933.

Schürer, E. *Geschichte des jüdischen Volkes im Zeitalter Jesu Christi*, I-III. 4th ed. Leipzig, 1901-1909.

Schwartz, J. "Note sur la famille de Philon d'Alexandrie." *Mélanges Isidore Lévy*. Annuaire de l'institut de philologie et d'histoire orientales et slaves, XIII. Bruxelles, 1955, 591-602.

Schwenn, [F.] "Korybanten," *PRE*, 22 (1922), cols. 1441-1446.

Siegfried, C. *Philo von Alexandria als Ausleger des Alten Testaments*. Jena, 1875.

Solmsen, F. "Greek Philosophy and the Discovery of the Nerves." *Museum Helveticum*, 18 (1961), 150-163, 169-197.

Staehle, K. *Die Zahlenmystik bei Philon von Alexandreia*. Leipzig and Berlin, 1931.

Stein, [A.] "Julius, 59." *PRE*, 19 (1918), cols. 153-157.

_____. *Die Präfekten von Ägypten in der römischen Kaiserzeit*. Bern, 1950.

Stein, E. *Die allegorische Exegese des Philo aus Alexandria*. Beihefte zur Zeitschrift fur die alttestamentliche Wissenschaft, LI. Giessen, 1929.

Tappe, G. *De Philonis libro qui inscribitur* Ἀλέξανδρος ἢ περὶ τοῦ λόγον ἔχειν τὰ ἄλογα ζῷα: *Quaestiones selectae*. Diss. Göttingen, 1912.

Taubenschlag, R. *The Law of Greco-Roman Egypt in the Light of the Papyri*. Rev. ed. Warsaw, 1955.

Tēr-Awetisian, S. *Katalog der armenischen Handschriften in der Bibliothek des Klosters in Neu-Djoulfa*, I-II. Vienna, 1970-1972.

Theiler, W. *Zur Geschichte der teleologischen Naturbetrachtung bis auf Aristoteles*. Zürich, 1925.

Thompson, D'Arcy W. *A Glossary of Greek Birds*. Rev. ed. London, 1936.

_____. *A Glossary of Greek Fishes*. London, 1947.

Thyen, H. "Die Probleme der nueren Philo-Forschung." *Theologische Rundschau,* 23 (1955), 230-246.

Turner, E. G. "Tiberius Iulius Alexander." *The Journal of Roman Studies,* 44 (1954), 54-64.

Verdenius, W. J. "L'*Ion* de Platon." *Mnemosyne,* III, 11 (1942), 233-262.

Voelke, A. -J. *L'idée de volonté dans le stoïcisme.* Paris, 1973.

Vogel, C. J. de. *Greek Philosophy,* I-III. Leiden, 1960-1964.

Völker, W. *Fortschritt und Vollendung bei Philo von Alexandrien, Eine Studie zur Geschichte der Frömmigkeit.* Texte und Untersuchungen zur Geschichte der altchristlichen Literatur, XLIX. Leipzig, 1938.

Volz, P. *Der Geist Gottes und die verwandten Erscheinungen im Alten Testament und im anschliessenden Judentum.* Tübingen, 1910.

Watson, G. "The Natural Law and Stoicism." *Problems in Stoicism.* Edited by A. A. Long. London, 1971, pp. 216-238.

Weaver, M. J. Πνεῦμα *in Philo of Alexandria.* Unpublished Ph.D. diss. University of Notre Dame, Ind., 1973.

Wellmann, M. "Aegyptisches." *Hermes,* 31 (1896), 221-253.

_____. "Alexander von Myndos." *Hermes,* 26 (1891), 481-566.

_____. "Juba, eine Quelle des Aelian." *Hermes,* 27 (1892), 389-406.

_____. "Leonidas von Byzanz und Demostratos." *Hermes,* 30 (1895), 161-176.

_____. "Pamphilos." *Hermes,* 51 (1916), 1-64.

_____. "Sostratus, ein Beitrag zur Quellenanalyse des Aelian." *Hermes,* 26 (1891), 321-350.

Wendland, P. "Philo und die kynisch-stoische Diatribe." *Beiträge zur Geschichte der griechischen Philosophie und Religion.* Festschrift Diels. Edited by P. Wendland and O. Kern. Berlin, 1895.

_____. "Philos Schrift περὶ τοῦ πάντα σπουδαῖον εἶναι ἐλεύθερον [Quod Omn]." *Archiv für Geschichte der Philosophie,* 1 (1888), 509-517.

_____. *Philos Schrift über die Vorsehung, ein Beitrag zur Geschichte der nacharistotelischen Philosophie.* Berlin, 1892.

_____. *Die philosophischen Quellen des Philo von Alexandria in seiner Schrift über die Vorsehung.* Programme, LIX. Berlin, 1892.

Westermann, W. L. "Tuscus the Prefect and the Veterans in Egypt (P. Yale Inv. 1528 and P. Fouad I 21)." *Classical Philology,* 36 (1941), 21-29.

Winston, D. "Freedom and Determinism in Greek Philosophy and Jewish Hellenistic Wisdom." *Studia Philonica,* 2 (1973), 40-50.

_____. "Freedom and Determinism in Philo of Alexandria." *Studia Philonica,* 3 (1974-1975), 47-70.

Wissowa, [G.] "Bacchanal." *PRE,* 4 (1896), cols. 1396-1397.

Wolfson, H. A. *Philo: Foundations of Religious Philosophy in Judaism, Christianity and Islam,* I-II. Rev. ed. Cambridge, Mass., 1968.

_____. "Philo on Free Will and the Historical Influence of His View." *Harvard Theological Review,* 34 (1942), 131-169.

INDICES

I

INDEX OF PROPER NAMES

Numbers refer to sections of the text

II

CLASSIFIED ZOOLOGICAL AND BOTANICAL INDEX

Numbers refer to sections of the text. Parenthesized references
are to sources from which the Arm. and Gr. equivalents are derived

MAMMALS

Ape (tamed)	կապիկ	καλλίας	a Cercopithecus	24 (transliteration; cf. Vita Cont 57)
Bear	արջ	ἄρκτος	Ursus arctos	51 (Leg All I 8, 22, 87; Abr 161; Dec 113)
Boar	խոզ	σῦς	Sus scrofa domesticus	51 (Spec Leg I 148; III 36)
Boar (wild)	վարազ	κάπρος	Sus scrofa	16, 51, 70
Bull	գուլ	ταῦρος	Bos taurus	66, 80 (Dec 76; Spec Leg III 44)
Bull (young)	զուարակ	μόσχος ταῦρος	see Bull	51, 64 (Abr 266; Spec Leg I 79, 135)
Camel	ուղտ	κάμηλος	Camelus bactrianus dromedarius	28 (Spec Leg I 135)
Cat	կատու	αἴλουρος	Felis domestica	22 (Dec 79 has կուկ)
Cattle	արջառ	βοῦς	see Bull	41 (Dec 77; Spec Leg I 135, 291; III 44, 46)
Cattle (herd)	անդիորդ	ἀγέλη	see Bull	41, 64 (Abr 45, 209, 220)
Deer	եղջերու	ἔλαφος	Cervus elaphus	33, 38, 68, 86-87 (Irenaeus)
Dog	շուն	κύων	Canis familiaris	70, 90 (Abr 266; Dec 79, 114-115; Vita Cont 40)
Elephant	փիղ	ἐλέφας	Elephas africanus indicanus	27-28, 51, 59, 89 (Vita Cont 49)

English	Armenian	Greek	Latin	References
Fawn	ուլ	ἔριφος	*Haedus caprae*	24, 90 (Irenaeus)
Fox	աղուէս	ἀλώπηξ	*Canis vulpes*	46, 66, 86 (Irenaeus)
Goat	այծ	αἴξ	*Capra hircus*	23, 38, 41 (Spec Leg I 135)
Goat (he)	բաղ	τράγος	*Caper hircus*	64, 66 (Leg All II 52; Dec 76-77; Spec Leg III 36, 46)
Goatherd	ծառայ	αἰπολίον	*see Goat*	41, 64 (Spec Leg I 136, 141)
Hare	նապաստակ	λαγώς	*Lepus timidus*	69, 86
Horse	ձի	ἵππος	*Equus caballus*	40, 56-59, 83, 86-87 (Leg All I 72; II 94, 99, 102-104; Abr 234; Dec 4; Spec Leg I 135; III 46)
Hound	որսորդ շուն	θηρευτῆς κύων	*see Dog*	45, 84 (Abr 266)
Leopard	ինձ	πάρδαλις	*Felis pardus*	70 (Dec 113)
Lion	առիւծ	λέων	*Felis leo*	16, 55, 70, 87 (Abr 266; Dec 78; 113; Vita Cont 8)
Monkey	կապիկ	πίθηκος	*Macacus inuus*	23, 46, 90
Mouse	մուկ	μῦς	*Mus musculus*	22, 81
Onager	ցիռ	ἡμίονος	*Equus onager*	90 (Spec Leg III 47)
Ram	խոյ	κριός	*see Sheep*	64, 66 (Dec 76-77; Spec Leg III 46)
Sheep (flock)	շատ	ποίμνη	*Ovis aries*	41, 64 (Abr 220; Dec 114; Spec Leg I 136, 141)
Whelp	կորիւն	σκύμνος	*Catulus leonis*	23
Wolf	գայլ	λύκος	*Canis lupus*	66 (Dec 79)

BIRDS

English	Armenian	Greek	Latin	References
Blackbird	կեռնեխ	κόσσυφος	*Turtus merula*	15, 98
Crow	ագռաւ	κορώνη	*Corvus cornix corone*	13-14, 98
Falcon	բազէ	ἱέραξ	*Gen. falco*	37 (Abr 266; Dec 79)
Fowl	հաւ	ὄρνις	*Gallina communis*	15, 37, 68, 80 (Irenaeus)
Kite	ցին	ἱκτῖνος	*Milvus ictinus*	80 (Vita Cont 8)
Parrot	պապկայ	ψιττακός	*Palaeornis cyanocephalus*	13, 98
Partridge	հաւրաւ կախաւ	πέρδιξ	*Perdix cynerea graeca saxatilis*	35, 80
Peacock	սիրամարգ	ταώς	*Pavo cristatus*	88
Pigeon	աղաւնի	περιστερά	*Columba palumbus*	66 (Irenaeus)
Plover (Egyptian)	որորիլnu	τροχίλος	*Pluvianus aegyptius*	60 (transliteration)
Starling	սարիկ	ψάρ	*Sturnus vulgaris*	15n (transliteration)
Stork	արագիլ	πελαργός	*Ciconia alba nigra*	61, 96 (Dec 116)
Swallow	ծիծառնունկ ծիծառ	χελιδών	*Hirundo rustica urbica*	15, 22, 81 (Leg All II 75)
Turtledove	տատրակ	τρυγών	*Turtur communis*	15

REPTILES

Asp	իժ	ἀσπίς	*Naia haie*	39 (Dec 78)
Cobra, Egyptian	իժ եգիպտացւոց	Αἰγυπτ' ἀσπίς	*Caluber haie*	52
Crocodile	կոկորդիլոս	κροκόδειλος	*Crocodilus niloticus vulgaris*	50 (Dec 78; Vita Cont 8)
Python	վիշապակ	δράκων	*Python cebae molurus*	52 (Irenaeus)
Reptile (gen.)	սողուն	ἑρπετόν	*Gen. reptilis*	80 (Leg All II 105; Dec 78)
Snake	աւձ	ὄφυς	*Gen. vipera*	80 (Leg All II 53, 71-72, 74, 76-81, 84, 87-88, 90, 92-94, 97-98, 106 39 (Vita Cont 49)
Tortoise	կրայ	χελώνη	*Testudo graeca marginata*	

FISHES, MOLLUSCS, CRUSTACEA, AND ECHINODERMS

Clam	զարդակուր	χήμη	*Gen. chama*	80, 84
Crampfish	նարկա	νάρκη	*Torpedo marmorata narce*	30 (transliteration)
Dolphin	դելփին	δελφύς	*Delphinus delphis*	67 (transliteration)
Oyster	խեցգետորթ	ὀστρακόδερμος	*Gen. mollusca*	31 (translation of a compound)
Pilot fish	պոմպիլոս	πομπίλος	*Naucrates ductor*	60 (transliteration)
Pinna	պիննա պիննոս	πίννη	*Pinna nobilis rudis*	60-61, 93 (transliteration)

English	Armenian	Greek	Latin	References
Pinna-guard	փիննապատնեկ / փիննապատնակ	πιννοτήρης	*Pinnoteres veterum*	60-61, 93 (transliteration of one and translation of the other part of a compound, the latter being a corruption of պահնակ, τηρητής)
Polyp	բազմոտանի	πολύπους	*Octupus vulgaris*	30 (translation of a compound)
Shellfish	խորժ / պետրական / ասող	κόγχη	Gen. *concha*	31, 31n.
Starfish	աստղ	ἀστήρ	Gen. *osterias*	30 (*passim*, in references to heavenly bodies)
Unidentified	սեմել[ա]	σέμελος/η		36 (transliteration)

INSECTS AND ARACHNIDA

English	Armenian	Greek	Latin	References
Ant	մրջիւն	μύρμηξ	Gen. *formicidae*	42, 91-92
Bee	մեղու	μέλιττα	*Apis mellifica*	20-21, 61, 65, 77-78, 91-92 (Spec Leg I 291) 61, 96 (Hes. *Th.* 599 quoted in § 61)
Drone	հշամեղու	κηφήν	see Bee	33
Fly	ճանճ	μυῖα	*Musca domestica*	33
Scorpion	կարիճ	σκορπίος	Gen. *scorpiones*	80 (Leg All II 84, 86)
Spider	սարդ	ἀράχνης	Gen. *arachnoidea araneida*	17-19, 77-78 (Vita Cont 51)
Spider (venomous)	պարգանդ / փարիժ	φαλάγγιον	see Spider	38 (transliteration)

PLANTS

Cabbage	կաղամբ [ը]	κράμβη	*Brassica cretica*	95 (transliteration)
Dittany	ողկուզուկ	δίκταμνον	*Origanum dictamnus*	38
Elm	նշրար	πτελέα	*Ulmus glabra*	94
Grapes (sour)	ազոխ	ὄμφαξ	*see* Grapevine	79
Grapes (sweet)	խաղող	σταφυλή	*see* Grapevine	79
Grapevine	որթ	ἄμπελος	*Vitis vinifera*	79, 94 – 95 (Quaes Gen I 28; Provid II 95)
Honeysuckle	մեղեզողվայ	μελίλωτος	*Trigonella graeca*	20 (transliteration)
Hop	խեցգետնի	ὄστρυα	*Ostrya carpinifolia*	38 (transliteration)
Ivy	բաղբեղ	κίττος	*Hedera helix*	94
Laurel	սարդ	δάφνη	*Laurus nobilis*	95
Marjoram	գուիրակ	ὀρίγανον	*Origanum heracleoticum*	39
Olive tree	ձիթենի	ἐλαία	*Olea europaea*	94 (Spec Leg I 141; Provid II 100)
Thyme	թիւմ	θύμον	*Thymbra capitata*	20 (transliteration)
Vine	գետնախախանձ	χαμαίζηλος	*Gen. vitis*	20 (Spec Leg III 1)

III

INDEX OF GREEK EQUIVALENTS

Equivalents are based on Philonic works extant in Gr. and Arm.: Abr, Dec, Leg All I-II, Spec Leg I 79-161, 285-345, III 1-63, Vita Cont, and the frags. of Anim, Provid II, Quaes Gen I-IV, and Quaes Ex I-II. Numbers refer to sections of the apparatus to the translation and/or the commentary.

IV

SUBJECT INDEX

Numbers refer to sections of the text

Abstinence from eating flesh, 62

Actions of animals reveal truth
about them, 44; of kindness, 63;
unjust, 70; rational, 95; involuntary, 96

Adultery, laws concerning, 49

Air, spider's web fills space of,
17; and water as natural riches,
48; hare full of, 69; fragrance
in, 79; escapes through wind
instruments, 99

Anarchy and ochlocracy, 65

Animals, *passim*; *see* Ind. II

Art of weaving, 17-19; of healing,
38; skill in, 77-78; knowledge
basis of, 77

Artisans depend on tools, 19

Assertions, critical examination
of, 3; of logicians, 46, 84;
of truth, 93

Attachment to offspring, 22, 86;
of plants to one another, 95

Attacks of malicious animals, 20;
of whelps, 23; of bulls, elephants, bears, and other wild
animals, 51; of boars, 51, 70;
of lions, 55, 70; of leopards
and dog, 70; *see also* Contests

Author, allusion to Alexander as,

1, 74-75

Authority, failure of exercising,
65; *see also* Leadership

Bacchanals and corybants, 73

Bakers and cooks, 47

Barbarians, wars with, 56; customs of, 62

Beautiful, the, 17

Believing the truth, 52; the revel-
ation of witnesses, 61; readily,
75

Betrothal agreement, 2

Body, posture of, 11; and soul, 20,
63, 86; impulses of, 58; the
things of, 62; strain on, 67;
parts of, 86

Care, precautionary, 22, 34; manifested by the vine, 79; foresighted, 80; *see also* Foresight

Centurions and tribunes, 53, 65

Chariot race, 58

Charioteer, monkey as, 23; animals
as, 54; eagerness of, 58; horses
as, 58

Chastity of animals, 49; of men,
62

Cheating of horse, 40, 83; *see
also* Fraudulance

Child, parental care for, 10;

307

Father, allusion to Alexander's,
8; care of, 10; affection of,
74
Fear of death, 39; horses gauded
by, 59; serving with, 65; contest of, 69; *see also* Cowardice
Feasts and winebibbling, 47
Females, disadvantaged, 11; copulate with males, 48-50; *see
also* Women
Fickle-minded people, 75; *see
also* Foolish, Stupid
Fishing, unrewarding pursuit, 36
Flesh, abstinence from eating, 62
Followers, animals as, 64, 87
Food, providing of, 22, 42-43, 80,
92; used in taming and training
animals, 23, 87; simple, 48;
desire for, 96
Foolish hunters, 16; those whose
eyes of the soul are blind, 25;
men and animals, 66, 95; *see
also* Fickle-minded, Stupid
Foresight of swallows, 22; of
other animals, 30, 34, 43, 80-
81, 92, 97; and judgment, 65;
see also Care
Fount of reason, 12, 96
Fowlers and partridges, 35; and
falcons, 37
Fraudulance of animals, 82; *see
also* Cheating, Hypocrisy
Friends of truth, 17

Games, panhellenic, 57
Gladiator, ape as, 24
Gladness, expressions of, 23
Gluttony of tortoises, 39
God, giver of truth, 10; giver of
common advantages, 16; esteems
animals, 25; and natural agencies
have endowed man with fundamentals of knowledge, 71; conception
of, 85; *see also* Divine
Growth of antlers, 33; of plants,
76; of creatures, 81; of child,
96
Guardian, responsibility of, 10

Happiness, aspiration for, 59;
see also Pleasure
Harmony in singing, 15
Healing, self-taught, 38-40, 80
Hearing and the mind, 13
Hedonism of Greeks, 62; *see also*
Pleasure
Help, giving of, 10; reciprocation
of, 63
History of animals, 97
Honor, exchanged for disgrace, 49;
love of, 56, 58; given to horses,
56; given to elephants, 59, 89
House management, 41, 65, 91
Hunger of animals, 32, 42, 90; for
truth, 76
Hunters, foolish, 16; attacked by
whelps, 23; of partridges, 35;
and falcons, 37; reports of 51

Hypocrisy of horse, 40; *see also*
Fraudulance

Image, divine, 16; mental, 29, 95
Ignorance of this subject, 71;
Alexander accused of, 82
Immature, seed given to, 49; infant is rational by nature, 96;
see also Child, Childish,
Mature
Industry of spider, 17-19; of
ants and bees, 20-21,42, 91-92
Infant, rational by nature, 96;
see also Child, Childish, Im-
mature, Mature
Injustice and justice, 61; of
men and animals, 66; of boars,
leopards, and lions, 70; storks
cannot be accused of, 96; of
granting equality to unequals,
100; *see also* Equality, Justice
Insight of mind, 86
Instinct of animals, 61; *see also*
Nature
Instruction, Philo's childhood,
73; *see also* Education, Knowledge, Learning
Intelligence of animals, *passim*
Interpreter, Philo as, 7, 74
Instruction. to be asked for, 6;
Alexander's personal, 8; oral
71; about animals, 73; plants
do not receive, 79; *see also*
Education, Learning
Instruments, musical, 17, 27, 98-
99

Interpreter, defined, 7; Philo as,
7, 74; and researcher, 73

Judgment and foresight, 65
Justice, commutative, 60-63; animals exhibit, 60-65; stork exhibits supreme, 61; and injustice,
61; and equality, 64; political,
64-65; *see also* Equality, Injustice

Kings hailed, 13-14; keep bees,
20; designated by bees, 65
Knowledge of truth, 10; self-
heard and self-taught, 13, 23;
of animals, 30, 41; men endowed
with fundamentals of, 71; basic
for art, 77; childish grasp of,
82; *see also* Education, Instruction, Learning

Law of nature, 48; concerning
adultery, 49; unalterable, 49;
conception of, 85
Leadership of animals, 64-65, 87
Learning, aid to, 3; those distinguished in, 7; of animals, 13,
23-28, 78, 81, 89; of truth,
25; unsound, 75; love for, 76;
see also Education, Instruction,
Knowledge
Leisure, want of, 4
Letters, elephant writes, 28
Lewdness of men and animals, 66
Light, affected by distortion of
evidence, 2; leads to knowledge
of truth, 10

V

INDEX OF PHILONIC PASSAGES

Parenthesized numbers refer to sections of the commentary

De virtutibus (Virt) 1-50 (51), 6
(48), 11-12 (13), 13 (10), 14
(47), 81, 116 (25), 119-120
(59), 125-160 (25), 129 (34),
137-138 (96), 146 (51), 172-173
(47), 180 (65), 203-205 (16)

De vita contemplativa (Vita Cont)
8, 10 (25), 12 (73), 17 (60),
35 (48), 40-47 (73), 48-63 (13),
59-62 (49), 70 (11)

De vita Mosis (Vita Mos) I 11 (90),
23 (25), 28 (48), 60-61 (16), 62
(25), 76 (3), 109 (52), 130 (70),
192 (80), 272 (85, 98), 279 (11),
327 (59), 328 (60)

II 13-14, 23, 48 (48), 65 (16),
81 (13), 82 (10), 104 (96),
127-130 (12), 140 (30), 162
(25), 163-164 (73), 181, 185
(30), 211 (13), 212 (3), 216 (30)

INDEX OF BIBLICAL PASSAGES

Parenthesized numbers refer to sections of the commentary

INDEX OF CLASSICAL PASSAGES

Parenthesized numbers refer to sections of the commentary

AELIANUS (Ael.)

De natura animalium (NA) Prologue (30); i. 9 (61), 21
(17-18), 32 (94), 59 (20);
ii. 10 (56), 11 (15, 27-28),
21 (27), 25 (42), 42 (37);
iii. 16 (35), 24-25 (22);
iv. 5 (94), 35 (83), 43
(42), 44 (83, 95); v. 10
(65), 11 (21, 34, 65), 13
(65), 26 (23, 46), 45 (51),
48 (94); vi. 1 (51, 54), 2
(62), 5 (33), 10 (ii) (83),
19 (13), 20, 23 (80), 24
(46), 32 (94), 43 (42), 44
(95), 48 (83), 50 (42, 60,
70), 59 (45); vii. 5 (64),
19 (66, 69), 21 (46), 23
(83), 26 (64), 48 (83); viii.
3 (83, 95), 22 (95), 32 (83);
ix. 3 (50), 12 (36), 30 (16),
58 (27); x. 48 (83, 95); xi.
14 (83); xii. 28 (13); xiv.
26 (46); xvi. 25 (40); xvii.
25 (46)

Varia historia (VH) i. 5 (36)

AËTIUS (Aët.)

Placita (Plac.) i. 7 . 33 (20);
iv. 11.1 (29)

ARISTOTELES (Arist.)

Analytica posteriora (APo)
76b24 (12)

De caelo (Cael.) 276a18-277b26
(61)

De anima (de An.) 410b22-23 (78),
25 (17), 413b2, 7-8 (78),
420b6, 29-31 (14), 427b15-16,
433b28-29 (78), 28-31 (29)

Ethica Nicomachea (EN) 1095a17-
18 (59), 1096b28 (25), 1098a16
(59), b12-14 (30), 23-30 (59),
1103a14-1109b26 (30), 1104b34-
35 (44), 1106a26-1108b13 (30),
1107b1-3 (51), 1109b30-1115a2
(80), 6-35 (51), 1117b20-
1118b1 (47), 1130b30-1134a16
(60), 1144a7-8 (30), 1151b32-
1152a3 (47), 1153a14 (44),
1177a11-1178a8 (59)

Historia animalium (HA) 487b9

325

326

(31), 488a10 (42), b12-26 (30), 15 (69), 20 (46), 547b15-33 (60), 555a27, b12 (38), 573b30 (66), 588a18-b14 (24-25, 29), 4-23 (78), 589a3-4 (97), 608a13-17 (30), b27-30 (64), 611a25-28, b8-18 (33), 34-612a8, 24-33 (38), 24-28 (39), b21-31 (22), 613b15-21, 31-33 (35), 620a33-b5 (37), 621a6-14 (36), 622b20-27 (42), 28 (38), 623a7-23 (17), b4-627b22 (20)

De interpretatione (*Int.*) 16a3-18 (14)

Metaphysica (*Metaph.*) 980b25 (83), 1074a31-40 (61)

De partibus animalium (*PA*) 681a10-682a2 (78), 686b22-25 (31), 689b12 (11)

Physica (*Ph.*) 252b26-27 (61)

Politica (*Pol.*) 1323a24-26 (30, 59)

Rhetorica (*Rh.*) 1369b33-35 (44), 1411b73 (25)

Topica (*Top.*) 108a11 (25)

ATHENAEUS (Athen.)

Deipnosophistae 89d (60), 104b, 278f, 280a (47), 605d (49)

CICERO (Cic.)

Academica (*Ac.*) i. 42 (17)

De finibus bonorum et malorum (*Fin.*) iii. 16 (33), 17 (17, 44), 18 (88), 33, 49 (96), 62 (34, 44), 62-63 (10), 62-68 (63), 63 (60); v. 11 (59), 17-18 (44), 65 (63)

De legibus (*Leg.*) i. 22 (63), 26, 30 (96), 45 (79)

Pro Murena (*Mur.*) 61 (30)

De natura deorum (*ND*) ii. 16, 18 (11), 22, 29-30 (61), 34 (44, 81, 92), 35 (79, 86), 47 (17), 57-60 (68), 120 (95), 122 (44, 81, 92), 123 (77), 123-124 (60), 124 (33), 126 (38), 127 (51), 129 (79), 133, 140 (11), 147 (82, 93), 154 (63), 156-157 (92), 158 (70), 160 (78)

De officiis (*Off.*) i.11 (22), 58 (10)

De senectute (*Sen.*) 52 (95), 53 (79)

Tusculanae disputationes (*Tusc.*) i. 16-17 (5)

DIO CASSIUS (D.C.)

xl. 32. 40-41 (42); lvi. 27.4 (27)

DIOGENES LAERTIUS (D.L.)

v. 42 (63); vii. 85 (25, 33, 68), 85-86 (44, 92), 92 (30), 93 (66), 100-102 (17), 107 (80), 116 (55), 128 (60), 129

(63), 131 (48), 135-136 (20),
138-139 (61), 140 (17), 142-
143 (61), 149 (44), 158 (99),
188 (48); ix. 37 (44)

EPICTETUS (Epict.)

Dissertationes (*Diss.*) i. 16.1-8
(10); ii. 11.3 (96), 22.1-37 (30)

EURIPIDES (E.)

Supplices (*Supp.*) 201-213 (11)

GALENUS (Gal.)
De placitis Hippocratis et
Platonis iii. 5 (13)

HERODOTUS (Hdt.)
iii. 36 (34); v. 55; vi. 123
(40)

HIEROCLES STOICUS (Hierocl.)
i. 38; iii. 19-52 (33)

HESIODUS (Hes.)

Opera et dies (*Op.*) 227 (60, 70)
Theogonia (*Th.*) 594-599 (61)

HOMERUS (Hom.)
Iliad (*Il.*) xi. 416, 474-475
(51); xx. 170-171 (54); xxii.
310 (69); xxiii. 276, 374 (57)

Odyssey (*Od.*) xi. 541-562 (59)

IAMBLICHUS (Iamb.)

De vita Pythagorica (*VP*) 168
(63)

ISOCRATES (Isoc.)

Antidosis (*Antid.*) 213 (51)

JOSEPHUS (J.)

Antiquitates Judaicae (*AJ*) i. 41
(98)

Contra Apionem (*Ap.*) i. 58, 116,
161 (56); ii. 168-169 (7),
201 (49), 210 (63), 213 (22),
273 (49)

JUVENAL (Juv.)
v. 153-155 (23)

MACROBIUS (Macr.)
Saturnalia (*Sat.*) ii. 4. 29-30
(13)

MARTIAL (Mart.)
i. 104 (23)

Spectacula (*Sp.*) 30 (23)

OPPIANUS APAMENSIS (Opp.)

Cynegetica (*C.*) i. 62-64 (37),
165 (69), 221-222 (40), 236-
238 (48); ii. 54-58 (51), 72-
82 (55), 181 (69), 187-205
(66), 605; iii. 45 (46), 139-
158 (66), 504 (69), 515-525
(66); iv. 77-211 (16)

Halieutica (*H.*) 144-147 (36)

ORIGENES

Contra Celsum (*Cels.*) iv. 48,
78 (85), 81 (85, 95), 81-
85 (92), 87 (33, 38); vi.
65 (12)

PHILOSTRATUS JUNIOR (Philostr. Jun.)

Imagines (*Im.*) ii. 28 (17)

PINDARUS (Pi.)

Isthmian Odes (*I.*) i. 40 (34)

Nemean Odes (*N.*) ii. 46 (34)

PLATO (Pl.)

Apologia (*Ap.*) 33A-B (7)

Cratylus (*Cra.*) 411E (47)

Epinomis (*Epin.*) 977D (30)

Euthydemus (*Euthd.*) 279A-B (30)

Gorgias (*Grg.*) 465A (77)

Ion 530A-542B (7, 77)

Laches (*La.*) 196C-197C (51), 196E (30)

Leges (*Lg.*) 652-674C (7), 697B, 743E (30), 814B (22), 823D-E (36), 836C, 838E (49)

Phaedrus (*Phdr.*) 227A-243E (1), 227B-C (3), 228A (4, 8), 228B (5, 9), 228C (3), 228D-E (9), 228E (3), 230E (6), 230E-234C (9), 234D (73), 235A (2, 8), 235C-D (7), 236A (3, 6), 236C (3, 5), 238D (2), 242C-243E (77, 100), 246A-256A (10, 47, 58), 259C (48)

Philebus (*Phlb.*) 48D-E (30)

Protagoras (*Prt.*) 321C (11)

Respublica (*R.*) 335B (57), 391E (11), 419-445E (30), 430B (51), 432A (47), 433E

(60), 439D (10), 443B (60), 514A-517A (10)

Sophista (*Sph.*) 263E (12)

Theaetetus (*Tht.*) 189E (12)

Timaeus (*Tim.*) 30D, 31A-B, 32C-33A (61), 33B (17), 40D (47), 44D (61), 47A-E (13), 64D (44), 69C-71D (11), 69E-71D (10, 30), 72E-18E (11), 75D-E (98), 90A (10-11), 91E (11)

PLINY

Naturalis historia (*NH*) i. (27-28); vii. 38, 48 (49); viii. 2 (27), 4 (27-28), 6 (28), 7 (27), 11-12 (59), 14 (27-28), 32, 35 (27), 49 (54), 50 (55), 96-101 (38), 98 (39), 103 (46), 112 (48), 115, 117 (33), 146 (83), 154-155 (56), 159 (40), 160-161 (58), 181 (51), 202 (66), 215-216 (23); ix. 55 (64), 146, 154, 183 (78); x. 23 (37), 80-124 (13), 100 (66), 103 (35), 104-105 (66), 120-121 (13), 171-172 (49), 203-208 (94); xi. 11-70 (20), 20 (21), 27-28 (61), 53-54 (65), 58 (21), 79-84 (17), 80 (18), 87 (80), 108-110 (42), 123 (68), 125 (24), 135 (11); xiv. 58; xvii. 239-240 (95)

PLUTARCHUS (Plu.)

Moralia (*Mor.*) 37C-48D (3), 39B-
D, 42F-43B (5), 45C-D (6), 47F-
48D (5), 75B-86A 441A (30),
493A-497E (22, 74), 494E (35),
777B-C (12), 949D (46), 959A-
965B (25), 959A-985C (31, 62,
73), 960A (12), 960B (1, 73),
960C-963A (17, 78), 960D-961B
(13), 960E-F (44), 960F (83),
961D (40, 44), 961F-962B (30),
962B (12), 962C-963B (29), 962D-
963A (31), 962E (66), 963B (30,
73), 963C (54), 963F (71), 964A,
D (10), 964E-F (62), 965C-E (73),
965F-966B (36), 966C (51), 966C-
D (16), 966D-E (22), 966E-967A
(17), 967D-968A (42), 968D (59),
968F-969A (46), 969A-B (45),
971C-D (35), 972F-973A (13),
973A (12), 973E (13), 974A-E
(38), 974C (62), 975A-B (64),
975D (73), 976C (64), 977B
(36), 980A-B (60), 982B-D (50),
984D (22), 986F-992E (30),
987C-988E (53), 988E (100),
988F-991D (48), 991B (70),
991B-D (62), 991E-F (38), 992A-
B (23), 992B (83), 992C-E (31),
992D (78), 992E (85, 100),
993A-999B (25, 62), 994A, 995A
(70), 997E (30), 999B (10),
1034C (30), 1038F (44), 1044C-
D (88), 1044F (48), 1050A (91),
1052D (18), 1060C (33, 68)

PORPHYRIUS TYRIUS (Porph.)

De abstinentia (*Abst.*) iii. 1-
15 (30), 3 (98), 9 (22), 10
(48), 16 (25, 66), 19 (63, 78),
20 (86), 23 (29, 66), 25 (63),
26 (62); iv. 14 (22)

PRE-SOCRATIC PHILOSOPHERS (Diels *VS*)

Orpheus 1B3 (58), 8 (47)

Xenophanes 21B7 (62)

Heraclitus 22A1§§7, 8; A8, 10;
B8, 10, 23 (61), 40 (62), 87
(75)

Philolaus 44B14 (58)

Democritus 68B145 (38, 44)

SENECA (Sen.)

De beneficiis (*Ben.*) ii. 18.2
(10); iv. 7.1-2 (20)

De clementia (*Cl.*) i.19. 3-4
(92)

Dialogi (*Dial.*) iii. 3.4 (88),
3.5-6 (95); vi. 7.2 (79);
viii. 5.4 (11); ix. 12.3 (92)

Epistulae (*Ep.*) xviii. 5-7; xx.
13 (48); lxvi. 12; lxvii. 10
(30); xcii. 13; xciv. 56 (11);
cviii. 14-16 (48); cxii. 1;
cxx. 14 (11); cxxi. 14 (63),
19-21 (33), 21-23 (92); cxxiv.
9 (96)

Naturales quaestiones (*Nat.*)
iii. 29.3 (96)

(30), 30 (17), 33 (30, 59),
38 (51, 85), 39-48 (30), 83,
92 (17), 136 (30, 59), 139
(30), 146 (33, 68), 154 (44),
156 (30), 169 (51), 178 (25,
33, 44, 68), 178-196 (92), 182
(33), 184 (63), 188 (30), 189
(17), 202 (30, 51), 207-208 (30),
211 (44), 227 (30), 245-254
(88), 255-260 (30), 262 (51),
262-263 (60), 262-264 (47),
262-265 (30), 264 (51), 266
(47), 268 (30), 274 (47),
277 (30), 280 (30, 51), 285-286
(51), 295 (30, 51), 301 (65,
77), 303-304 (30), 306 (10),
308 (60), 312 (30, 79), 323
(48, 65, 91, 95), 336 (92),
337 (48, 52, 95), 339-342 (63),
340 (34), 340-348 (100), 343
(96), 362-363 (75), 367-376
(63, 100), 368 (85, 92, 95),
369 (60), 372 (30, 78, 88),
392 (30), 406 (13), 459 (10),
493 (80), 505 (88), 512 (30,
80), 522 (3), 527, 536, 548,
557-566 (30, 567 (41), 573

(10), 609 (90), 624 (65), 707
(48), 709 (47), 728, 744, 753
(48), Diog. 29 (12), 29-30 (10),
Ant. 43 (17), 47 (78), Apollod.
5 (17)

THEOPHRASTUS (Thphr.)

De causis plantarum (*CP*) ii.
18.4 (95)

Historia plantarum (*HP*) iv. 16.
6 (95)

THUCYDIDES (Th.)
i. 20 (40); iv. 62 (34); vi.
52-59 (40)

VARRO (Var.)

De lingua Latina (*L.*) vi. 56
(99)

Res rusticae iii. 16.8 (61),
16.9 (21)

VERGIL (Verg.)

Georgica (*G.*) iii. 234, 241
(51); iv. 67-87 (21)

XENOPHON (X.)

Memorabilia (*Mem.*) i. 10 (7);
iii. 9 (30)